TELEVISION AND SOCIAL BEHAVIOR:

BEYOND VIOLENCE AND CHILDREN

*A Report of the Committee on
Television and Social Behavior
Social Science Research Council*

Edited by

STEPHEN B. WITHEY
University of Michigan, Ann Arbor

RONALD P. ABELES
*National Institute on Aging,
Bethesda*

LEA LAWRENCE ERLBAUM ASSOCIATES, PUBLISHERS
1980   Hillsdale, New Jersey

Lawrence Erlbaum Associates, Inc., Publishers
365 Broadway
Hillsdale, New Jersey 07642

Library of Congress Cataloging in Publication Data

Main entry under title:

Television and social behavior.

"Sponsored by the Social Science Research
Council."
  Bibliography: p.
  Includes index.
    1. Television broadcasting—Social aspects—
Addresses, essays, lectures.   2. Television—
Psychological aspects—Addresses, essays, lectures.
I. Withey, Stephen Bassett, 1918–
II. Abeles, Ronald P., 1944–   III. Social
Science Research Council.
PN1992.6.T39   302.2'3   79-29684
ISBN 0-89859-014-0

Printed in the United States of America

Contents

Preface *vii*

1. **Beyond Violence and Children** 1
   Ronald P. Abeles
   Entertainment and Television 2
   Televised Portrayal of Ethnicity 3
   Television as an Industry 4
   Children 6
   Conceptual and Methodological Problems 6

2. **An Ecological, Cultural, and Scripting View of Television and Social Behavior** 9
   Stephen B. Withey
   An Ecological View 10
   Cultural Approaches 11
   Script Models 13

3. **On the Nature of Mass Media Effects** 17
   Jack M. McLeod and Byron Reeves
   Types of Mass Media Effects 17
   Complexity of Evidence Required 27
   Complexity of Media Stimuli 30
   Varying Strategies of Inference 33
   Peculiar History and Current Structure
     of the Field 35
   Media Effects Research and Public Policy 41

**4. The Audience for Television—
and in Television Research**   55
Herbert J. Gans

Introduction   *55*
The Audience—and Audience Research   *56*
Viewer Involvement: Research and Policy
   Implications   *67*
The Suppliers of Television and the Audience   *70*
The Larger Society and the Television Audience   *75*
The Television Researchers and the Audience   *78*

**5. An Organizational Perspective on Television
(Aided and Abetted by Models from Economics,
Marketing, and the Humanities)**   *83*
Paul M. Hirsch

Economic and Organizational Aspects of the
   Television Industry   *84*
Television Viewing: A New Typology   *87*
Television as a Consumer Good   *89*
Market Research Findings and Issues for
   Social Scientists   *91*
Social Science Research on Television and
   Mass Communication   *98*

**6. After the Surgeon General's Report:
Another Look Backward**   *103*
Leo Bogart

The Terms of Debate   *104*
Premises and Parties in the Debate   *106*
Television Content and Television Research   *108*
Television and Film   *112*
"Measuring" Violence   *114*
Questioning the Evidence   *117*
Network-Sponsored Research   *118*
The Public Relations of TV Violence   *122*
Does He Who Pays the Piper Call the Tune?   *126*
From Violence to Sex, or Worse   *128*
Two Concluding Notes   *130*

7. Social Influence and Television   *135*
Hilde T. Himmelweit

The Annan Committee   *136*
A Conceptual Model of the Interdependence of
   Broadcasting and Society   *143*
Social Science and Television   *151*
Needed Research   *153*

8. **The Influence of Television on Personal Decision-Making**   161
Irving L. Janis
Potential Power of Television   *161*
Availability of Images and Personal Scripts   *164*
Cumulative Effects of Exposure to Recurrent
    Themes   *173*
Acquisition of Personal Scripts   *177*
Effects of Content Themes Bearing on
    Decision-Making Procedures   *181*

9. **When I Was a Child I Thought as a Child**   191
Aimée Dorr
What, Really, Are Those Things in the Box?   *194*
How Come Those Programs Are There?   *201*
What's the Story About Anyway?   *204*
What Does It Mean That Things Are So Much
    the Same on Television?   *209*
What Do the Pictures Mean?   *211*
What Do I Think About How People Behave?   *213*
The End?   *221*

10. **Television and Afro-Americans:
Past Legacy and Present Portrayals**   231
Gordon L. Berry
Television as a Medium and a Mediator
    in Society   *231*
Psychosocial Legacy from the Past   *232*
Television and Its Portrayals of Blacks   *235*
Implications for Research   *243*
Conclusions: Shall the Legacy Survive?   *245*

11. **Social Trace Contaminants:
Subtle Indicators of Racism in TV**   249
Chester M. Pierce
How a Black Watches Television   *251*
Advertising Behavior   *256*
Conclusions   *257*

12. **Psychological Effects of Black Portrayals
on Television**   259
Sherryl Brown Graves
Introduction   *259*
Television Content and Black Portrayals   *262*
Television Viewing Patterns   *263*
Effects of Content Featuring Blacks   *264*
Summary   *277*

13. **An Aerial View of Television and
Social Behavior** *291*
Stephen B. Withey
An Aerial View *294*

APPENDIXES
 I: Members and Staff of the Committee on Television
and Social Behavior *303*
 II: A Profile of Televised Violence *305*
III: Report of the Study Group on the Entertainment
Function of Television, March 4, 1976 *323*
IV: Report of the Study Group on Television and
Ethnicity, March 2, 1976 *331*
 V: Description of Television Programs Cited *337*

Author Index *347*

Subject Index *355*

Preface

Shortly after the publication of the Report of the Surgeon General's Scientific Advisory Committee on Television and Social Behavior (1972), the National Institute of Mental Health—which had sponsored the $1.8 million of research commissioned by the Advisory Committee—convened a workshop in order to solicit advice on how to follow up on the work and report of the Advisory Committee. A major topic of discussion was the necessity for a profile of televised violence that would measure the extent and nature of violence portrayed on television. Among the suggestions that emerged from the workshop, which strongly urged NIMH to take the initiative in the conduct of further research, was the participation of a nongovernmental group such as the Social Science Research Council (SSRC) in the planning and instigating of further research on television and social behavior. In particular, the consultants to the NIMH believed that an SSRC committee might be a good vehicle for considering the advantages and disadvantages of various measures of televised violence and for suggesting research directions beyond the then-fashionable emphasis on children and violence (Cater & Strickland, 1975; NIMH, 1972).

SOCIAL SCIENCE RESEARCH COUNCIL

As a not-for-profit organization devoted to the advancement of research in the social and behavioral sciences and with a long history of stimulating research on a wide variety of topics (Sibley, 1974), the Council was receptive to this suggestion. Supported by grants from private foundations and governmental agencies, the Council's multidisciplinary research planning and appraisal committees are appointed to consider means of advancing research in selected areas of current significance—areas that are deemed ready for focal attention because of rapid expansion of research, the availability or need for new research techniques, the convergence of interests of social scientists from various disciplines, or the applicability of new theoretical perspectives. The members of these committees are selected on the basis of their interest and expertise and serve voluntarily without financial compensation. Each committee is assisted by one or more members of the Council's professional staff, who are themselves social scientists.

The Committee on Television and Social Behavior was appointed by the Council in December 1972, and a proposal for a 3-year program of activities was submitted to NIMH. (See Appendix I for a list of committee members.) In general terms, the committee was established to assess the current state of research in the area and to plan and stimulate new research. The committee was charged with the immediate task of conceptualizing and giving scientific overview to the research required for the development of a multidimensional profile of violence in television programming. The committee was also charged with: (1) examining and clarifying the major theoretical models underlying research on the long- and short-term effects of television on children; (2) examining methodological issues involved in this research; (3) planning and stimulating needed research on the short- and long-term effects of televised violence on children; (4) planning and stimulating research on socially and behaviorally important effects of television viewing on children, other than the linkage of televised violence and children's aggression; and (5) exploring the feasibility of research on the institutional context in which the content of television programming is determined. These goals were to be pursued through periodic meetings, commissioned papers and reports, and topical work groups or conferences organized by the committee.

While the committee was considering as its first task the pros and cons of a violence profile (see Appendix II), there was early agreement that its program of activities should not be limited to the influence of television on children and certainly should not be confined to the influence of televised violence. Strong interest was expressed in considering television's influence on the behavior of adults and of such special audiences as the elderly and minority groups. This early decision led to the sponsoring of a workshop on the entertainment functions of television (see Appendix III and Tannenbaum, 1980) and to another on televised portrayals of blacks (see Appendix IV) as well as to exploratory efforts toward the study of how programming decisions are made.

Another early decision resulted in a narrowing of the committee's agenda. It soon became apparent that it would not be necessary for the committee to undertake a general literature review and assessment of the field because such a project, by George Comstock and his colleagures at the Rand Corporation (Comstock, 1975; Comstock & Fisher, 1975; Comstock & Lindsey, 1975), was already underway. In addition, a major conference assessing general needs for research—planned jointly by the Ford Foundation, the John and Mary R. Markle Foundation, and the National Science Foundation—resulted in a large conference in November 1975 at Reston, Virginia (Andersen, Comstock, & Dennis, 1976; Comstock, 1976). Thus, rather than duplicate these efforts, the committee decided to devote its attention first to the violence profile and then to other selected topics, building where possible on the activities and products of such programs as those of the Rand Corporation and the Reston Conference.

The present volume is a reflection of these two early decisions. It consists of a series of essays on topics other than violence and children (although children are not ignored), and it presents the reports of the study groups organized by the committee. Together with Percy Tannenbaum's volume on *The Entertainment Functions of Television,* it summarizes the various conceptual, substantive, and methodological issues that dominated the committee's attention.

The committee owes a debt of gratitude to the many consultants who gave it wise counsel. In particular, the committee wishes to express its appreciation to George Comstock, George Gerbner, Larry Gross, Paul Hirsch, Elihu Katz, and Robert Liebert for their contributions and participation in its discussions of the television violence profile and the institutional context of television. In addition, the Council and the committee would like to express their appreciation to the National Institute of Mental Health and to David Pearl and Joel Goldstein of the Institute for their support and encouragement.

STEPHEN B. WITHEY
RONALD P. ABELES

REFERENCES

Anderson, K., Comstock, G., & Dennis, N. Priorities and recommendations. *Journal of Communications,* 1976, *26*(2), 98–107.

Cater, D., & Strickland, S. *TV violence and the child: The evolution and fate of the Surgeon General's report.* New York: Russell Sage Foundation, 1975.

Comstock, G. *Television and human behavior: The key studies.* Santa Monica: Rand, 1975.

Comstock, G. Setting the stage for the conference on priorities. *Journal of Communications,* 1976, *26*(2), 95–97.

Comstock, G., & Fisher, M. *Television and human behavior: A guide to pertinent scientific literature.* Santa Monica: Rand, 1975.

Comstock, G., & Lindsey, G. *Television and human behavior: The research horizon, future and present.* Santa Monica: Rand, 1975.

National Institute of Mental Health, *Report of special consultation on the development of measures of tv violence.* Washington, D.C.: June 2, 1972 (mimeographed).

Sibley, E. *Social Science Research Council: The first fifty years.* New York: Social Science Research Council, 1974.

Surgeon General's Advisory Committee, *Television and growing up: The impact of televised violence.* Rockville, Md.: Department of Health, Education, & Welfare, 1972.

Tannenbaum, P. H. (Ed.). *The entertainment functions of television.* Hillsdale, N.J.: Lawrence Erlbaum Associates, 1980.

1 Beyond Violence and Children

Ronald P. Abeles
National Institute on Aging (Bethesda, Md.)

People are attracted to the subject of television and social behavior for a variety of reasons. Some are challenged by the promise of the medium and point to its potential for expanding our experiential horizons through our almost instanteous and simultaneous exposure to people and events around the world. There are others who express concern about the possible harmful effects of television. They raise the specter of its potential for constricting our intellectual and social horizons by converting active human beings into passive and highly influenceable "vidiots." However, both critics and proponents share the perspective that we are all embroiled in mass media environments that require improved understanding if we are to comprehend complex societies and if we are to propose sensible public policies for the mass media.

The present volume is a reflection of this common perspective and is offered in the hope that it may stimulate others toward a better understanding of television and social behavior. It is not meant to be a comprehensive overview of the field but rather a set of selected contributions that have been informed by and grown out of four years of discussions among the members and consultants of the Social Science Research Council's Committee on Television and Social Behavior. For the most part, the following chapters are speculations about the problems of doing research on television, about the contributions that particular perspectives might make to the study of television, and about the directions that television research might most productively take.

The major theme to emerge from the committee's deliberations was a strong interest in moving beyond the traditional focus on violence and children. Although continued research on this topic was considered to be still valuable, it was also believed that the time was ripe, if not overripe, for expanding the range of independent and dependent variables and for enlarging the heterogeneity of the populations studied. Thus, beyond the consideration of the advantages and disadvantages of a profile of televised violence (Appendix I), the topic of violence and children is almost completely absent from the program of the committee and from the chapters comprising this volume. Instead a variety of other substantive and conceptual issues were considered as examples of the array of topics awaiting the media researcher's attention. In the following pages, synopses of these substantive and conceptual issues are offered as a means of introducing the chapters comprising this book and as a means of summarizing the committee's other activities.

ENTERTAINMENT AND TELEVISION

Although most of the research dealing with the mass media generally, and television in particular, has focused on direct or mediated learning from communications messages, it has overlooked one of the more salient facts of media consumption: Most of the deliberate exposure of most people to television is motivated *not* by a desire to "scan the environment in order to seek information" but rather in search of something vaguely referred to as "entertainment." Given this, it is surprising that there has been so little work on the meaning of entertainment in everyday life, especially in regard to the factors leading to exposure to television and hence its potential influences on vast numbers of individuals.

Consequently, Percy H. Tannenbaum (University of California, Berkeley) organized a study group to develop and to share their thoughts on this issue. Among the questions that the group was asked to consider were: What is meant by "entertainment" and what is to be included under and what excluded from this rubric? What is it about entertaining materials that provides positive incentives for people to seek it out, often at the expense of other desired activities? Is there anything special about television that enhances these positive functions (other than the obvious fact that it provides a less costly means of access)? What are the by-products of "being entertained"? Are they a set of conditioned responses we produce to certain patterns of stimuli; reactions experienced, labeled for the moment, and then promptly set aside; or do the effects of "entertainment" tend to linger on and affect one later? In what ways, if any, does the purely entertainment function help or hinder the mediation of other effects of the message and subsequent behavior?

Obviously these are questions that could not be answered in a single meeting. The purposes of the sesssions were to stimulate thinking on the topic and, it was hoped, to encourage empirical research. One direct product of the discussion was the preparation of a set of commissioned papers, which will be published as a companion volume to this book. A summary of the study group discussions is included in the present volume as Appendix III.

TELEVISED PORTRAYAL OF ETHNICITY

In American society, television programming may be a major source of information about other racial and ethnic groups. Unfortunately this information is often incomplete and stereotyped and hence may lead to misconceptions about one's own and about other racial or ethnic groups. In addition, television presentations may influence the self-concepts of children and adults whose racial or ethnic groups are depicted in simplistic or stereotyped fashion. These attitudes and self-concepts may or may not promote the mental health and positive social functioning of television viewers. Although there is a growing literature on the televised portrayal of minorities and on the consequence of that portrayal, little is as yet known about the functional importance of television in the formation of such attitudes and self-concepts or about the way in which people deal with such information, much of which is presented through subtle cues within a television program.

With these issues in mind, Aimée Dorr (University of Southern California), Irving L. Janis (Yale University), and Chester M. Pierce (Harvard University) convened a study group of researchers and television industry members to schematize the types of cues that may influence self-concepts and attitudes; to explore the ways in which viewers recognize (or fail to recognize) and process these cues; to consider content characteristics, viewer characteristics, and relationships between the two that may predict differential outcomes; and to suggest strategies for promoting research on this topic. In order to limit the discussions to some degree, the study group confined its considerations to the forces within the television industry shaping the portrayal of *blacks* and to the potential consequence of such portrayals for black and white Americans. (A summary of the meeting is presented as Appendix IV to this volume.)

However, additional funding was made available to the study group participants and to others to convene smaller meetings under their guidance. These meetings were intended to broaden the participation in these discussions and to encourage further consideration of these issues in regard to other minority groups. Such smaller meetings were eventually convened by Gordon L. Berry (University of California, Los Angeles), James Blackwell

(University of Massachusetts, Boston), Mary S. Harper (National Institute of Mental Health), and Domingo Nick Reyes (Institute for the Study of the Hispanic American in U.S. Life and History, Alexandria, Virginia). In addition, as a result of his participation in the committee's study group, Berry convened a two-day conference with funds obtained from NIMH.

The committee's interest in television's portrayal of blacks is also reflected in three chapters of the present volume. The triad begins in chapter 10 with Gordon Berry's summary of the historical background of blacks in the United States, his description of their portrayal on television over recent decades, and his consideration of some implications for research resulting from his analysis.

The middle chapter (11) in this triology is by Chester Pierce and can be read in three ways. From one point of view it describes a black psychiatrist's acute sensitivities to nuances in the television treatment of blacks. Whether or not large numbers of black viewers perceive these nuances at a conscious level is unimportant. From another point of view the chapter reflects the difficulties of television executives, producers, and directors in negotiating their way from the climate of the early portrayals described by Berry to one that attempted to avoid offending the sensibilities of a highly perceptive viewer. Finally, the chapter may be read as a set of hypotheses about the subtle ways in which television fare may influence its viewers. From this last perspective the chapter suggests how models drawn from medicine may clarify the study of the effects of television.

The final chapter on television and blacks offers a good springboard for anyone coming into the field. In chapter 12 Sherryl Browne Graves (New York University) describes thoroughly the current research on the psychological effects of black portrayals on black and white children and adults. Although her discussion addresses the images of blacks presented on television and their effects on black and white viewers, it is probably valid to draw lessons and ideas from this material for the understanding of the effects of television portrayals of other American minorities.

TELEVISION AS AN INDUSTRY

Early in its deliberations the committee emphasized that the study of how television content comes to be what it is represents a critical part of the larger problem of determining television's social and cultural effects. The committee's major concern was how the organization of the television industry influences the nature of television programming. The principal assumption underlying its discussions was that the organizational arrangements peculiar to American television have resulted in particular types of programs (e.g., action–adventure, situation comedies), in particular formats (e.g., 30-minute presentations with commercial breaks), in particular thematic content (e.g., violence), in particular target audiences (e.g., mass appeal), and in particular goals (e.g., entertainment over education).

Given this concern and this assumption, the committee's discussions addressed two broad topics. The first centered on the institutional structure of the television industry in terms of its relationships to other media (e.g., radio, movies) and to other organizations (e.g., governmental regulatory agencies, citizen pressure groups). The second focused more on the internal organization of the industry (e.g., the organization chart of a television network). In both instances the key question was how decisions are reached about what types of programs will appear on television.

Despite great interest in this topic, it proved extremely difficult to organize a sustained, productive activity focusing on the television industry. A few informal and exploratory contacts were made between the committee and television industry personnel, including a productive discussion as part of the Study Group on Television and Ethnicity (see Appendix IV). However, it became evident that the investigation of such a sensitive topic as program decision-making would require the development, over a long period of time, of a sense of trust in the research community by members of the television industry. This is probably best accomplished through the establishment of individual relationships between researchers and television personnel. Unless members of the committee were either willing themselves to undertake a research program or could identify others who were, it seemed counterproductive to pursue the initial overtures. That is, if the committee appeared to have been "investigating" the television industry, this might have aroused the suspicions and fears of individuals in that industry and made access by other researchers more difficult.

Consequently the committee opted not to pursue the topic through such a formal mechanism as a conference or study group. Instead, it invited Elihu Katz (Hebrew University of Jerusalem) and Paul M. Hirsch (University of Chicago) to discuss their research on organizational aspects of television. In addition it arranged Hirsch's attendance at an informal discussion with two television executives that was sponsored by the committee. The formal products of this line of interest are four chapters in this volume.

In chapter 4 Herbert H. Gans (Columbia University and the Center for Policy Research) probes the relationships among television, its audience (not just individuals) in the larger society, its suppliers, and its researchers. This fresh examination of contexts, settings, and interactions adds to the skepticism surrounding some of today's currently held conclusions but also adds provocative perspectives on where research on television might go. In the following chapter (5) Hirsch examines the organizational and economic context in which the television industry operates and arrives at several implications for research. Much of his treatment of audiences complements those of the preceding chapter as does his general humanistic approach to the study of the television industry.

Leo Bogart (Newspaper Advertising Bureau, New York) goes deeply into the television industry in action by tracing the course of events related to the multisided debate over television violence since the publication of the surgeon general's 1972 report. Chapter 6 offers a detailed account of these events, and also provides insights into the general functioning of the television industry. In the final chapter (7) on this subject, Hilde T. Himmelweit (London School of Economics and Political Science) presents a conceptual model of the interdependence among the television industry, other social institutions, and its audience. This model is developed and examined within the context of the British television industry, which offers a cross-cultural context for considering the consequences of different institutional arrangements for television broadcasting.

CHILDREN

Although its emphasis was on adults, the committee also considered the continuing need to study the interplay between children and television, but within the context of behavioral domains besides violence. Two of the chapters reflect this concern with broadening the field of study to include cognitive and social–emotional aspects of behavior.

In chapter 8, Irving L. Janis (Yale University) draws our attention to two areas: the influence of television on young people and the "picture of the world" provided by entertainment programs. He considers in some detail the possible influence of television on both the process whereby people make decisions and the options posed and the decisions made in such major choices as careers, marriages, and life styles. The major question addressed is whether or not television programs provide models or scenarios for how to make decisions and for the desirability of certain options over others.

In chapter 9 Aimée Dorr addresses the limits to a child's understanding of television and the implications of those limits for the possible effects of television. By taking us into the mind of the child viewer, she explores the child's growing sophistication and understanding of the medium and its messages. In so doing, she outlines the state of the art in this area as well as suggesting the direction of current research.

CONCEPTUAL AND
METHODOLOGICAL PROBLEMS

A recurring theme in committee discussions was the problem of conceptualizing "television effects." It is quite easy to develop a large list of potential effects, ranging from changes in the behavior of individual persons to alterations in the structure and functioning of whole societies. The difficulty lies in conceptualizing *how* these effects might operate. Are they direct or mediated by other variables such as ethnicity, viewing situations, or personality? Are there complicated feedback loops between television and a person's behavior such that it does not make sense to think in simplistic cause-and-effect terms? Both at a conceptual level and a research operationalization level, how are the potential effects of television as a medium to be separated from the effects of the messages it transmits (i.e., McLuhan's "the medium is the message")?

These types of questions quickly lead to concerns about how to study such a phenomenon as television. That is, when a stimulus is everpresent, how can its effects be weighed? Children born today in the United States have the potential of being exposed to television almost from the moment of their birth, of viewing television daily for the rest of their lives, and of interacting regularly with other confirmed viewers of television. Consequently, serious questions must be raised about the possibility of ever teasing out the effects of television, because it is confounded with many other influences on the person. Experimental attempts to assess the influence of television either by limiting people to particular types of television content or by depriving them of viewing altogether lead to epistemological problems. In the former case, for example, is the effect due to viewing a particular diet or to being deprived of another viewing diet? In the latter case, because total deprivation is extremely unlikely to occur naturally, how generalizable are any results? In short, the problems faced by television researchers in teasing out effects are quite similar to those faced in the study of the effects of schools or of parents on children. It may well be beyond our current methodological abilities to disentangle the complicated web of interactions.

These conceptual and methodological discussions are reflected in some depth in the present volume. Stephen B. Withey (University of Michigan) proposes three perspectives from other fields of inquiry that might serve as stimuli to new ways of conceptualizing research on television. His chapter (2) suggests the use of ecological, cultural, and "scripting" concepts as general orientations to television research. Jack M. McLeod and Byron Reeves (both of the University of Wisconsin at Madison) systematically lay out in chapter 3 the methodological issues confronting would-be researchers of the effects of television. They also answer the common question of why, after all the research and public clamor about television's effects, we have only probabilistic answers.

In the concluding chapter of this volume, Withey presents a summarizing, aerial view of research on television and social behavior. He notes conceptual, substantive, and methodological areas in which research has been relatively heavy, points to others deserving of more attention, and simultaneously suggests the strengths and weaknesses of current and possible future research.

In conjunction with the preface, the preceding material was intended to serve as a brief summary of the concerns and activities of the Committee on Television and Social Behavior. As such, I hope it provides a context for the following chapters, which were written by members of the committee and by a few consultants to the committee. The chapters have been grouped into two sets of six. The first (chapters 2–7) deals with relatively broad and abstract issues such as general perspectives for understanding and studying television, and the second (chapters 8–12) focuses upon relatively narrower and more concrete issues such as the consequences of particular portrayals for particular audiences. In reading the chapters it will become apparent that the volume is not a unified "final report" that presents the consensus and position of a committee of experts. It is not intended to be such a report. Rather, it should be read as a collection of essays that have been influenced by a common experience. They are being offered in this volume as a means of sharing that experience with a wider audience and in the hope that they may stimulate further research into television and social behavior.

2 An Ecological, Cultural, and Scripting View of Television and Social Behavior

Stephen B. Withey
University of Michigan

It would seem that in one way or another, proximally or distally, directly or deviously, significantly or inappreciably, everything in this world is related to everything else. However, no scientific inquiry can assess very much or take very many factors into account. Any study tears out of reality a selected group of variables and assumes or hopes that these aspects are more significantly related to each other than they are to the multitude of unconsidered variables.

Changes in theoretical orientations are most often evolved by challenging assumptions about the adequacy of a selected set of variables. Sometimes the challenge is to redefine or realign the variables. More often the challenge is to change the boundary of the set, severely restricting it or including other, unconsidered variables that were heretofore regarded only as more or less irrelevant components of the stable environment for the system of variables isolated for study.

Studies of television and viewers have with great frequency restricted their inquiry to variables of television content (the message) and their relations to viewers' perceptions, feelings, recall, and behavior. Much of the so-called effects research follows this pattern. Expansions on this restricted approach have included lengthening time periods for effects to surface. Others have added variables from the viewing situation or added to the differentiating characteristics of viewers. A few have looked at behavior precipitating conditions outside the viewing occasion. Occasionally researchers have asked how televisions fare comes to be. Such questions explode the area of inquiry far beyond the home viewing room. Similar enlargements occur when studies look at social consequences that are immeasurable in projects focusing on a few laboratory viewers.

Most of the chapters in this volume turn our attention to various sets of variables isolated from or overlapping with the areas selected by others. Differing selections of significant variables make for differing interpretations, models, and theories of what is going on. But it is challenging to have these various conceptual systems and approaches. Their integration is far from obvious at this stage, but there is a nagging and tantalizing suggestion that some moderately encompassing theory of media communications and their social effects might be evolving. If that idea is too distant in realization, at least different and competing perspectives or approaches can stimulate proponents.

In an effort to stimulate my own thinking I took a different tack. Instead of developing ideas about what is related to what or, synonymously, what set of variables relevant to television and social behavior belong in a researchable system, I turned the task on its head. I looked for models about social behavior, not used heretofore in television research, that might well be applicable and perhaps fruitful if tried as a conceptual approach to studying television and social behavior. The following three approaches are candidates for experiments on their usefulness.

AN ECOLOGICAL VIEW

Ecology deals with the interrelationships between organisms and their environments. As such, its models offer an invitation to any student of the complexities of environmental impacts—including the media environment. An ecological perspective provides an example of an approach to a subject matter in which there are chains of effects, such as food chains and water cycles, and complex patterns and interdependencies. The buildup of damaging pollution is matched by the buildup of supportive systems that encourage growth, health, and bounty. Present conditions not only are sustained by complex networks of processes and interactions and dependencies but also are related to prior circumstances, not only in simple lines of cause and effect but also in degree of restriction both of freedom and in the setting of boundaries for options and choices. Interesting concepts such as dosages or diet supplements may be more than an analogy. Certains inputs may be harmless under brief exposure but deleterious if they are sustained and if exposure occurs a certain number of times within a duration of much longer length. Such a perspective is tried in Chapter 11 by Pierce. Similarly certain beneficial inputs may have no potency until they achieve certain strengths, frequencies, proportions, or durations. Ideas of niches with special environments and the concept of potentiated inputs, which are harmless separately but harmful in conjunction, are also suggestive metaphors.

One area in which ecological models would seem relevant is the moderately well-researched study of diffusion of innovations through society. Such studies have stressed the central importance of communication and the role of opinion leaders. The fact that such leaders may be pioneering people has been stressed, but the idea that they are trying to maintain their position by means of various organizational and institutional behaviors has not. The emerging orientation is an ecological systems approach that regards as intertwined research, technological development, organizations, institutions, production, and the "market." Roles, communication, individual behavior and so forth become aspects of the dynamic processes involved with no one factor viewable as an independent or dependent variable (Saint & Coward, 1977).

Ecological approaches are not characterized only by a recognition of complex interdependencies and tangled chains or sequences of effects. The most productive (Odum, 1977) perspective seems to be one in which some single concept or common variable is traced or followed, be it food, habitat, oxygen levels in the ocean, chemical contaminants, or territory. One can imagine tracing the content and meaning of a new event through the channels of communication or tracking a program idea through the networks. Or one might examine the media "food" for a particular minority or the competition for television territory.

One twist on ecological orientations has involved economic models of resource allocation processes. Settings are viewed in terms of the economics of consumption and production and of mechanisms for bringing into balance producer–consumer activities (Rapport & Turner, 1977). One can regard television programs as the product and the audience as consumers or the audience as the product and the advertisers as the consumers. The struggle for primary rating looks like the optimizing behavior of economic models, but for the audience a more suitable notion may be that of "satisficing." A satisficer examines alternative courses of action and then chooses the first one that meets a set of *minimal* requirements. This simplifies behavior in complex, uncertain, and variable situations (Simon, 1957). Such an idea may characterize the behavior of both producers and program consumers in many circumstances.

Models in one discipline are not necessarily suitable or adaptable to different subject matter. But the ecological model seems to be more than a handy metaphor for television. Public television is struggling for "species survival." The telecommunications industry is affected by new technologies, and the whole system adapts. Industries and other institutions compete with each other for some balance. Parts of chapter 5 by Hirsch adopt a somewhat ecological orientation.

CULTURAL APPROACHES

Culture involves the total pattern of human behavior and its products, including meaning, symbols, and artifacts. Sometimes "culture" is regarded as the dominant pattern in a society, and sometimes the meaning of the word has been restricted to the idea of "most cultivated." Television is certainly an artifact of our society, and television fare certainly deals with the meanings, symbols, and patterns of our behavior. When one considers the cultural impacts of television or television as a component of our culture, it is relevant to conceptualize the aspects of culture that are affected. Culture might be a concept that encompasses most television fare. Pop culture, haute culture, and folk culture or a sports subculture do not superficially seem to have much in common. What constitutes a cultural group is also not obvious without some arbitrary definition involving government, language, territory, interests, or customs.

Bennett (1976) proposes an idea that suddenly makes the concept of culture less diffuse and, from a research point of view, more manageable. The idea is that culture can be regarded as the *precedents* (the ways people think they and others will behave) and *aspirations* held and shared by a group. Such an approach allows one to define a cultural group by the degree to which its aspirations and precedents are similar or complementary. A group defined some other way might turn out to display cultural disparity or conflict. Such perspectives on culture could include the law, art, ethics, a society's incentives, reward priorities, notions about equity and justice, and dreams of success. The core idea is "images." What actually happens in courts, factories, playgrounds, offices, and homes contributes to these images, but images are only partially congruent with the empirical reality of what has become custom and what can realistically be accomplished. Forms of popular adaptation to common situations and problems are the substance of a shared culture. The aspirations can supply the driving force and the impetus to change; the precedents can act conservatively.

Such concepts make the mass media and particularly television, with its direct and covert messages about the way things are and the way things could be, a dispenser of popular culture in a much broader sense than the usual connotation of "pop" culture. What is regarded and reported as news, what is suggested in the commercials, what is portrayed in fiction and what is given prime display—all communicate aspects of culture from this point of view and do so in the direct form of "images".

Firth (1955) has proposed that there are temporary organizations of these images (and their implied roles, behaviors, etc.) that change quickly and frequently to cope with equally labile circumstances. He also suggests *structures* of these images that are more deeply rooted and longer established and that evolve slowly and show considerable stability. Thus one can speak of a changing but shared current culture as well as a relatively stable current culture. Television and its allied media may show rapid and readily discernable influence in the former yet slow and complex impacts on the latter more stable structures.

SCRIPT MODELS

A provocative approach to analyzing programs is offered by Goffman (1974). He analyzes dramas, spectacles, contests, and ceremonials; material that is the substance of television fare; material that includes both the world of fiction or fantasy and the semireal world of arranged, staged, and produced events. In his model a drama frames and structures an aspect of life by creating a portrayal that is a way of looking at a piece of the world. The flow of events in everyday life is often hard to interpret and difficult to understand. The dramatist selects what is significant, what creates the plot, and what provides a sense of outcome or resolution. The drama declares that this topic is socially significant. It merits presentation and it deserves an audience. What happens in the play is a series of happenings in an unusual and meaningful relation of cause–effect of sequential development. This sense of meaningfulness is not generally characteristic of real life events, in which the meaning of changing circumstances is usually quite ambiguous. Thus a program of fiction points to what is significant in a culture (the outcome) and elaborates what is meaningful in the course of events (what influences what) toward a significant crisis and denouement.

In this sense the popular fiction of television and the other media have an effect in providing ideas of what is significant and meaningful out of the less structured stream of life's events. (See chapter 8.) Similarly a ceremony, a ritual, a demonstration, or even a replay or news report frames and organizes a real event by providing significance and giving meaning to its context and content. Many deceptions, Goffman claims, work because they create the frames of significance and meaning without any validity. Contests provide a frame like a drama or a ceremony, but the scripting is less set and rehearsed; but unlike average life they are produced within a time frame during which the issues will be settled and a criteria is provided by which to judge success—the rules of the game and the scoring that settles the outcome.

All television products are framed in this sense, and what they accomplish is not just a portrayal of this or that scene and event and the display of this or that set of characters. The frame, the format, the style, the time limit, and so forth all communicate the significance of the components and the meaning of their relationships, sequence, outcomes, and interactions. It is this structuring that adds communicated content to the episodes and elements of television products. It may well be that these organizations of content are more important characteristics of the messages of television than the components taken alone. Violence, justice, love, survival, or winning are examples of themes running through many televised programs. Aspects of these themes are given significance. Falling in love or getting married ends the story. Winning the contest or capturing the criminal closes the program. Meanwhile, what is meaningful in the course of developments is woven into the story and action line. What leads to violence or love or success? What are the consequences? How can the events be understood? These meanings and significances are also social products of television broadcasting. The strength of this model is that it comes to grips not just with a way of analyzing "staged" performances from news broadcasts to sports events and dramas but with a means for judging their impact.

Aspects of this model are now showing up in thinking about attitudes. Attitudes have usually been seen as evaluations (like–dislike, support–oppose, approach–avoid, etc.) and associated affects or feelings, of some object, act, person, group, condition, and so on. Messages or experiences are seen as possibly changing attitudes, but communication effects get bogged down in the swamps of situational contexts in the message itself and in one's exposure to it. Attitudes, in turn, have been seen as directing behavior but attitude effects are also constrained by situational characteristics in which behavior occurs. Attitudes that are learned by being told or shown something lack something in experiential context, and they do not have the same influence on behavior as do attitudes that are learned from direct participatory experience. Certain repeated personal experiences are crushingly overlearned. Even the limited spectator involvement in a picture such as *Jaws* can create an attitude of fear or anxiety better than just being told something about the behavior of sharks. What happens is that something, an event or several events, is scripted into a scenario, and Abelson (1976) proposes this idea as the more appropriate organizational form for the consideration of attitudes. Like Goffman's frames, Abelson's scripts are structures of bits of information linked into vignettes or episodes that are "written" into a script with significance and meaning.

He proposes three processes, the first being the organization of actors, objects, scenes, and so forth into an experience or incident. Several of these can be organized into similarity groupings and placed in categories based on one or another criterion of perceived similarity. Features of incidents help not only to group similar experiences but also to differentiate contrasting ones. With a growth of experiences an individual is able to organize these various stored scripts into hypothetical options and alternatives in a sort of decision tree or "program" through which he or she is able to exercise some control over what script is to be played in a particular situation or context. One develops a kind of repertoire for dealing with aspects of the world. A repertoire is a more complex idea than attitude.

A similar structure is proposed by Powers (1973), who sees a perceptual hierarchy in which objects and people are perceived in the context of acts, motions, and changes that are in turn seen in the context of an event scripted into a beginning, development, and ending. The event is seen in the context of a staging in which causes, consequences, and other relationships provide a temporal, spatial, and circumstantial context that are all the ingredients of the next hierarchical context of "programs" about what might happen in one or another contingency. Looked at this way, there is also some room for Attribution Theory, which concerns itself with how people find meaning in their world through the causes and consequences they attribute to their and others' behaviors (Kelley, 1973).

For instance, to turn again to overused examples of violence, a gun may be fired and someone is hurt. This mini-event is seen in terms of provocation and consequences and other contexts that make it either similar to others or unique. In this process "violence" (a larger category than just gun-shooting) can be seen as sequentially necessary (meaning) in obtaining certain conquest outcomes (significance). The criminal's violence is lumped with and reinforced by the policeman's violence. On the other hand, a different script would put the policeman's act in the category of social control if he was subsequently kind and helpful.

The idea of scripting is not an abandonment of the concept of attitudes but rather a recognition that they function in a context of scenarios about events and behaviors. *This idea of how attitudes are organized* may prove more fruitful than the concept of attitude as a vector or powerful force determining behavior by itself. The simple concept of attitude requires more consistency than tends to be characteristic of human behavior.

Although the concept of scripting, like writing, seems very linear, like grammar it is equally hierarchical. Many have pointed out various levels at which scripts can be simultaneously appreciated. Objects, actors, and behaviors are given meaning in the context of an event. The events "stage" the characteristics of participants. But an event is itself characterized by its perceived or implied context within the plot, and the plot is seen in the context of other plays and real life experiences.

Certainly scripting and the composition of scenarios is what people do individually. It is what a culture does in prescribing precedents and proposing acceptable aspirations. It is what television writers, directors, and producers do. It is what policy makers and legislators do in their attempts to restrict or facilitate the direction of public behavior. The methodology of script analysis is not well developed, and the obstacles to further development may be severe. The appeal of the idea of such an analysis is that it gives the elements of communications a behavioral meaning and context.

What is being proposed here is that although research on television and social behavior has been extensive, it has tended to be limited in its product. Media influences have been assumed to be rather immediate and direct, so that violence in programs has been assumed to precipitate symptoms of violent behavior. Research has focused on the explicit communicated message individually interpreted, rather than on subtle messages infiltrating social networks. Thus the impacts of the message were assumed to be directed at individuals rather than at social institutions. The technology and organization of the industry with their consequences were considered a given rather than products of social systems and market structures. Some of the current and pending issues within and regarding the telecommunications industry would be illuminated by still further research as well as by some clarification of the values, preferences, and priorities that fuel policy-making in the complex and changing area.

REFERENCES

Abelson, R. P. Script processing in attitude formation and decision making. In J. Carroll & J. Payne (Eds.), *Cognition and Social Behavior.* Hillsdale, N.J.: Lawrence Erlbaum Associates, 1976.

Bennett, J. W. Anticipation, adaptation, and the concept of culture in anthropology. *Science,* 1976, *192.*

Firth, R. *Journal of the Royal Anthropological Institute of Great Britain and Ireland,* 1955, *85*(1).

Goffman, E. *Frame analysis: An essay on the organization of experience.* Cambridge: Harvard University Press, 1974.

Kelley, H. H. The process of causal attribution. *American Psychologist,* 1973, *28,* 107–128.

Odum, E. P. The emergence of ecology as a new integrative discipline. *Science,* 1977, *195,* 1289–1293.

Powers, W. T. *Behavior: The control of perception,* Chicago: Aldine, 1973.

Rapport, D. J., & Turner, J. E. Economic models in ecology. *Science,* 1977, *195,* 367–373.

Saint, W. S., & Coward, E. W. Agriculture and behavioral Science: Emerging orientations. *Science,* 1977, *197.*

Simon, H. A. *Models of Man.* New York: Wiley, 1957.

3 On the Nature of Mass Media Effects

Jack M. McLeod
Byron Reeves
University of Wisconsin (Madison)

An uncomfortable question is often asked of those professing expertise in the field of mass communication: "Why, after all this research and public clamor about television effects, can't we say with greater clarity and certainty whether the medium does or does not affect the behavior of children and adults in harmful or beneficial ways?" An adequate answer to that question, as is usual in science, is uncomfortably complex. Even a partial answer must consider the following: the number and types of potential effects; the complexity of media stimuli; the special problems in documenting effects; the varying strategies of making inferences from evidence; and the peculiar history and current structure of the communication research field.

TYPES OF MASS MEDIA EFFECTS

A common idea of a mass media effect is that some aspect of content has a direct and immediate impact on members of the audience. In the vocabulary of the philosophy of science, this implies that the content is viewed as a necessary and sufficient condition for some effect. Unfortunately, such simple models of causation seldom fit the reality of any area of human behavior, and the study of communication is no exception. We are more likely to find media effects if we understand that the consequences of exposure to media content are likely to be varied and complex.

Television particularly seems to be a topic about which we all have opinions. Perhaps as a result of this, literally hundreds of possible effects of television and other media have been suggested, even if few of these assertions have been backed by solid research evidence. Lurking within this long list are dimensions along which these supposed effects vary and that illustrate the complexity of what we call effects. We can classify variances according to *who* is affected, *what* is changing, *how* the process takes place, and *when* the impact is evidenced.

Micro vs. Macro

Most experimental studies of communication effects restrict their attention to individual audience members and are often criticized for doing so. At least their data and their inferences are consistently at the micro level. For nonexperimental field studies, however, the "who" of media effects is often ambiguous. Quite often, for example, effects are measured in terms of individual audience members, whereas the inferences from those effects are made with respect to the larger society. That is, the micro data gathered from individuals are simply summed to come to macro societal conclusions. For example, if some members of the audience are found to become more informed by using media content, it is sometimes assumed that such information gain must be functional for the society. But societal consequences cannot be inferred solely from estimates of the number of changes. The social location (e.g., social class) of those gaining information must be considered in assessing system consequences. The same problem applies to the term "public opinion," which is often used as a grand reification of individually measured opinions having little connection to their mode of organization in the community or society.

Certain other potential effects are clearly not identifiable from changes in individual behavior alone. For example, research investigating the hypothesis that the media contribute to a "knowledge gap" between the more advantaged and less advantaged groups depends on an analysis of the *relative* gain in information for each status level (Robinson, 1972; Tichenor et al., 1970, 1973). Two communities might have the same average level of knowledge, but yet be very different in the way information is distributed across subgroups in these communities. Similarly, other assessments of effects such as the diffusion of information through a population become meaningful only when plotted against time and in comparison between two social systems differing in macro characteristics such as population density, degree of stratification, and so on. Still other types of research problems seem to focus entirely on the analysis of effects at the macro level—the effects of concentration of media ownership on the quality of news coverage, for example (Gormley, 1976). Other more whimsical illustrations suggest the depletion of societal resources—for example, the idea that intensive television use drains electrical energy, or that the newspaper industry depletes the wood pulp supply, or that the sudden demands of half-time toilet flushing during the Super Bowl seriously affect the municipal water levels.

Direct vs. Conditional

The popular idea of media impact carries with it an implicit assumption that effects are equally probable for everyone in the audience. The model implies an immediate response without either delay or alteration by the emotional states, cognitive processes, or social behavior of the recipient. Such a simple model of media effects would be considered naive by most investigators of communication effects. The bulk of recent research has indicated that media message effects do not appear to have direct or across the board impact. Rather, the thrust of research has been to identify various conditions under which effects are present or not present or present with varying probabilities. Unfortunately, theoretical development and practical understanding of such complications are hindered by a lack of uniformity and clarity as to the labels and meaning of the role of various third "conditional" variables, as we call them, affecting the relationship between exposure to media and effect of that exposure.

One set of conditional variables influencing the relationships between media exposure and effects are those originating prior to media exposure. If the control for the third variable identifies some subgroup or situation in which the effect does *not* take place, we can say that the conditional variable represents a *contingent* condition. For example, there is evidence that suggests that the highly touted agenda-setting effect of the newspaper in "telling the public not what to think but what to think about" (Cohen, 1963; McCombs & Shaw, 1972) is limited to those who consider newspapers to be their major source of political information (McLeod, Becker, & Byrnes, 1974). Thus, newspaper reading is a necessary condition for the operation of agenda-setting. Another type of prior situation is where the third variable acts as a *contributory* condition making the effect of media exposure more likely. For example, the prior angering of members of an audience can make the instigation of aggressive behavior more likely after they have seen a violent film (Berkowitz, 1962).

Contributory and other types of conditional variables can also have impact *after* exposure to media content. A variety of labels are used for third variables operating in this way: hypothetical construct, intervening variable, and mediating variable among others. In psychology a distinction has been made between a hypothetical construct as a theoretically postulated but unmeasured variable and an intervening variable as the measured and/or experimentally manipulated concept clarifying the relationship between a stimulus and effect (MacCorquodale & Meehl, 1948). The intervening variable is often conceived of and measured at a less abstract level of discourse, as, for example, a physiological state in research at the individual cognitive level or a psychological or cognitive state in studies of communication in social systems. The term "intervening variable" is, however, more loosely used in more sociologically oriented research as an alternative to the term "mediating variable." It is sometimes used to mean a social process set off by communication exposure and at other times simply as a third variable affecting the exposure–effect relationship. An example of its use as a social process is to be found in the recent research on the 1976 presidential debates where greater debate exposure appeared to simulate interpersonal discussion. Discussion, in turn, had much greater impact on the political process than did the initial exposure (McLeod, Durall, Ziemke, & Bybee, in press). A large part of the impact of viewing the debates, therefore, can be said to be an indirect effect operating through the stimulation of discussion. Because discussion effects might be expected to be delayed a day or more after exposure, studies of the immediate direct effect of debate watching might well miss the stronger indirect effect. Other research on the 1976 debates found that the perceptions of who won these encounters were not well formulated until several days afterward, when respondents presumably had a chance to read press evaluations (Lang & Lang, in press; Morrison, Steeper, & Greendale, 1977).

Conditioning third variables may also operate at the same time as the communication variable, or their time order may be undetermined. To the extent that media exposure and the third variable operate independently to produce an effect, we can say that each has a main effect (Kerlinger & Pedhazur, 1973).

An example from recent communication research may help to illustrate the necessity of specifying the way in which conditional variables operate. As one explanation for the rather modest strength of the direct relationship between children's exposure to television and the various effects of television, it has been suggested that the relationship is dependent on or its strength is proportionate to the child's perceived reality of the content (Feshbach, 1972; Greenberg & Reeves, 1976; Hawkins, 1977). Perceiving program content to be realistic is assumed to make television information more socially useful and more likely to be assimilated equitably with information from nontelevision sources. Thus all or most of the impact of television exposure operates through the perception of reality as a conditional variable. There are, however, several different ways of specifying *how* perceived reality operates as a conditional variable, and each of these specifications suggest quite different theoretical interpretations of media effects.

Figure 3.1 shows six examples of how perceived reality may operate as a conditional variable. All of the examples show the relationship between the frequency of exposure to television and the magnitude of effects at two levels of perceived reality (I = high perceived reality; II = low perceived reality). The high reality condition in each of the examples is shown with the same difference in effects at the two levels of exposure (as indicated by the distance "a"). The low reality condition is shown with different changes in the

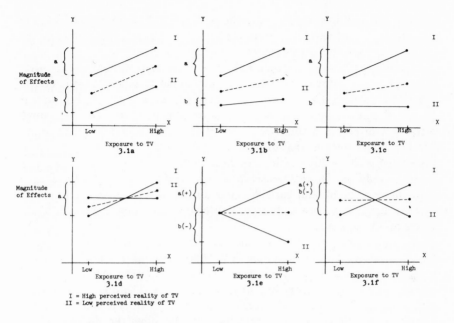

I = High perceived reality of TV
II = Low perceived reality of TV

FIG. 3.1.   Hypothetical example of the perceived reality of television as a conditional variable in the relationship between exposure to television (X axis) and the effects of television (Y axis).

magnitude of effects between low and high exposure (as indicated by the distance "b") to illustrate changes in the theoretical interpretation of the role of perceived reality as a conditional variable in each of the examples. The dotted line in each example indicates the change in effects from low to high TV exposure that would be found if the conditional variable were ignored.

Figure 3.1a shows perceived reality as a conditional variable that operates independent of children's exposure to television. Although the high perceived reality condition results in greater *levels* of effects, there is no difference between the two conditions in the estimation of the change in effects from low to high TV exposure (as indicated by the fact that a = b). If perceived reality were not considered in this case, the estimation of the difference between low and high exposure would not change (as indicated by the dotted line), although an estimate of the level of effects at both low and high levels of TV exposure would not represent either condition. In this case, then, the conditional variable is additive and noninteractive. It is important to note that most studies in which conditional variables are hypothesized make the assumption that the effects of the third variable—that is, perceived reality— are additive in relation to exposure. This is a mathematical assumption, for example, of multiple regression techniques in which no interaction term is introduced. Third variables are typically added to a prediction equation as main effects without consideration of possible interactions.

Figure 3.1b shows an interaction between perceived reality and exposure. The high perceived reality condition results in the greatest change in effects from low to high TV exposure (as indicated by a > b). In this case, if perceived reality were not considered as a conditional variable, some difference in effects at the two levels of exposure would likely be found (dotted line); however, that difference would underestimate the differences associated with the high reality conditon and overestimate those in the low reality condition. In this sense, perceived reality could be viewed as a contributory conditional variable. Accounting for perceived reality would *increase* the ability to estimate accurately changes in media effects at different levels of TV exposure.

A similar overestimation of the low reality condition would exist in Fig.3.1c; however, in this example there is no relationship between exposure and effects in the low reality condition (as indicated by the absence of change in the level of effects from low to high exposure). There is a relationship between exposure to television and effects only for those in the high reality condition. In the sense that perceived reality limits the domain of the exposure–effect relationship, perceived reality becomes a contingent variable. If perceived reality were not measured in this case, there might appear to be a slight relationship between exposure and effects (as indicated by the dotted line); however, that relationship might not be significantly different from zero.

It would also be possible to generalize this example to a situation in which several levels of perceived reality were considered. With 10 levels of reality, for example, it could be that a significant positive relationship between exposure and efffects exists only for the 10th or highest level of perceived reality. In this situation, the overall relationship would almost certainly not be significantly different from zero (the average of nine conditions showing no change in effects and one condition showing a change), although a near-perfect relationshp may exist at the 10th level of perceived reality. Overall, this may not be an interesting finding, because the relative number of individuals in the 10th category would be small. However, in cases of significant rare effects of media (e.g., commission of felony crimes such as murder or rape that are modeled from TV portrayals), the conditional variable may be a crucial piece of information in the explanation of effects.

Another version of this contingent condition is shown in Fig. 3.1d. There is still no relationship between exposure and reality for the low reality condition; however, in this case the relationship between the two conditions is disordinal (i.e., the lines for the two conditions intersect). Although the relationship between exposure and effects in this case would be exactly the same as those in Fig. 3.1c (because the dotted line in each example indicates the same change in effects from low to high exposure), the interpretation and use of the data could be very different. In Fig. 3.1d there exists a critical point at which those in the low perceived reality condition are affected more than those in the high reality condition (as indicated by the exact point of intersection). This would mean that for the low level of TV exposure, those who perceived television to be most real would be affected *less* than those in the low reality condition. A similar problem would exist in this example if perceived reality were ignored as a conditional variable. It is possible that the combination of the conditions would make the relationship between exposure and effects appear to be nonexistent or at best uninteresting.

The final examples, Figs. 3.1e and 3.1f, demonstrate that ignoring a conditional variable may actually preclude understanding the relationship between exposure and effects. In Fig. 3.1e the high reality condition produces an increase in the level of effects from low to high exposure (as indicated by the fact that a is positive), and the low reality condition produces a decrease in the magnitude of effects (as indicated by the fact that b is negative). Perceived reality makes a difference as a conditional variable only for the high level of exposure—augmenting the effects in the high reality group and diminishing effects in the low reality group. In Fig. 3.1f, which is the disordinal version of this example, the two conditions again show either a positive or negative change in effects; however, these effects are found in the low reality condition at low exposure *and* in the high reality condition at high exposure. The important point to consider in these final examples is that traditional correlational procedures would totally miss the relationship, *even if the conditional variable was measured and evaluated in the three-variable relationship.* The change in the magnitude of effects in both cases would average to zero (as indicated by the dotted lines). Although it is difficult to imagine these results actually being operative for the perceived reality of television, they are nevertheless possibilities that may fit other conditonal variables.

From these examples it is clear that merely including conditional variables in analyses of media effects is potentially misleading unless the specific relationships of these variables to exposure and effects are also studied. Unfortunately, once conditional variables are identified as important in the effects process, they are often studied only descriptively or as dependent or independent variables. Perceived reality, for example, has been theoretically hypothesized to be an intervening variable; however, most studies using perceived reality have concentrated on the dimensions of perceived reality; levels of perceived reality for various program and character types; or the ability to predict different levels of perceived reality by accounting for variables such as childhood experiences with the real-life counterparts of television, time context of the program, and overall exposure to television

(Reeves, 1978). In studying other conditional variables that may help explain individual or subgroup differences in how media have effects, it seems important at least to measure and simultaneously evaluate all three variables. Studying conditional variables descriptively or as independent and dependent variables in two-variable designs could lead to errors in conclusions about television effects or to a misallocation of effort, should the variable be of little value in explaining effects. Although we explore conditional variables independent of exposure and impact, we should also study them within the process they are expected to have a role in explaining.

Other communication concepts have undergone a transition similar to that of perceived reality in being seen first as conditional variables and then as effects in themselves. The gratifications sought from media were first treated as correctives to a simple exposure to effect model but later came to be treated as important phenomena without specified effects and most often without any cause (Blumler & Katz, 1974).

Conditional variables may be important beyond their role in clarifying media effects. For example, they may prove to be more likely targets for policy change or social action than is the body of content found in the mass media. Parental intervention strategies, educational programs, and warning messages all may be more viable than any direct governmental control over television programming.

Content-Specific vs. Diffuse–General

It is natural to look for media effects that bear a one-to-one relationship to the specific content of the medium. We seek to measure the aggressive responses of children exposed to violent content and to assess the stereotypical cognitions of those paying closest attention to programs portraying biased sex or occupational roles. There is also an unfortunate tendency for many observers to skip the step of actually assessing audience reactions and to infer effects from the content alone. At the time of the Surgeon General's report on effects (Comstock & Rubinstein, 1972) of televised violence, for example, one magazine editor wondered aloud why all that money had been spent to document what was obvious from the content that he could see on his own television screen. From such logic we might see a parallel to the expression "we are what we eat" and assert instead "we become what we see."

The great majority of media effects research also focuses on content-specific effects, albeit with more complicated models than the exposure to effect causation implied by the editor. Not all of the effects attributed to the media, however, are so directly tied to the content. For example, erotic as well as aggressive film content can enhance subsequent aggressive behavior among previously angered subjects (Zillmann, 1972). It is argued that the observed effects may not be so much the consequence of exposure to the content per se as they are the function of the excitatory potential of the communication. Such nonspecific arousal can also enhance more posi-tive effects such as music appreciation (Cantor & Zillmann, 1973). Nonexperimental research has shown that aggressive behavior can be predicted as well from the *form* of a given television program (e.g., unpredictability of audio and visuals, location and characters, mode of presentation) as from the frequency of violent content alone (Watt & Krull, 1977).

A very different set of examples of effects not tied to specific content can be found in the various studies of the displacement effects of media. For example, Parker (1963) found that the advent of television in a community was associated with a decline in the circulation rates of fiction in public libraries but not with the level of use of nonfiction materials. The heavy dosage of fictionalized content of television apparently displaced fiction books in serving whatever needs such materials fulfill, but the effect was not a direct effect of television per se. A host of studies have dealt with either the replacement of one medium with another, or with displacement in terms of lowering the time devoted to the original medium, and some have examined the displacement effect of television on children's play (Himmelweit, Oppenheim, & Vince, 1958; Lyle & Hoffman, 1972; Schramm, Lyle, & Parker, 1961). Speculation about other displacement effects have included more elusive and less researched criteria such as language skills, reading behavior, impaired eyesight, declining college entrance test scores, and general apathy. What these speculations have in common is that they assert that it is the activity of watching television per se and not its specific content that generates the effect. Finally, we should add to this list of proponents of diffuse noncontent effects the name of McLuhan (1964) and his assertion that it is the form of the medium and not its message that is the critical element in understanding the consequences of media use.

Attitudinal vs. Behavioral vs. Cognitive

The history of media effects research is very nearly the history of attitude change research. For 40 years various source, message, personality and situational characteristics have been studied in relation to their effectiveness in shifting audience attitudes. Much less attention has been given to the more natural function of the media in conveying knowledge, changing various types of cognitions, and altering overt forms of behavior. This imbalance would be less serious if we could assume a strong causal flow, say, from media exposure to knowledge gain (or other cognitive change) to attitude shift to behavior change. Unfortunately, none of these links is supported by strong evidence. The last step, from attitude shift to behavior change, has been given close scrutiny in recent years, and it has been concluded that supportive evidence is lacking (Festinger, 1964; Siebold, 1975). It is obvious that much more systematic research into nonattitudinal effects of media exposure is needed along with how such changes relate to attitude change.

Alteration vs. Stabilization

Another basic distinction among types of effects is between the facilitation of change and its prevention or stabilization of existing attitudes or behavior. Although there has been a substantial amount of work done on immunization against persuasive messages (McGuire, 1964; Tannenbaum, 1967), the overwhelming proportion of studies has dealt with enhancing attitude change. Although nonexperimental survey research on media effects has also concentrated on persuasive attitude change, a dominant inference from the research of the past 30 years has been that the major effect of media exposure has been to "reinforce" preexisting attitudes (Berelson, Lazarsfeld, & McPhee, 1954; Hyman & Sheatsley, 1958; Lazarsfeld, Berelson, & Gaudet, 1948; Star & Hughes, 1950). Unfortunately, extremely gross measures of change were used such that only those who showed large shifts (e.g., conversion from one party's candidate to another's) were counted as changing, whereas all other members of the audience were assumed to be "reinforced." More recent research continues to show evidence of stabilization of attitudes, but by using more varied and sensitive measures of effect other more nonreinforcing destabilizing types of changes also have been shown (Becker, McCombs, & McLeod, 1975; Blumler & McLeod, 1974; Chaffee, 1975; Kline, Miller & Morrison, 1974).

Other Dimensions of Effects

In addition to the five foregoing dimensions there are many other ways in which the various alleged effects of the mass media might be classified. For example, most experimental studies deal with relatively short duration of effects following exposure to a message and do not address the long-term consequences of the change directly. This is true more often than not for all types of research strategies, but it is also the case that many proposed media effects such as the diffusion of information through a population require a longer time perspective of effect. Other dimensions include cumulative versus noncumulative effects and learning of novel versus previously learned behaviors.

The basic point is that each of these dimensions represents possible types of effects that can be put together in a complex matrix of possible effects. If each of the five dimensions we have described were to be treated as dichotomies, 2 to the 5th power or 32 possible types of effects are possible. Among these possible effects, however, only a few dominate the existing research of the field. The combination micro-direct-content–specific-attitudinal-alteration occupies the foreground whereas most other types are virtually ignored. Answers to questions of the extent of media effects may depend on which combination of these cells we are referring to. It follows that adequate answers to media effect questions require more systematic attempts to investigate effect combinations other than those that have dominated attention up to this time. The marketplace of ideas, unfortunately, is likely to keep the focus of attention on only certain types of effects. Equally sobering is the realization that many specific effects simply are undetectable given the current crudeness of measurement available within the social sciences.

COMPLEXITY OF EVIDENCE REQUIRED

We have seen that the diversity of potential effects has contributed to the confusion of evaluating mass media impact. Another major reason for the scarcity of definitive answers about media effects is the complexity of evidence needed to document and to make inferences about such effects. These requirements may differ in degree but probably not in form from the necessities for testing any kind of effect in the social sciences or in other scientific fields: knowledge of the stimulus material, control of its application, assessment of effect, and an understanding of the mechanism or process underlying the effects. In the analysis of media effects we need to accomplish the following:

(1) Assess the media *content* in relation to the expectations about how media have an impact. This is often approached by simply analyzing what is most quantifiable or obvious about the media content under study. What is actually needed is a coding scheme whose categories are isomorphic to (have a one-to-one relationship with) the dimensions and categories used to measure the effect on the audience and the process by which the message is received. Put simply, we need evidence that the audience is reacting to the same things as is the content analyst. The problem is no less for the experimental analysis of messages, if the manipulation is to have an effect or at least one that is interpretable.

(2) Control the *exposure* of the audience to the content. In natural settings this is a major problem, for much of media use is done under conditions of low attention. People frequently watch television without much specific motivation or rational choice about content, and their viewing is often lacking in focused attention (Bogart, 1972; Gans, Chapter 4, this volume; Goodhardt, Ehrenberg, & Collins, 1975). This has some serious implications for the conduct of media effects research. First, the lack of conscious selection leads to imperfect recall by audience members about what programs they

watched. Second, the variation in motivation and in attention makes measures of mere exposure frequency relatively weak indicators of the actual strength of the media stimulus. As a result of these problems, the reliability and validity of measurement of exposure may be lessened, and any coefficient may understate the actual strength of the relationship between exposure and effect. Laboratory experiments, of course, do not suffer from these problems, because they manipulate the message content and maximize attention through control of extraneous conditions. They pose the problem not of threats to internal validity from these sources but rather of external validity in generalizing to the less than perfect conditions outside the laboratory.

(3) Assess the *effect* of media content. As discussed earlier, a great number of types of effect are possible. The question is then to determine which of these can be identified as functionally related to the message. The closer the particular effect to various conditional processes as well as to exposure to the specific content of the message, the more fully the relationshp can said to be documented.

(4) Elaborate the *conditional processes* that help interpret and specify the relationship. Research linking media exposure to effects frequently invokes as an explanation for the relationship an unmeasured process that is assumed to have been stimulated by the exposure to the message. If research findings indicate that children who frequently watch situation comedies with highly traditional family roles also have more highly stereotyped views of sex roles than other children, the investigator is apt to assume that the child is learning such views through the process of identification with one of the leading characters. Direct measurement of identification would help to test this inference against various alternative explanations.

There are abundant numbers of other processes and concepts that have been suggested as modifying or interpreting media exposure to effect relationships. Variables such as the gratifications sought from content should be analyzed as both additive increments altering effects beyond those of exposure and as potential factors interacting with exposure to produce nonadditive effects (McLeod & Becker, 1974), not simply as isolated variables. For the most part, the most interesting communication theory results from the unraveling of these conditions and interactive relationships, not from the simple assertion that the media set public agendas or that children learn from television.

An assumption about a conditional process—namely, that children will learn antisocial or prosocial material from television to the extent they perceive the video portrayals as real—serves to illustrate the inadequacies of incomplete evidence for media effects. Because the perception of reality is both measurable and experimentally manipulable, in recent years the concept has been the subject of several studies that have examined its dimensionality (Hawkins, 1977); its antecedents such as age (Lyle & Hoffman, 1972), socioeconomic background (Dominick & Greenberg, 1970), and real life experience (Greenberg & Reeves, 1976); and its consequent effects (Feshbach, 1972). Despite all the research now in print and the theoretical predictions that perceived reality operates as a conditional variable enhancing the effect of exposure to television content, very little research has actually addressed the three-variable question or attempted to integrate two-variable studies into a comprehensive statement about how perceived reality works (Reeves, 1978).

Our basic point is that evidence for full documentation of media effects requires coordinated codification and measurement from all four sources of variables: content, exposure, effect, and conditional processes. A larger proportion of media research examines only one or two of those variable sources, sometimes making unsubstantiated assertions about change. This leads to inaccurate interpretation and to the disparity of current views about media effects.

Nonexperimental studies of media effects face challenges to the validity of their inferences, because media use is an activity embedded in other activities and its presumed effects can be and often are caused by other activities. A simple positive correlation between media exposure and some effect is a necessary but not a sufficient basis for inferring that the media exposure caused the effect. Many other alternative explanations for the correlation can be advanced. These alternatives fall into two classes: third variable causation and reverse causation. Third variable explanations assert that some other antecedent or concurrent factor may have caused both the media exposure and the effect. For example, children of lower social class status may watch more television and also exhibit more aggressive behavior. We are more confident that our inference is not spurious if we can show that controlling for social status does not make our original exposure–effect correlation disappear (McLeod, Atkin, & Chaffee, 1972).

The second alternative explanation, reverse causation, asserts that the proposed effect may actually have caused an increase in media exposure rather than the reverse. This is difficult to test in the usual cross-section study conducted at a single point in time. Longitudinal designs measuring the audience at two or three points in time represent greater opportunities for testing the direction of causation. Cross-lagged correlation analyses that examine the strength of the across-time correlations of one variable with another can be useful to the causal direction problem (Chaffee, 1972a, 1972b; Kenny, 1973). An opportunity to test for media effects with panel designs is provided when the exposure variable is measured in regard to a particular time-bound set of events such as the Watergate hearings (McLeod, Brown,

Becker, & Ziemke, 1977) or the presidential debates (McLeod, Durall, Ziemke, & Bybee, in press). Here the pre-event levels of the effect variable along with prior measures of potential third variable challenges can be entered as preliminary control in regression analyses, and the test of media exposure to the events is made with respect to the remaining variance in the effect variable not accountable to prior levels or to the other prior variables (Kerlinger & Pedhazur, 1973). It forms a stronger measure of change in the effect variable not possible in static, one-time designs.

COMPLEXITY OF MEDIA STIMULI

A good test of effects depends on the presence of strong stimulus conditions. The ideal characteristics of good stimulus conditions include the following: There should be a stimulus unit signifying the nature of the stimuli and their beginning and end; the stimulus measurement should be well specified and precise or its manipulation sufficiently strong such that we can reasonably expect detectable effects; and the stimulus should be independent of other extraneous stimuli so that effects can be properly attributed. Unfortunately, research on media effects in natural field situations seldom meets any of these criteria.

Natural Stimulus Units

With the exception of highly salient events like the media coverage of the Watergate hearings, the presidential debates, or the Super Bowl, media exposure is most easily describable as a habitual ongoing behavior taking place over an extended period of time. We expect that effects will come largely from the viewing of typical content over a relatively long period of time. As a result, both the nature and the level of abstraction used to describe and measure media exposure is varied and uncertain. Time spent with television is the most frequently used measure of exposure, and, unfortunately, it suffers from being unreliably measured (for reasons described earlier) and also from being dependent on a rough correspondence between viewing time and the "typical" content of the media. It is not surprising that few media effects have been uncovered using exposure time or time spent measures. Beyond time, the selection of a unit of time requires a choice among different media, program types, specific programs, characters, the interaction sequence within programs and so forth. At least in part, the selection of a unit of media exposure is an empirical question that can be ascertained through research. But at present there is no standardization of units in the field, a fact that contributes to the disparity of reported media effect findings.

Strength of the Stimulus

There is an old dictum in experimental research that urges the investigator to "start strong" in manipulating the differences between conditions. The nonexperimental counterpart of this dictum is to find sufficient variance in the natural conditions of the stimulus such that significant relationships with effects might be obtained. This is a potential problem for mass media research where the level of exposure is being measured. For example, there might be insufficient variation in the large amount of violent television content viewed by young adolescents. This would markedly lower the likelihood of finding any strong association with effect variables.

It can be argued that customary nonexperimental research designs are weak in that they depend on natural variation in media behavior within a given community or area within a single country. Admittedly, the strongest differences may be found within countries or communities before and after the introduction of television, but studies of these are rare and may be contaminated by the prior availability of radio and other media (Cramond, 1976; Furu, 1971; Harrison & Ekman, 1976; Himmelweit, Oppenheim, & Vince, 1958; Schramm, et al., 1961; Werner, 1971). A more viable possibility exists in designs that intentionally sample and compare communities on the basis of the "richness" of the available media ranging, say, from New York city with three daily newspapers and a multitude of television and radio stations to a small community with only a local weekly and single radio station (Chaffee & Izcaray, 1975; Chaffee & Wilson, 1976; Kraus, Davis, Lang, & Lang, 1975). Apart from the characteristics of individuals in those communities, the context provided by the media systems could well have important consequences. The detection of these consequences, of course, depends on the ability to control the many social factors likely to be associated with the media richness of the community.

There is a potential problem of insufficient stimulus strength even in field experiments in which a set of messages is disseminated differently to experimental and control communities to study effects. If any manipulated set of messages about a topic on which subjects have preexisting attitudes is nonnovel in that they merely add to the customary level of information flow on the topic, then the expected effects may be only an incremental fraction and any effects may escape unnoticed. The alternative is to stay with novel messages about previously unknown or low salient topics; however, in the latter case the dilemma is that it may be impossible to generate enough salience in the topic adequately to test the effect of the message. A final problem of stimulus strength is that it may not bear a one-to-one relationship to the number of repetitions of the message. Density as measured by the number of repeated messages per unit time may be a more appropriate measure where a very dense set of redundant messages may actually lower the effects of mere exposure presumably because of their irritating quality (Becker & Doolittle, 1975).

Independence of Stimuli

It is important to the interpretation of media effects to be able to specify as precisely as possible what aspect of media content led to any effect found in research. This is very difficult to do at the gross level at which most media effects research has been conducted. Perhaps the most basic question is whether the effect is due to something the media added to or subtracted from "reality" or conversely, whether the media simply acted as a conduit in expediting what the audience would have obtained from other sources. If it is the latter, do we wish to label an effect as a media effect or merely as an effect transmitted by the media?

The production of network news and public affairs programming serves to illustrate the wide variation in the amount of discretion and judgment involved in presenting various types of content. At one extreme is the large amount of selectivity involved in producing the approximately 18 news stories contained in the average evening network newscast. A given story has been chosen from among hundreds of possible items, and each has been condensed to fit the tight time average of a bit over 1 minute per story. A voice-over interpretation has been added in most cases. Clearly differences between viewers and nonviewers of such programs might be attributable to the news judgments of the reporters and editors. At the other extreme with much less reportorial discretion are political party conventions and presidential press conferences, which represent live performances with most of the judgment falling to the party managers and political leaders. In the second instance it is debatable just how much of any effect can be attributable to the media. The research on the effectiveness of television versus classroom teaching serves further to illustrate this problem. Schramm (1962) reports that 83 studies showed television to be superior; another 55 studies found the reverse, with classroom teaching better; and another 255 could draw no conclusions. We suspect that not only is the strength of the contrasting teaching conditions likely to have been weak but that the various studies were apt to have been confounded by variables other than the mode of instruction.

Another problem of stimulus specification derives from the fact that much of the natural audience exposure to media content is embedded in a complex of other social and media behaviors. The analytical problem becomes one of controlling these other sources of potential influence in order to evaluate the media exposure of immediate interest. The 1976 presidential debates offer a good example. During the period of the debates, voters were also learning about the campaign from other information sources such as the daily accounts of the campaign, political advertising, and so forth. Any net gains shown in knowledge, vote preference, or other effect variables might be due to debate viewing; but they might also be due to influence from these other sources. Before strong inferences can be made regarding debate effects per se, correlations between debate viewing and an effect variable require control for other information sources that might be correlated with both viewing and effect changes. Examination of the debates relative to the more customary information sources also is needed to assess any *unique* contribution of the debates.

The presidential debates also illustrate another problem in specifying media stimuli. We might analyze the direct impact of debate-viewing by examining the strength of its relationship to various effect variables measured after the debates controlling for predebate levels of these measures. But this may be too narrow a view of the complexity of the debates as stimuli in that it ignores the likelihood of their initiating a diverse set of social processes that themselves may have effects. Because the content of these processes may go well beyond the issues covered in the broadcast debates, it is likely that some of these social-process-induced effects will not be detectable from the content of the debates alone. Investigators taking the narrower view of effects obviously will be likely to come to less strong conclusions about debate impact than will those taking a broader view that includes such processes as discussion of the debates, following accounts of them in the media, and perceiving one or the other candidate as winning the debates. Research taking the broader view has found these subsequent variables to have greater effects than did the level of debate exposure (McLeod et al., in press).

VARYING STRATEGIES OF INFERENCE

Up to this point we have been discussing the consequences that various research decisions have on the nature of evidence regarding mass media effects. Differing criteria and research strategies will produce very different evidence. Even within any particular body of evidence, however, there is considerable room to make very different inferences about the extent of media effects. The strength and nature of the inferences that investigators make depend in turn on what kinds of risks they are willing to take regarding the evidence. We can draw a parallel to the testing of statistical data where the investigator must make a decision based on research evidence whether to reject the null hypothesis of no difference or to decline to reject that hypothesis. Whichever decision is made, the investigator risks making an error. The question is which type of error will be more damaging or costly, given the purpose of the research.

An analogy can be made to the drawing of inferences from evidence. Here it is an even more general problem than in the testing of hypotheses, because research inferences and conclusions are made not only by the investigator conducting the research but even more often by secondary analysts who have their own values and priorities. Historically, public knowledge of mass media effects has been set more by oversimplified summaries (Berelson & Steiner, 1964; Klapper, 1960) than by more detailed research findings. Two quite different images of media effects have emerged from various reviews of research findings, reflecting which type of error the reviewer worries about most. At the risk of oversimplification, we have called these two types of analysts Type One and Type Two worriers.

Type One worriers fear the possibility of making too strong inferences and consequently tend to accept a null position of no media effects. The evidence they cite tends to be from field studies of persuasion using gross measures of both media exposure and attitude change. The lack of major change and the apparent stability are attributed to selective exposure and selective perception, and the major impact is thought to be reinforcement of preexisting attitudes. If they should encounter a correlation coefficient of 0.30 between media exposure and effect, they are likely to point out that (by squaring the coefficient) this accounts for less than 10% of the effect variance. It is interesting that much of the evidence cited was obtained before the advent of television (Berelson et al., 1954; Katz & Lazarsfeld, 1955; Lazarsfeld et al., 1948) and that secondary citation of the original evidence has tended to move toward limited inconsequential effects or no effects at all.

Type Two worriers hold a diametrically opposing set of concerns in making research inferences. They worry most about overlooking any media effect and frequently cite the difficulties that beset attempts to find even "obvious" effects. At worst they are willing to accept any change taking place at the same time as some media event (e.g., debates, sports events) as evidence of media effects. They seek results among a much more diverse set of effects and may overlook the possibility of obtaining chance findings in such a large number of comparisons. They worry less about the proportion of effect that media variables account for, noting that even a small effect—say 2% in an election campaign—can make a very great difference.

Type Two worriers note too that communication variables tend to be less reliably measured than demographic and other variables and hence that correlation coefficients understate considerably their "true" predictive power. They are not likely to hestiate counting as effects the media acting as mere conduits of public events, as discussed earlier. Their interest includes indirect as well as direct effects. Their definition of media as stimuli tends to break down media exposure into fine units and includes motivation to viewing and social processes set off by media exposure as sources of effect. Finally, they seem to assume that the public spends so much time with television and other media that there must be some kind of effect in at least some segment or subpopulation of the audience. Theirs is an almost unbeatable faith.

Our point is that variance in discussion of media effects comes from more than contrary evidence or the level of conservatism in setting statistical alpha levels in testing hypotheses. The disparity relates to the nature of what is evidence and what kinds of errors are most likely to be harmful. Obviously, we have overdrawn the two types of worriers in order to make our point clear. But rather different sets of inferences are apparent in the literature of the field, and anyone wanting to understand more about media effects should be aware of this. Neither type of worry is completely foolish or unwise. Perhaps one way out of the dilemma is to use the broad research strategies of the Type Two worrier in combination with the basic caution of the Type One worrier.

PECULIAR HISTORY AND CURRENT STRUCTURE OF THE FIELD

It is important for anyone who wishes to gain an adequate understanding of mass media effects to learn some of the peculiar history of the communication research field. The history is indeed peculiar but probably not unique in the social sciences. A rough parallel can be seen, for example, between the development of social psychology as a field and similar stages in mass communication research some 15 to 25 years later.

Changing Models of Effects

Polemical writing about the effects of mass media was frequent in the post-World War I era when massive propaganda impact was attributed to the persuasive content of the media. The mass model of society prevailed in these writings, and the dominant influence mechanism was thought to be simple learning through repetitive messages (Bauer & Bauer, 1960). It was in reaction to this view that the so-called "limited effects" model of media influence developed largely through the voting and functional media use studies at Columbia University in the 1940s (Berelson, 1949; Berelson et al., 1954; Lazarsfeld et al., 1948; Wolfe & Fiske, 1949). This view has become the dominant one in reviews of the field in various social science texts despite the lack of methodological sophistication in its original evidence and a dearth of fresh evidence from the posttelevision era. For example, communication was of little concern in the voting studies from 1952 to 1972, but that situation has begun to be corrected in studies such as those of the 1974 and 1976 election campaigns funded by the John and Mary R. Markle Foundation and the National Science Foundation, under the auspices of the Social Science Research Council's Committee on Mass Communication and Political Behavior.

Although the limited effects model still dominates academic reviews of the field, the public, along with many public action groups, holds to a view that is much closer to a mass persuasion model. This split has resulted in a wide gap between the basic academic research and applied concerns. At least until recently, the limited effects view has also discouraged "visitors" from "outside" social science departments from doing research in this area. Gans (1972) reviewed the mass communication research activity within sociology and other standard social science departments and judged that a "famine" existed. The situation has changed recently, particularly in political science where it has become apparent that the standard party affiliation and socioeconomic variables are declining in their power to predict voting behavior. They reason that the potential for television and other media influences may well be growing (Nie, Verba, & Petrocik, 1976).

It is clear that the earlier models of mass media effects did little to stimulate and develop communication research. Although no single model has emerged to replace them, unquestionably much greater attention will be paid to models that include the specification of conditions under which media exposure will produce effects on certain audience members. These are more complicated multivariate models, but they are likely to be much closer to the realities of mass media influence than are simple models overstating or understating effects on the total public.

Emphasis on Applied Research

A second historical factor limiting the integration of mass media effects findings is the fact that the field arose largely from the study of applied problems rather than as a response to intellectual concerns. Evidence-based research began in the 1930s with sample surveys that attempted to identify the audiences of various mass media, but particularly that of radio. World War II provided an unparalleled stimulus to communication research through government sponsorship of experimental studies of the effectiveness of film and other media in persuading audiences. Sample surveys studying problems relevant to the war effort also touched on issues of media effectiveness. The war thereby served as a common meeting ground for social scientists from a variety of academic disciplines who were out of necessity involved in communication research problems. Some of them retained an interest in these problems when they returned to academic settings, but their allegiances tended to return to their previous academic fields. The emergence of communication as a separate discipline did not take place until a decade later.

Applied problems continued to dominate communication research activities after the war. The perception that America's world responsibilities included aiding less developed nations led to a ready acceptance of the idea that the mass media could foster economic growth and political development by disseminating needed information, by stimulating achievement motivation and empathy, and so forth (Lerner, 1958; Schramm, 1964; Rogers, 1972). The empirical difficulties of inducing social change appear to have scaled down this optimistic hope, at least to a point where the mass media are no longer seen as sufficient causes (i.e., media exposure producing change all by itself) of the various forms of national development (Rogers, 1976).

The domination of applied research in the study of mass media effects has continued to the present day. In 1969, for example, pressure from the public and from Congress led to the formation of the Surgeon General's Scientific Advisory Committee on Television and Social Behavior and to the earmarking of about one million dollars for research recommended by this committee. The funding resulted in the publication of 60 reports from investigators representing a wide variety of fields (Comstock & Rubinstein, 1972). Although the major focus was on the possible effects of televised violence on the aggressive behavior of children, the topic did not represent the dominant research interests of most of the investigators. Their tangential interests and the relatively short duration of the research contracts (12 to 18 months) undoubtedly contributed to a lack of theoretical orientation among the reports. The history of communication research is replete with other examples of funding being infused into the field on a short-term basis for the study of a variety of applied problems: health communication (e.g., heart disease, cancer campaigns); dissemination of birth control information; television advertising effects; cable television and other technological advances; and effects of televised courtroom proceedings. No comparable funding is available for basic research, and communication researchers must apply to agencies whose scientific panels are labeled according to established academic fields. The relative adequacy of funding for applied research in combination with the scarcity of support for basic research has helped to make the mass communication area highly relevant to public policy issues but at the same time theoretically impoverished.

Late Emergence as an Academic Discipline

The applied focus has also helped to retain communication research as "an academic crossroad where many have passed, but few have tarried" (Schramm, 1963 p. 2). To the present day, a large share of the research relevant to mass media effects emanates from social scientists with allegiances to a variety of fields other than communication. Few would identify themselves as communication researchers. Not only does this tend to impede theoretical integration of research, but it also makes it very difficult for any given person to locate the scattered studies relevant to mass media effects.

There has been a gradual shift away from the temporary crossroads situation described by Schramm and toward the emergence of communication as an academic discipline. This was seen first in the mid-1950s with the founding of several communication research centers at universities like Illinois, Stanford, Wisconsin, and Michigan State. From these arose academic programs including doctoral training and degrees in communication and mass communication. The resulting research has had major impact on existing academic organizations such as the Association for Education in Journalism and the Speech Communication Association and has probably contributed to the founding of new organizations like the International Communication Association. Not surprisingly, these developments have served to develop orientations that view communication research as a new discipline with unique concepts and styles of theorizing rather than as an applied derivative of existing social science fields. The growing numbers of self-conscious communication researchers helps to account for the birth of several new communication journals within the past few years (e.g., *Communication Research, Human Communication Research*) and the growing amount of evidence-based articles in the older publications of the field.

Communication as a "Variable Field"

Communication research represents what has been termed a "variable field" as opposed to a "level field" (Paisley, 1972). Level fields within the social sciences are those whose concepts and theories tend to operate at a common level of discourse. For example, psychologists study individual systems using appropriate cognitive, affective, and behavior concepts, whereas sociologists examine the structure and processes of various types of social systems (e.g., societies, communities, organizations, primary groups) at a more abstract level. Of course, there are many exceptions among psychologists using physiological variables and sociologists having recourse to psychological concepts and explanations. But in general the consistency in abstraction is much greater in these level fields.

Variable fields, on the other hand, focus on a particular class of variables that have consequences at more than one level of abstraction. Economics and political science are examples of relatively old and established variable fields, but level fields tend to have longer research traditions. Business and educational research are also variable fields, as are the new hybrids of social psychology, cybernetics, and communication research. The older variable fields have tended to resolve their inherent differences in abstraction by coexistence of two traditions such as macroinstitutional versus microbehavioral with different research problems and methodological strategies. These arrangements are still to be worked out in communication research, and the shifting levels of abstraction have implications for what we know and don't know about the effects of the mass media.

Variable fields tend to utilize research strategies that are common to the level field at any given level of abstraction. Experimental research dominates at the less abstract levels, whereas nonexperimental field studies are more common in research in the more abstract social systems level fields. As a result, standards of evidence vary, and there tends to be little agreement about strategies of research. Theoretical development in the newer variable fields tends to be sporadic and characterized by fads and "small islands" of theory rather than by traditional "grand" theories.

Communication research appears to closely fit the model of the variable field. There is little standardization of research approaches. Not only does the field admit both experimental and nonexperimental evidence, but knowledge claims are also based on legal and historical scholarship, literary criticism, and other strategies. In the attempt to study media process and effects at all levels of discourse, communication research has no broad encompassing theories that cover the field. Rather, most theories are borrowed from other level fields, particularly from psychology and sociology and from the more established variable fields. Unfortunately there is often a lag between the prominence of a theory in another field and its appearance in communication research, so that the latter is often in a position of pursuing a dead theory.

The dispersion of communication interest across levels has also led to looking to methodological innovation in lieu of unifying theory. Sequentially, over the past 30 years, various techniques have been seized upon as solutions to communication problems: factor analysis, the authoritarian f-scale, the semantic differential, Q-methodology, coorientation, smallest-space analysis, path analysis, and gallilean coordinate systems. Although most of these techniques became useful tools in communication as well as in the other social sciences, none proved to be the panacea some hoped for. There remains, however, a tendency to propose technical solutions to substantive and theoretical problems.

Variable fields tend to have as part of their origins a close connection with public or professional institutions. Economics and political science (originally called political economy) have traditional ties to financial institutions and government as have business and education research to their respective institutions. Communication research is no exception to this pattern, at least if we consider those investigators who profess allegiance to the field rather than those from other fields who occasionally do research others might label communication research. Given the already existing tradition of audience research, it was not surprising that communication research found a home in schools of journalism in the post-World War II era. The resulting tie to the professional field of journalism and to the mass media industry has had important consequences for research priorities and questions. The more direct ties to media institutions have maintained the focus on applied problems and reduced the proposition of esoteric research. This contact also may have made even the more basic research more realistic in the sense that the economic and procedural realities of media production constraints are taken into account. On the other hand, much of the research emanating from academic journalism has tended to be too narrowly applied to have much general value. Confusion over whether the client is the media institution or the public also has consequences for research.

Communication research is done today not only in schools of journalism but also in departments of speech communication (or equivalent title). The latter field originates in large part from the humanities rather than from the social sciences and includes such areas as rhetoric–public address, speech correction and audiology, radio–TV and film, interpersonal communication, and sometimes theater and drama. In both journalism schools and speech communication departments, those interested in mass media effects are apt to be a small minority (sometimes one person) at a given school. As a result, investigators are spread thinly across the country and, for quite different reasons, across the rest of the world as well. The isolation of researchers is alleviated only partially by the research divisions within professional associations and more informally by "invisible colleges" of like-minded scholars. Autonomous schools and departments of communication have arisen at Michigan State and Pennsylvania, for example, but the dominant picture is that of isolation with little standardization of definitions, goals, or standards of training. This also tends to spread research findings across the journals of the various fields, although the recent founding of new publications not tied to any one field may lessen their problems.

Media-Centric vs. Effects-Centric
Approaches to Research

In examining the more recent work on mass media effects, we can identify two rather different perspectives guiding research. Individual researchers tend to follow one or the other but not both of these perspectives. The first, a "media-centric" approach (a term coined by Chaffee, 1977), begins with a concern for the structure and content of media and often involves a monitoring of media content. The focus is frequently on the media content and its origins. As a result, communication variables are often treated as dependent on various economic and production factors, whereas effects on the audience are sometimes tacitly assumed from the content. Research within this perspective is more likely to be done by mass communication researchers allied to journalism or speech communication settings than by other social scientists and by others trying to "prove" that the media have effects. Researchers in this mode are in business as long as the content of the media keeps changing or as long as they are able to find new content that could have effects. The tradition of "critical events" analysis (e.g., the Lang and Lang [1960] study of the MacArthur homecoming parade, studies of the Kennedy assassination, presidential debates) illustrates the strengths and weaknesses of the approach. The results of such studies provide rich detail about the operation of the media and their audiences in a particular situation; however, the conditions are so specific as to prevent generalization to other situations. Media-centric research not tied to any critical event has similar strengths and weaknesses. In order to make a lasting contribution to the understanding of media effects, the media-centric perspective must be integrated with other studies to explain effects more fully. This is particularly important in considering the large majority of effects in which the media are likely to be only one among many sources of influence.

The second perspective, which can be called an "effects-centric" approach, is likely to emanate from a concern for the dependent effect rather than for the communication variables. At least prior to the 1970s, research on the relationship between media violence and aggressive behavior showed much more concern with the latter and used filmed violence largely because of experimental convenience of providing a common stimulus to subjects. Research in this tradition is apt to study communication variables relative to other influences on the particular effect of interest. The result may tend to be the opposite of that of the media-centric approach. The effect criterion is well understood and measured, but the mass media stimulus may be overly simplistic. The social context of media use is particularly likely to be ignored in effects-centric research. Although the historic segmentation of the field has tended to produce and continue this dualistic set of research approaches, it is clear that communication research would be improved if the strengths of the media-centric and effects-centric approaches were effectively combined. This might be facilitated through revising graduate training, with programs such that communication students with predominately media-centric orientations would receive more intensive experience regarding particular classes of effects and those with effects-centric orientations would concentrate on gaining greater awareness of the nuances of media perspectives. It also might be useful to mix both media-centric and effects-centric researchers in teams investigating applied problems or developing and evaluating media programs designed for children, the elderly, and other specific audiences.

MEDIA EFFECTS RESEARCH AND PUBLIC POLICY

The foregoing material should convince even the most optimistic reader that identifying or answering questions about media effects is extremely complicated. Several concerns were mentioned that make media effects questions difficult, including the number of different types of possible media effects; the complexity of media stimuli; the special problems in studying and documenting media effects; the varying strategies of making inferences from evidence; and the peculiar history and organization of the field of communication research. Despite these difficulties, however, it is still important to confront the application of research to important social problems. The fact that most media research has a traditionally applied orientation may be a weakness for the development of communication theory, but it may be considered a strength when considering the potential for research to contribute to media policy.

Researchers in the field of communication are not narrowly entrenched in the esoteric concerns of level fields (i.e., psychology and sociology). They are more likely to have ties with the media, perhaps even professional experience, and have a more thorough understanding of the context of media policy. We should not, however, expect immediate answers. The utilization of research is complicated by a very real discrepancy in the understanding of media effects among policy makers, researchers, and the general public.

Public interest in the mass media seems to be guided by an implicit assumption that media have powerful and pervasive effects. Although media researchers have developed through various conceptions of media effects— from direct effect models to limited effects models to concern for the conditional aspects of media effects—it seems that the popular conception of effects has been universally one of undifferentiated concern for the all-powerful media. Current policy related to media, however, does not reflect the popular image of image effects. Policy actions, which are perhaps largely a function of the Constitutional and economic limitations of regulating media, seem unrelated to public concern or evidence from social research. The orientations of the public, policy makers, and the research community are currently very much out of synchrony.

Despite the apparent gap between social research and public policy, most people at least recognize the potential for empirical research as an input to policy. The translation of data into action, however, is possibly more complex and involves more issues than at first appear. To suggest that media research should be an important part of public policy logically assumes that we have answered or at least considered several questions: (a) Is the field of communication research ready to offer policy suggestions based on empirical research? What standards of evidence are appropriate in deciding which research should apply to which problems? (b) Do we know what the goal of media policy should be? Should policy guide media in the direction of social change, or should media be encouraged to reflect reality as closely as possible, or should media be left to serve commercial interests at whatever cost to the quality of content? (c) Do we know where we want to direct policy? Should we consider only the content of media and those people responsible for the content, or are there other elements in the process of communication effects that would make appropriate targets for social policy? (d) What role and amount of responsibility should researchers, policy makers, or other intermediaries be willing to assume in the translation of research into public policy?

How Should We Evaluate Communication Research?

A first policy consideration asks what communication researchers are able to say about media policy, given the complexity of the process they are trying to explain. A careful analysis of the many different types of media effects, and especially of the problems in designing empirical tests that identify unique media effects, could lead to an unnecessarily dismal view of the policy potential of media research. We do have persistent evidence that media have some effects on audiences that at least justifies the consideration of research in the formulation of policy. There is overwhelming evidence, for example, that children can learn from television and that voters at least partially depend on media for political information. Although there is a great deal of variance in the confidence we place in the documentation of several other media effects, there is a corresponding variance in the risk of enacting the policy counterparts to research.

Two considerations seem important in first determining the need for policy action: (1) the magnitude of the impact of media and (2) the costs involved in originating and applying policy related to the effects. First, the media issues addressed by research involve effects that could vary greatly in terms of social significance. The potential of media to influence elections, alter the crime rate, or narrow the knowledge gap between disadvantaged and advantaged children may be of greater significance than media effects on adolescent dating behavior or audience knowledge of sports activities. Although setting an agenda of the importance of various media effects involves value judgments as much as examination of data, media messages do vary according to the negative threat or potential positive gain they may cause. Unfortunately, this is difficult to infer from either experimental studies that show that media are sufficient conditions for effects or from correlational studies that examine the degree of relationship between exposure to media and media effects. It is still unclear how much change in criterion behaviors (e.g., cutting the incidence of violence) is to be gained from a given amount of change in media content.

A second criterion that could determine the appropriateness of policy is the cost of originating and applying the policy solutions. Policy will require various degrees of change from current standards both in direction and in relation to the regulatory group assigned to monitor its progress. Certainly policy decisions that require tampering with constitutional guarantees will involve greater risks than those included in voluntary codes. Similarly, government regulation of media, whether by congressional action or through federal commissions, could involve greater costs than policy that is initiated and controlled through industry self-regulation, public action groups, or educational systems.

The standards of evidence required to evaluate a research input into policy need not be any more consistent than the costs of various policy solutions.

Perhaps we should not be willing to become *either* Type One *or* Type Two worriers but rather be willing to adopt a degree of conservatism in evaluating research that corresponds to the amount of changes and costs required to enact policy. Type One standards should apply to high cost policy, for example, requiring by law that networks cut the amount of violence on television by 50% over the next five years. Adopting an addition to a self-regulating voluntary code that suggests the elimination of gratuitous violence between 7 and 9 p.m., however, could more appropriately be based on a Type Two approach to research.

Policy to What End? What Is the Goal?

Despite the attempts by social scientists to be guided to the greatest extent possible by standards of objectivity, it is often difficult, especially for researchers dealing with media effects, totally to ignore personal biases and outside pressures in designing and conducting social research. Even by the selection of certain media effects questions to the exclusion of others, we are making statements that at least identify the effects we feel are most important. It is, however, much easier to *conduct* social research without subjective evaluations of the system being studied than it is to *translate* research into social policy. The necessity for media policy logically assumes a goal for the system, and if research is to be utilized in policy, its application must be with respect to the standards and goals of the system.

Frequently, however, the goals of different groups involved in policy decisions are in conflict. When confrontations between television networks and public pressure groups take place, for example, the two groups may conceive of the social function of media very differently, although their conceptions are seldom a major part of their arguments. One group may believe media should operate to promote social change whereas the other may feel media exist mostly as a commercial venture. In formulating policy, research would be applied very differently to policy questions in these two systems, and therefore it seems essential that we consider the end goal of policy before attempting to translate research into viable policy solutions. Identical evidence could be used in opposing ways, depending on the values of the system. This would apply to different values within the American media system and especially to systems in different countries.

If we consider, for example, the policy translation of a recent finding about the effects of television on children's perceptions of occupational sex roles, the difficulties in utilizing research become clear. The study found that the frequency of portrayals of women in various occupations traditionally held by men (e.g., police officer) was related to children's perceptions of the ratio of men and women who held these jobs in real life (Miller & Reeves, 1976). Imagine now that we have been asked to determine, in a policy statement, the number of females that must be portrayed as doctors and lawyers on prime-time television. Should we say that television has a responsibility in social change and insist that 50% of the portrayals include women? Or should we insist that television portrayals reflect real occupational statistics so that media do not become responsible for creating a falsely reassuring picture of the world for young people? Or should we say that as a commercial venture, media have the responsibility to attract the largest possible audience despite incidental effects of their programs. Several other alternatives are probably also possible. The major point is that the same data could support several policy alternatives. We must agree on policy or at least be aware of the differences in our conceptions of media systems before research can be meaningfully applied to policy actions.

Policy About What?

The first objects of social policy that most people consider are the sources of media messages. Furthermore, most policy proposals usually include laws, regulations, rulings, and codes. Although policy could effectively deal with communication sources, we should not consider them the only outlet for solutions, no more that we should consider their messages the only required piece of evidence to document media effects. Policy could especially consider the conditional aspects of the effects process and focus on those variables that determine when and for whom media effects will occur. This would include, for example, teaching children about unreal aspects of television and educating voters about the role of media in elections.

There could be several advantages to focusing policy on conditional variables. First, those people most affected by the media would be the primary objects of the policy. If it is true that several of the media effects we are most concerned about are operative only for certain subgroups of the audience, then policy directed primarily at those subgroups would preserve the content for those not affected. For example, those who approach television drama with a willing suspension of belief would not be penalized by policy that attempted to educate those who perceive television drama to reflect the real world. Policy related to conditional variables may also require less risk, because most of the complex legal and economic questions surface only when control of media content is suggested. Finally, by not having to control content, we could be more assured that content changes would not invoke a new set of incidental effects.

There are also different aspects of the type of media research that is used for policy that may determine what policy is about and the risks that policy-makers are willing to take. Utilization of effects-centric research may be less likely to suggest high cost policy, because media effects would be reported in relation to other information sources that could be as responsible—and in many cases more responsible—for the effects. Given the complexity of the process by which media have effects, reliance on only effects-centric research for policy may also result in a less precise description of the portion of the effects that can be uniquely attributed to media. Media-centric research, although more limited in its coverage of the causes of effects, may actually provide more useful information that could be translated into policy. A tolerance for less comprehensive but more detailed research, whether oriented toward media or toward effects, may ultimately result in more accurate and beneficial policy directives.

It may even be reasonable to suggest that there should be media- and effects-centric policy-makers, just as there are media- and effects-centric researchers. Consider, for example, two government commissions that have both examined the violence issue. The Eisenhower Commission on Violence, operating from an effects-centric perspective, approached the general issue of violence and considered several possible causes of violence. The Surgeon General's Scientific Advisory Committee on Television and Violence, operating from a media-centric perspective, considered only a single possible cause of violence—television. From a reading of both reports, it would seem much easier to derive policy directives concerning the media from the surgeon general's report. However, the more general conclusions of the Eisenhower Commission offer important information about other aspects of violence that may be crucial to an understanding of the social context of the effects of televised violence. The major point here is that although both perspectives are useful, policy may to some extent be determined by the orientation of the research that is used in its formulation. Rather than relying on a random matchup, we should make a deliberate effort to correlate policy goals and research perspectives.

Policy by Whom?

Even a sketchy introduction to media effects research suggests to most people several possible linkages between research and policy. If this is so, what role and amount of responsibility should policy-makers and researchers take in the translation process? This has traditionally been a problem of policy-makers mixing in the objectivity of research versus researchers mixing in the subjectivity of policy. Both must recognize, however, that at least to some extent their roles are not mutually exclusive. Researchers, especially those interested in applied research, must depend on policy makers to define problems accurately and issue priorities, and policy-makers must rely on researchers to insure that results are interpreted correctly and accurately applied to policy concerns.

There are at least two aspects of the current organization of the communication field that may contribute to the gap between research and policy. A major problem may be simply geographical. A substantial portion of the research in communication and in media effects is conducted in the Midwest, far from the administrative headquarters of media in the East and the production facilities in the West. There are few established communication links between universities, Hollywood, and New York. Second, as the brief history of the development of communication as a unique area of theory and research indicated, we are a new field, unknown even to many academicians. Certainly a majority of the communication research currently being done comes from new departments of communication, from communication and mass communication research centers, or from schools of journalism and departments of speech. Many policy-makers view communication only within the context of psychology and sociology, which is probably correct as far as the content of most communication research is concerned; however, this view ignores the rather peculiar organization of the communication field.

There is at least one other important consideration in determining the role of researchers and policy-makers in this process. Although policy decisions should be based on evidence from social research, this should not be done in isolation from other considerations or expertise from other information sources. The violence issue, for example, is to a large extent a legal and economic concern. From a policy perspective, the constitutionality of legislative action should be considered in relation to research evidence about the effect. Why should we expend large efforts relating media research to policy issues only to find afterward that control is legally or economically impossible? All information that is relevant to policy should be collated and considered in the appropriate sequence.

The roles of persons with various types of input and evidence relating to policy questions, however, should not be confused. Scientists attempting to translate systematically collected data into policy should not be confused with experts offering a legal opinion. To some extent, legal standards of evidence, which are currently responsible for most policy decisions, may be incompatible with what media researchers have to offer to policy. Although different types of input need simultaneous consideration, we should recognize the uniqueness of each input and attempt to weight the various policy formulas to maximize the benefit from each source.

In addition to differences between contributions of social, legal, and economic information to the formation of policy, there are important differences in the roles that individual researchers play in the translation of social research into policy. Recognizing these differences illustrates the number of unique ways in which research can be utilized and also points out that although an attempt is made to conduct research under the highest possible standards of objectivity, the use of research in policy necessarily implies a consideration of values that may determine and shape final policy as much as or more than the research itself.

Before considering the different roles that researchers can play and have played in the translation of media research into policy, it is important to note that the *type* of information supplied in each of these roles can vary as much as the situation in which the information has been offered. Each type of information requires a different amount of commitment to issues by the researcher and also draws on different forms of evidence. First, researchers are likely to make data-based contributions either by generating original data or summarizing data already available. The Surgeon General's Commission on Television and Violence, for example, required researchers to propose and carry out original research that would recommend policy. Reviews of research, however, are just as likely to have contributed to policy decisions; a good example is the recent review of research on the effects of television advertising on children compiled for the National Science Foundation for the purpose of providing research information to policy-makers as well as for guiding future policy-related research (Adler, 1977). The problem with such literature reviews is that there is considerable slippage in precision in moving from the original work. Findings and qualifications are often rounded off to fit an overly simple and consistent summary of research results.

A second type of contribution is opinion. Regarding researchers' expertise in certain areas, policy-makers are often anxious to have these experts help resolve policy conflicts by offering opinions based on a unique perspective on an issue. A frequent recipient of this type of information, for example, is the Federal Trade Commission, which is increasingly soliciting opinions from researchers as well as consumers on the public's probable understanding of potentially deceptive advertising (Brandt & Preston, 1977). The transition from presenting evidence to giving opinion is a particularly abrupt one for both the researcher and the policy-maker. Hearings are conducted in a quasi-legal format that dictates opinion about a specific case whereas the researcher is accustomed to dealing with research findings aggregated across a large number of cases. The researcher is nervous about extrapolating to a specific case and policy, and the policy-maker or commission attorney is impatient to be rid of all the qualifications and timidities of the social scientist.

Regardless of the type of information researchers bring to the formation of policy, the nature of the interaction between researchers and policy-makers has a great potential to determine how research will be used. We have identified seven researcher roles that could affect the use of research in policy (Broom & Smith, 1978).

First, researchers often find themselves in the role of *legitimizer*. In this situation, researchers and research are typically used to confirm an already established position or policy. Regardless of the type of evidence submitted, the researcher in this role serves to enhance the credibility of current practices by selectively linking evidence with policy after the policy has been established. Reviews of research defending network policies on the portrayals of violence, for example, often fall within this perspective.

In what is often their most influential role, researchers may contribute information via a one-to-one relationship with the highest sources of policy-making authority within an organization or agency. This is sometimes referred to as the *Merlin* role (Sussman, 1966). In this case, researchers offer information without the restrictions often imposed by formal decision-making processes and the complexities of bureaucracies. The potential for more accurate translations of research may actually be increased in this role because of the absence of political and legal constraints and by the fact that consensus may not be as important because both parties have relatively little to lose within their private relationship.

Researchers also play the role of *information broker*, relating the concerns of the research community to institutions and policy-makers. Although single individuals are seldom placed in this role, organizations have been created and are often staffed by individuals from the research community for the purpose of providing a clearinghouse of research information. To some extent, this role is played by the professional associations within the general field of communication research (e.g., the International Communication Association, the Association for Education in Journalism, and the Speech Communication Association); however, this role is the prime focus of private nonprofit agencies such as the Aspen Institute.

Very often researchers are called on after a policy conflict has already been identified. In these situations the researchers can act either as mediator or arbitrator. A researcher as *mediator* will typically make contributions aimed at resolving a policy conflict, while at the same time relying on the policy-makers themselves to make the final policy decisions. It is even possible in this situation for the researcher to initiate the interaction between policy-makers and to make empirical research the focus of that interaction. This role is exemplified by a recent joint effort by British political parties, media representatives, and political communication researchers to restructure British political campaigns (Blumler, Gurevitch, & Ives, 1977).

The researcher as *arbitrator* is usually presented with a policy conflict and then asked, on the basis of data or expertise, to select the right alternative. One of two outcomes is almost inevitable in this situation: Either the researcher will alienate one of the two sides or the research contributions and policy recommendations will be compromised with the goal of pleasing both sides and with the risk of making the policy too innocuous to be effective. The dilemma posed to the arbitrator is exemplified by requests of researchers to decide who is right concerning the question of television and violence—the networks or those interested in stricter regulation.

One of the more difficult roles for researchers is *policy-maker*. There are several instances in which researchers have been a part of commissions or other decision-making groups that are ultimately responsible for policy. Although researchers are typically part of this group because of familiarity with empirical evidence relating to policy, it becomes very difficult to make contributions to decision-making based on social research while at the same time having to consider politics, law, economics, and personal and system values that are also relevant to the final decisions. Separating the roles of researcher and policy-maker, especially within a single individual, is very difficult and could involve serious conflicts of interest.

A final role involves the researcher as a *passive participant* in policy. In this role, researchers maintain that the objectivity of research would be tarnished if the researcher were involved in any way in the translation process. Researchers are often anxious to operate in this role, especially in situations in which the translation of research into policy obviously involves taking a stand on an issue based on subjective values. Although it is somewhat difficult to imagine "no role" as a role, it is just this orientation to policy that critics frequently accuse researchers of playing—with policy the likely loser, because it will be based on factors that are possibly less appropriate than social research.

Comparing these roles, it is evident that the role of researchers in policy formation may differ along the following important dimensions: the type of information contributed to policy; access to policy-makers; the stage of the decision-making process in which researchers are consulted; researchers' concern for the process of decision-making versus the content of policy decisions; the degree of advocacy required; and the ownership of ideas. What is important to remember in the context of the larger discussion of media effects research is that finding logical conclusions for research, even if the research is formulated and designed as applied research, very much depends on the type of information and type of role relationships that are structured to link the research community with policy-makers.

Although the optimal arrangement of communication researcher roles with respect to policy is not yet clear, it is evident that present arrangements should be readjusted in at least two major ways. First, more sensitive communication policy decisions are likely to be made to the extent that research findings can be introduced more quickly into the planning process. This means that research must play a more central role in the actual formulation of policy, not just in evaluating decisions or programs after the fact. This has been referred to as formative as opposed to summative use of research evidence (Scriven, 1967). The production of "Sesame Street" serves as a model for this approach (Lesser, 1975).

A second imbalance, arising from the origins of communication research as an applied field, is an overly narrow view of policy research. If research investigates variation only within the mass media constraints currently operating on a given media system, research may only ameliorate present limitations. If policy-makers alone have the responsibility of deciding what questions to ask about media effects, this is likely to be the case. Increased attention to the support of basic communication research to some extent independent of the directness of the apparent linkages between the research question and the social problem would seem appropriate to encourage creative solutions to policy issues.

REFERENCES

Adler, R. P. *Research on the effects of television advertising on children.* Washington, D.C.: U.S. Government Printing Office, 1977.

Bauer, R. A., & Bauer, A. America, "mass society" and mass media. *Journal of Social Issues,* 1960, *16,* 3-66.

Becker, L. B., & Doolittle, J. C. How repetition affects evaluations of and information seeking about candidates. *Journalism Quarterly,* 1975, *52,* 611-617.

Becker, L. B., McCombs, M. E., & McLeod, J. M. The development of political cognitions. In S. H. Chaffee (Ed.), *Political communications: Issues and strategies for research.* Sage Annual Reviews of Communication Research, Vol. IV. Beverly Hills: Sage, 1975.

Berelson, B. What "missing the newspaper" means. In P. F. Lazarsfeld & F. N. Stanton (Eds.), *Communication Research 1948-49.* New York: Duell, Sloan & Pearce, 1949.

Berelson, B. R., Lazarsfeld, P. F., & McPhee, W. N. *Voting: A study of opinion formation in a presidential campaign.* Chicago: University of Chicago Press, 1954.

Berelson, B., & Steiner, G. A. Human Behavior: An inventory of research findings. New York: Harcourt, Brace, 1964.

Berkowitz, L. Violence in the mass media. In L. Berkowitz (Ed.), *Aggression: A social psychological analysis.* New York: McGraw-Hill, 1962.

Blumler, J. G., Gurevitch, M. & Ives, J. The challenge of election broadcasting. Leeds, Centre for Television Research, University of Leeds, LS2 9JT, 1977.

Blumler, J. G., & Katz, E. (Eds.). *The uses of mass communication.* Sage Annual Reviews of Communication Research, Vol. III. Beverly Hills: Sage, 1974.

Blumler, J. G., & McLeod, J. M. Communication and voter turnout in Britain. In T. Leggatt (Ed.), *Sociological theory and survey research.* London: Sage, 1974.

Bogart, L. *The age of television.* New York: Ungar, 1972.

Brandt, M. T., & Preston, I. L. The Federal Trade Commission's use of evidence to determine deception. *Journal of Marketing,* January, 1977, *41,* 54-62.

Broom, G. M., & Smith, G. C. *Toward an understanding of public relations roles: An empirical test of five role models' impact on clients.* Paper presented at the meeting of the Association for Education in Journalism, Seattle, August, 1978.

Cantor, J. R., & Zillmann, D. The effect of affective state and emotional arousal on music appreciation. *Journal of General Psychology,* 1973, *89,* 97-108.

Chaffee, S. H. Memorandum to the Committee on Mass Communication and Political Behavior, Social Science Research Council, August 30, 1977.

Chaffee, S. H. *Longitudinal designs for communication research: Cross-lagged correlation.* Paper presented at the meeting of the Association for Education in Journalism, Carbondale, August 1972.(a).

Chaffee, S. H. Television and adolescent aggressiveness. In G. A. Comstock & E. A. Rubinstein (Eds.), *Television and social behavior. Vol. 3. Television and adolescent aggressiveness.* Washington, D. C.: U.S. Government Printing Office, 1972. (b)

Chaffee, S. H. (Ed.). *Political communication: Issues and strategies for research.* Beverly Hills: Sage, 1975.

Chaffee, S. H., & Izcaray, F. Mass communication functions in a media-rich developing society. *Communication Research,* 1975, *2,* 367-395.

Chaffee, S. H., & Wilson, D. *Media rich, media poor: Two studies of diversity in agenda-holding.* Paper presented at the meeting of the Association for Education in Journalism, College Park, Md., August 1976.

Cohen, B. C. *The press, the public and foreign policy.* Princeton: Princeton University Press, 1963.

Comstock, G. A., & Rubinstein, E. A. *Television and social behavior* (Vols. 1-4), Washington, D.C.: U.S. Government Printing Office, 1972.

Cramond, J. Introduction of TV and effects upon children's daily lives. In R. Brown (Ed.), *Children and television.* Beverly Hills: Sage, 1976.

Dominick, J., & Greenberg, B. S. Mass media functions among low-income adolescents. In B. S. Greenberg & B. Dervin (Eds.), *Use of the mass media by the urban poor.* New York: Praeger, 1970.

Feshbach, S. Reality and fantasy in filmed violence. In J. P. Murray, E. A. Rubinstein, & G. A. Comstock (Eds.), *Television and social behavior* (Vol. 2). *Television and social learning.* Washington, D.C.: U.S. Government Printing Office, 1972.

Festinger, L. Behavioral support for opinion change. *Public Opinion Quarterly,* 1964, *28,* 404-417.

Furu, T. *The function of television for children and adolescents.* Tokyo: Sophia University Press, 1971.

Gans, H. J. The famine in American mass-communications research: Comments on Hirsch, Tuchman, and Gecas. *American Journal of Sociology,* 1972, *77,* 697-705.

Goodhardt, G. J., Ehrenberg, A. S. C., & Collins, M. A. *The television audience: Patterns of viewing.* London: Saxon House, 1975.

Gormley, W. T. *The effects of newspaper-television cross-ownership on news homogeneity.* Chapel Hill: Institute for Research in Social Science, University of North Carolina, 1976.

Greenberg, B. S., & Reeves, B. Children and the perceived reality of television. *Journal of Social Issues,* 1976, *32* (4), 86-97.

Harrison, R., & Ekman, P. TV's last frontier: South Africa. *Journal of Communication,* 1976, *26,* 102-109.

Hawkins, R. P. The dimensional structure of children's perceptions of television reality. *Communication Research,* 1977, *3,* 299-320.

Himmelweit, H. T., Oppenheim, A. N., & Vince, P. *Television and the child.* London: Oxford University Press, 1958.

Hyman, H. H., & Sheatsley, P. B. Some reasons why information campaigns fail. In E. E. Maccoby, T. M. Newcomb, & E. L. Hartley (Eds), *Readings in social psychology.* New York: Holt, 1958.

Katz, E., & Lazarsfeld, P. F. *Personal influence.* Glencoe: Free Press, 1955.

Kenny, D. A. Cross-lagged and synchronous common factors in panel data. In A. S. Goldberger & O. D. Duncan (Eds.), *Structural equation models in the social sciences.* New York: Seminar Press, 1973.

Kerlinger, F., & Pedhazur, E. *Multiple regression in behavioral research.* New York: Holt, 1973.

Klapper, J. T. *The effects of mass communication.* New York: Free Press, 1960.

Kline, F. G., Miller, P. V., & Morrison, A. J. Adolescents and family planning information: An Exploration of audience needs and media effects. In J. G. Blumler & E. Katz (Eds.), *The uses of mass communication.* Beverly Hills, Sage, 1974.

Kraus, S., Davis, D., Lang, G. E., & Lang, K. Critical events analysis. In S. H. Chaffee (Ed.), *Political communication issues and strategies for research.* Sage Annual Reviews of Communication Research, Vol IV. Beverly Hills: Sage, 1975.

Lang, K., & Lang, G. E. The unique perspective of television and its effect: A pilot study. In W. Schramm (Ed.), *Mass communications.* Urbana: University of Illinois Press, 1960.

Lang, G. E., & Lang, K. Immediate and mediated responses: First debate. In S. Kraus (Eds.), *Great Debates, 1976—Ford vs. Carter.* Bloomington: Indiana University Press, in press.

Lazarsfeld, P. F., Berelson, B., & Gaudet, H. *The people's choice.* New York: Columbia University Press, 1948.

Lerner, D. *The passing of traditional society.* New York: Free Press, 1958.

Lesser, G. S. *Children and television.* New York: Vintage, 1975.

Lyle, J., & Hoffman, H. R. Children's use of television and other media. In E. H. Rubinstein, G. A. Comstock, & J. P. Murray (Eds.), *Television and social behavior. Vol. 4. Television in day-to-day life: Patterns of use.* Washington, D.C.: U.S. Government Printing Office, 1972.

MacCorquodale, K., & Meehl, P. E. On a distinction between hypothetical constructs and intervening variables. *Psychological Review,* 1948, *55,* 95–107.

McCombs, M. E., & Shaw, D. C. The agenda-setting function of mass media. *Public Opinion Quarterly,* 1972, *36,* 176–187.

McGuire, W. J. Inducing resistance to persuasion. In L. Berkowitz (Ed.), *Advances in experimental social psychology* (Vol. 1) New York: Academic Press, 1964.

McLeod, J., Atkin, C., & Chaffee, S. Adolescents, parents and television use: Adolescent self-report measures from Maryland and Wisconsin samples. In G. A. Comstock & E. A. Rubinstein (Eds.), *Television and social behavior. Vol. 3. Television and adolescent aggressiveness.* Washington D.C.: U.S. Government Printing Office, 1972.

McLeod, J. M., & Becker, L. B. Testing the validity of gratification measures through political effects analysis. In J. G. Blumler & E. Katz (Eds.), *The uses of mass communication.* Beverly Hills: Sage, 1974.

McLeod, J. M., Becker, L. B., & Byrnes, J. E. Another look at the agenda-setting function of the press. *Communication Research,* 1974, *1,* 131–166.

McLeod, J. M., Brown, J. D., Becker, L. B., & Ziemke, D. A. Decline and fall at the White House: A longitudinal analysis of communication effects. *Communication Research,* 1977, *4,* 3–20, 35–39.

McLeod, J. M., Durall, J. A., Ziemke, D. A., & Bybee, C. R. Reactions of young and older voters: Expanding the context of debate effects. In S. Kraus (Ed.), *Great Debates, 1976— Ford vs. Carter.* Bloomington: Indiana University Press, in press.

McLuhan, M. *Understanding media.* New York: McGraw-Hill, 1964.

Miller, M. M., & Reeves, B. Linking dramatic TV content to children's occupational sex-role stereotypes. *Journal of Broadcasting,* 1976, *20,* 35–50.

Morrison, A. J., Steeper, F., & Greendale, S. The first 1976 presidential debate: The votes win. Paper presented at the meeting of the American Association for Public Opinion Research, Buck Hills Falls, May 1977.

Nie, N. H., Verba, S., & Petrocik, J. R. *The changing American voter.* Cambridge: Harvard University Press, 1976.

Paisley, W. *Communication research as a behavioral discipline.* Stanford: Institute for Communication Research, 1972. (mimeographed)

Parker, E. B. The effects of television on library circulation. *Public Opinion Quarterly,* 1963, *27,* 578–589.

Reeves, B. Perceived TV reality as a predictor of children's social behavior. *Journalism Quarterly,* 1978, *55,* 682–689, 695.

Robinson, J. P. Mass communication and information diffusion. In F J. Kline & P. F. Tichenor (Eds.), *Current perspectives in mass communication research.* Beverly Hills: Sage, 1972.

Rogers, E. M. *The communication of innovations.* New York: Free Press, 1972.

Rogers, E. M. Communication and development: Critical perspectives. *Communication Research,* 1976, *3,* 99–106, 213–240.

Schramm, W. Mass communication. *Annual Review of Psychology,* 1962, *13,* 251–284.

Schramm, W. Communication research in the United States. In W. Schramm (Ed.), *The science of human communication.* New York: Basic Books, 1963.

Schramm, W. *Mass media and national development.* Stanford: Stanford University Press, 1964.

Schramm, W., Lyle, J., & Parker, E. B. *Television in the lives of our children.* Stanford: Stanford University Press, 1961.

Scriven, M. The methodology of evaluation. In S. Tyler, R. M. Gagne, & M. Scriven (Eds.), *Perceptives on curriculum education.* Chicago: Rand, 1967.

Siebold, D. T. Communication research and the attitude-verbal report—overt behavior relationship: A critique and theoretical reformulation. *Human Communication Research,* 1975, *2,* 2–32.

Star, S. A., & Hughes, H. M. Report of an educational campaign: The Cincinnati plan for the United Nations. *American Journal of Sociology,* 1950, *55,* 389–400.

Sussman, M. The sociologist as a tool of social action. In Arthur B. Shostak (Ed.), *Sociology in action.* Homewood, Ill.:Dorsey Press, 1966.

Tannenbaum, P. H. The congruity principle revisited: Studies in the reduction, induction, and generalization of persuasion. In L. Berkowitz (Ed.), *Advances in experimental social psychology,* (Vol. 3). New York: Academic Press, 1967.

Tichenor, P. J., Donohue, G. A., & Olien, D. N. Mass media and differential growth in knowledge. *Public Opinion Quarterly,* 1970, *34,* 158–170.

Tichenor, P. J., Rodenkirchen, J. M., Olien, C. N., & Donohue, G. A. Community issues, conflict and public affairs knowledge. In P. Clarke (Ed.), *New models for mass communication research.* Beverly Hills: Sage, 1973.

Watt, J. H., & Krull, R. An examination of three models of television viewing and aggression. *Human Communication Research,* 1977, *3,* 99–112.

Werner, A. Children and television in Norway. *Gazette,* XVII (3), 1971.

Wolfe, K. M., & Fiske, M. The children talk about comics. In P. F. Lazarsfeld & F. N. Stanton (Eds.), *Communication research, 1948–49.* New York: Harper, 1949.

Zillmann, D. The role of excitation in aggressive behavior. *Proceedings of the Seventeenth International Congress of Applied Psychology, 1971.* Brussels: Editest, 1972.

4 The Audience for Television— and in Television Research

Herbert J. Gans
Columbia University and Center for Policy Research

INTRODUCTION

This chapter is, in a way, a summary report of my participation in the SSRC Committee on Television and Social Behavior, for it puts together some questions and hunches that I have developed over the years that the Committee has met and worked together. In fact, the chapter consists largely of questions and hunches—although the latter may often sound like assertions. My primary concern is with the television audience, but it cannot be studied in isolation. Consequently, the chapter also deals with three other sets of participants in television. One set includes the "suppliers" of television fare, by which I mean the "creative" people such as scriptwriters, business people, and "the industry" as a whole. A second set is "the larger society," a fuzzy term to describe not only the macroeconomic and social structural conditions that affect the television industry but also the events in society from which scriptwriters generate story and character ideas and the conditions in that society from which viewers may seek diversion. A third set of participants is us, the television researchers, because we are also, if more indirectly, involved with the audience.

The chapter deals with each of these participant sets in turn, centering mainly around my primary question: What are the relationships between television and its viewers and how should they be studied? The basic hypothesis, which comes up at various times in the chapter, is that for most viewers, involvement with what they see on television is low. Involvement is ✓ defined as the extent and intensity of people's interaction (emotional and intellectual) with the small screen; it could, I presume, be measured by the

55

amount and degree of attention paid to television fare. By low involvement I mean simply that viewers neither care strongly about nor are touched deeply by what they see; television fare often goes in one eye and out the other. Or as Leo Bogart (cited in Katz, Blumler, & Gurevitch, 1974) once put it: "Most mass media experiences represent pasttime rather than purposeful activity, very often (reflecting) chance circumstances within the range of availabilities rather than the expression of psychological motivation or need [p. 21]." To be sure, pasttime activity can sometimes be purposeful, too; and low involvement does not always mean low effects, for when viewers are uninvolved it is possible that they also lower their guard and are thus affected without being aware of it. My basic hypothesis has a variety of research and policy implications, which I spell out in the pages to follow.

THE AUDIENCE—AND AUDIENCE RESEARCH

I want to discuss the audience by raising some questions about the ways in which it has been studied, both in the past and present, beginning with content analysis and then considering effects and uses and gratifications research.

Content Analysis

Content analysis is not entirely concerned with the audience; indeed, many content analyses describe and analyze what is going on in American "culture," a symbolic system that is assumed to exist apart from the audience. Such studies identify values (and, more recently, constructions of reality) expressed in the content without much concern as to who, if anyone, holds these values or reality constructions. Some studies assume that everyone holds the values, or that the values are suprasocietal, surrounding us in much the same way as the atmosphere surrounds the Earth. Other content analyses are, however, concerned with the audience, and they are often indirect effects studies, for they make inferences about audience effects from content analysis. In this kind of study the content analyst generally posits attitudinal or behavioral implications that follow from the content and assumes that all or some of the viewers feel, think, or act in accordance with these implications.

The shortcoming of this approach is that content analysts are trained professionals whose "use" of television fare is totally unlike that of viewers. As analysts, they view the content so often and at such a high level of involvement that they see many themes in it which the audience, which watches less closely, cannot possibly see. This does not invalidate content analysis, of course, but it suggests that inferences about audience behavior and attitudes—other than as hypotheses for audience research—cannot be made from content or content analysis. It also suggests that if we are to understand the relationship between content and audience, *professional* content analyses must be complemented by *audience* content analyses. Viewers must be asked to content-analyze the television fare they watch so that we can learn not only what they see in the content but also with what concepts and categories they approach it. Even this is not entirely satisfactory, for audience members who become content analysts will also be viewing television fare differently than when they were simply audience members. Indeed, everyday viewing is itself a form of content analysis, in which people select and categorize what they see in order to make cognitive and emotional sense for themselves out of what they see. Such a content analysis is neither systematic nor comprehensive, and if the viewers' involvement is low, they will leave out much of the actual content. It is probably impossible, however, to duplicate this kind of audience content analysis for research purposes.

Effects Research

Effects research faces the insuperable task of isolating the impact of the mass media from the impact of other institutions, and obviously, methodological problems abound. These have received so much attention in the literature that I shall not discuss them, but there are two other shortcomings of effects research that deserve to be mentioned.

First, a good deal of effects research has been done through interview studies, which have in turn focused largely on measuring the effects of television (and other media) on attitudes or on behavior as reported by interviewees. The differences between what people say they do and what they do can perhaps be resolved by fieldwork studies, as I shall argue later; the emphasis on attitudes is a fundamental theoretical issue about which psychology and sociology have long been divided. If the issue is put, over-simply, as the causal relation between attitudes and behavior, psychology believes that, all other things being equal, attitudes shape behavior. Sociology, on the other hand, believes that behavior shapes attitudes or that the two are not always related, for people can hold and report attitudes at odds with their actions. Not all sociologists and psychologists accept these beliefs, of course, and the issue is obviously more complex than I have suggested. This is not the place to debate the interdisciplinary dispute, however; furthermore, some issues in the dispute are amenable to empirical testing, and considerably more research needs to be done on the relationships between attitudinal and behavioral effects of television.

The second shortcoming is that so far an equally significant amount of effects research, particularly on the effects of violence, has been laboratory research, the findings of which are difficult to transfer to the world outside and especially to at-home viewing. For example, although enough studies have been done to indicate that laboratory viewing of television or film violence will raise aggression levels among children, we do not know whether the same increases occur at home—and even if they do, whether children subsequently act aggressively at home when no experimenter is present to encourage them to do so and when, conversely, parents or siblings are present to discourage them.

Nor can we generalize from the distinctive social situation of the laboratory to the naturalistic context in which children carry out aggressive behavior. In such contexts, aggression is almost always an outcome of group processes; it comes to the fore when children have value conflicts or are engaged in power struggles in a peer group or between rival peer groups. Presumably they resort to force only when they cannot deal with the situation otherwise or when aggression brings some reward to the aggressor—for example, higher status within the peer group. Moreover, aggressive acts often develop on an ad hoc basis, and such situations cannot be duplicated in a laboratory. Nor am I sure they are necessarily affected by how much or what violence children have previously seen on television.

Many past laboratory studies have also assumed that all violent content has the same effects, although recent studies replicate the discovery, by Himmelweit, Oppenheim, and Vince (1958), that "realistic" television violence is more potent than stylized versions. Moreover, laboratory studies have not yet dealt sufficiently with sociological and other variations in the population. Thus effects research must still consider further the implication of findings that television violence sometimes has greater effects on boys than girls, and on poor and working-class children than on middle- and upper-middle-class ones. But if these sociological variations are relevant, then television fare per se plays a smaller role in real life aggression than has been assumed, and conditions in the larger society must be brought further into the analysis. Are girls less affected by television violence because females are less violent; and what will happen as sexual equality reduces some of the current sex role differences? And if non-middle-class boys are touched more by television violence than others, is this because they see more real violence in their neighborhoods? But many poor and working-class children grow up without seeing more violence than middle-class children. Or are less affluent children angrier at specific individuals or at general social conditions, and do they find insufficient the verbal and other nonphysical forms of aggression used by affluent children to conquer deficiencies in their social environments? And where does race enter in? Are poor black children more likely to react aggressively to television violence between white characters, or to violence by whites against blacks, or vice versa?

It is hoped that laboratory studies which vary the television fare and the demographic makeup of experimental subjects, field studies and natural experiments, and longitudinal surveys will one day answer the question of how much aggression is generated by television violence and how much by other factors. Nevertheless, it remains to be seen whether or not television violence per se can be shown to have more than short-term effects. Presumably, effects cannot follow unless there is a high degree of involvement with what appears on television, but if my primary hypothesis, that most people are not intensely involved with television, is valid, it may turn out that the effects of television have been overestimated.

The hypothesis raises two fundamental questions that need themselves to be empirically resolved. First, my concept of involvement may be too simple; as Paul Hirsch has suggested in personal communication, involvement may actually be a continuum. In such a continuum, low-involved "random viewers" who watch television rather than specific programs are on one end of the pole; "television addicts," who are so intensely involved (presumably with specific program types) to justify labeling them with a concept usually associated with alcohol and drugs, are at the other end. In addition, it is possible that for the majority of viewers there is low involvement with some programs but higher involvement with "favorite" programs.

The second and more crucial question concerns the relationship between involvement and effects, and that relationship may take two directions. One hypothesis is that low involvement generates few effects, conscious or other, because people are not paying much attention in the first place. The other hypothesis is that precisely because people are not paying much attention, they lower the guards of predispositional screening and selective perception, letting themselves be affected without knowing it. Thus Hilde Himmelweit (Himmelweit et al., 1958, p. 37) has argued that when viewers are not deeply involved, television can nevertheless produce what she calls the "drip effect," a cumulative reaction to similar stimuli (or programs) over time. Her hypothesis is well illustrated by commercials, which sometimes have a drip effect on enough people to produce huge sales increases for the product, even though viewers are not paying much attention to the commercials (Krugman, 1965). Similarly, low involvement can "sneak in" at least attitudinal effects, for as public opinion polls suggest, the attitudes people hold on many foreign events reflect what they have learned about them from the news media. If viewers are personally uninterested in the latest change in White House Latin American policy, they are apt to feed back to pollsters what they have seen or read about it.

Which hypothesis is accurate is, in the end, an empirical issue, although it is also a function of what one means by effects. If viewers do not really care what happens in Latin America, or whether they buy Brand X rather than Brand Y, their low involvement may produce an effect, but it is a temporary one that wears off quickly. My own hunch, therefore, is that low involvement is likely to produce temporary, especially attitudinal, effects but that more permanent, and behavioral, effects can be produced only when involvement is high.

Involvement does not, however, take place in a vacuum; it is generated on the one hand by the backgrounds and predispositions people bring to television and on the other hand by the content of television. If predispositions and content coincide, then content is personally relevant; if they do not, then content is personally irrelevant, but in that case involvement remains low and effects remain temporary.

The relationship between media content and media effects can, it seems to me, be looked at fruitfully with an adaptation of the classic Katz and Lazarsfeld (1955) two-step model. Television, like other media (including schools), sends out messages, which sometimes try explicitly or implicitly to educate or to change and reinforce attitudes and behavioral patterns and which sometimes do so even when they do not try. Still, the messages are filtered in two ways: first, by whether they are personally relevant or irrelevant for viewers; and second, by the viewers' level of involvement. These filters, which are both individual and socially created, help determine whether or not attitudes or behavior are affected, although a change in attitude may not necessarily produce a change of behavior.

The relationship between media content and effects is, however, also part of a feedback loop, for the people who supply media fare are, as I suggest later in more detail, attempting to guess what the audience wants or will accept. Further, they are themselves members of society, so that they share, at least to some extent, the attitudes and values of the viewers. Television violence can thus excite them in the same ways as viewers. Consequently, media effects may be shared by suppliers and users rather than being imposed by the former on the latter, although suppliers impose on some viewers what they believe the mass of viewers to want. Moreover, whatever suppliers or users want is further qualified by the economic, political, and other incentives and restraints under which suppliers work as employees of firms in the television industry.

In addition, television does not affect only people but also institutions; in fact, my hunch is that its most important behavioral effects are institutional rather than personal. Whatever the reasons, institutions seem to be more involved with television than viewers. For example, television's coverage of election campaigns has not, so far, affected the outcome of many viewer voting decisions (although perhaps enough to swing close elections), but it has surely affected how candidates and political parties act in election campaigns. Consumer goods firms also react institutionally to television; for example, in 1977 one airline, noting the high ratings of "Roots," sought to persuade ethnic viewers to schedule trips to their country of origin. If this institutional effect was also personal, however, is unknown; even if more ethnics did buy plane tickets to their ancestors' homelands, one would have to interview them to find out whether or not their purchase decision was affected by seeing "Roots." Conversely, toy manufacturers now regularly develop new toys based on television fare (and heroes and villains), but the extent to which youngsters buy "Superman" and "Star Wars" dolls suggests that in this instance the effect is both institutional and personal.

This raises another issue, empirical and conceptual, with respect to effects. In practice, television has been accused, or praised, when only a few viewers are affected by its fare, and researchers have been more concerned with the existence or absence of effects than with the numbers of people, or the percentage of the viewing cohort, who are affected. If one-tenth of 1% of the viewers buy Brand Y in response to a series of new television commercials, the effect is different than when 5% or 10% or more do so. Still, the numbers per se cannot determine the evaluation of effects. For one thing, glasses that are half-full are concurrently half-empty; more important, if 5% of the viewers buy Brand Y, the result is of interest mainly to the sponsor, but if 5% of the viewers change their vote as a result of television election coverage, or if 0.05% imitate a grisly murder or a terrorist act, then numerically small personal effects may have far wider implications and ramifications.

Because of the Commitee's concern with television violence, it may be useful to apply some of the preceding observations to that topic. My hunch is that despite the automatic arousal that programs containing violence generate (Tannenbaum & Zillmann, 1975), such shows have only a temporary effect on most viewers because their involvement, beyond automatic arousal, is low; moreover, for most viewers, violent television fare contains little personally relevant content. Television addicts, who are deeply involved, may be affected by television violence, but then one would still have to investigate why some people become television addicts. More important, because that addiction seems to be more prevalent among low income viewers and because for them crime and violence are in some cases personally relevant—either because they live in high crime districts or because they have little economic hope and many therefore entertain a temporary or permanent career in crime—it is possible that the effects of television violence on low income viewers may result from a combination of high arousal and the personal relevance of the programs. Consequently, one must then go one step further and ask whether low income and the conditions associated with it do not play as important a role in creating the effects of television violence as the medium itself.

In other words, the increase in actual crime and violence and in fear of crime over the last decade may help to explain the popularity of television violence. In fact, my hunch is that whatever the effects of television violence, actual violence encourages its simulated equivalent on the small screen. Thus people who have been victimized by crime, or by fear of crime, may enjoy watching television shows in which criminals eventually get their just deserts—precisely because they do not in the real world. (It would be interesting to compare the television preferences of matched samples of viewers, one victimized by crime, the other not; a further one fearful of crime, the other not.) If my hunch is correct, the effects of television violence may be a latent dysfunction of actual violence. In that case, however, public policies to reduce television violence may have little or no impact on actual violence.

My skepticism about the independent effects of television violence is also supported by the historical record. Americans have now been using the mass media since the beginning of the century, and if the years of steady exposure to radio, comic book, movie, and now, television violence could generate violent acts, we should now all be at each others' throats, and rates of violence, however defined, should have risen steadily ever since the first movie cowboy killed the first movie Indian. Rates of violence have fluctuated over this century, however, and, more important, they seem to have declined considerably since the 19th century, before the mass media were invented.

Uses and Gratifications

The various uses and gratifications approaches have gone considerably further than effects research to emphasize the interaction between viewers and content, but sometimes they have also made assumptions about the audience that need further study. As I understand the approach, uses are equivalent to goal-oriented intentions, and viewers make use of television much as they make use of voluntary associations or supermarkets. Gratifications, however, are a mixture of intended and unintended rewards that fulfill "needs." The conceptual problems of the approach have received considerable attention in the literature (Blumler & Katz, 1974), and I want to raise only those that pertain to my low-involvement hypothesis.

That people use television intentionally has been empirically verified in studies of television news, especially election campaigns (e.g., Blumler & McQuail, 1969; Levy, 1978), but these are situations in which people clearly need—and intend to get—information, particularly now that television is replacing the political party as the main medium for disseminating campaign information.

I wonder, however, whether or not these assumptions hold equally for entertainment television and thus for the main body of television viewing. Intentional usage, other than of the random viewing variety, again implies high involvement, as does, in a somewhat different way, the gratification of "needs." The problem is that we still do not know much about actual audience behavior. For one thing, we have little information about how people watch television. We assume, I think, that they sit in front of the set—but do they? Among the working-class families I knew when I was studying the West End of Boston and suburban Levittown, the set was always on, and often people used it as a radio to provide background "company." Moreover, even when they sat in front of the television set, they did not always watch. The West Enders usually entertained friends at the same time, and everyone watched the first two minutes of each program, two more minutes just before what was then the middle commercial break, and the final two minutes to see how the half-hour programs of that era ended (Gans, 1962, p. 189).

Nor do we know how many people watch the programs of their choice, because in one-set families the person who controls the knob makes the choice, and all others may be involuntary viewers. Presumably, in two-set families of small size all family members get to watch the programs of their choice, but if the typical American family of four (or three) has two sets, someone could still be without choice.

Second, we know little about how people choose their television fare. Some people watch the channel on which they get the best reception, regardless of programming; other choose a favorite channel and then stay with its prime-time fare. Television programmers now select programs with "channel loyalty" in mind, thus further encouraging viewers to make only a single choice every evening. How do viewers make this choice? Do they pick a favorite program and then also watch the remaining programs on that channel? Or do they review the evening program offerings and pick the least objectionable choices, as Paul Klein (1971) has argued? Presumably, some choices are made on a long-term basis, for some viewers have favorite programs which they watch every week. Yet even long-term choosers sometimes make new short-term decisions, as did, for example, the huge audiences who watched "Roots," and then evidently stayed with ABC to sample some of its other programs.

Nor do we yet know what components of a program viewers take into account when making a choice. Every program contains many "bits" of information—and entertainment value—and different viewers may choose the same program for different reasons. Some programs are surely chosen as genres—for example, detective mysteries and suspense programs—but the question remains: How do viewers decide which of these to watch? Many programs must be chosen for their characters, for ever since television became a mass medium, most programs have focused around one or two major characters. Even then one can ask what aspects of a character people choose; for example, do viewers gravitate to Archie Bunker because of his age, sex, socioeconomic status, his attitudes, his poor education, or because of the way he relates to the other characters in his family? And how many choose "All in the Family" for Edith Bunker and its other characters?

Third, and for me most important, too little is known about the qualitative aspects of program use. To begin with basics, we do not know how much of what people see on television they comprehend, (Katz, 1977, pp. 61–63). Nor is there good data on how much attention viewers pay to what they watch, how much they care about what they see, whether or not they become personally involved in programs, and, if so, for how long and at what level of intensity. I have already suggested that some research methods may assume a higher level of involvement than actually exists. If television viewing exists for diversion, asking whether programs are used, whether people derive gratifications from them, whether they have effects, or functions and dysfunctions, may be to take the viewing experience too seriously.

Not only are adequate data lacking, but there is a widespread and growing popular literature that takes the viewing experience very seriously. In recent years, television programs that have become huge ratings successes and movies that have broken box office records have encouraged a good deal of speculation about their social significance and their effect—often negative—on American culture and the American psyche. In 1977, for example, such speculation focused on "Roots" and "Star Wars"; before that it centered on films like Love Story, The Godfather, The Exorcist, and Jaws. But audience size has no necessary relationship to audience involvement or, for that matter, to effects; thus I question that very popular television programs and films generate involvement merely by virtue of their popularity. (I also doubt that because audiences for high culture fare are small, they are therefore involved.) In any case, although every new ratings or box office success encourages new speculation about the American psyche, the viewers themselves soon lose interest in what they streamed to see earlier; sequels and even revivals, both in the movies and on television, are often failures. Moreover, speculation about the social significance of very popular television programs and movies is sometimes based on news media reports of intense audience involvement, but these report the most intense reactions by a handful of people; then, too, it is possible that some of the reports are supplied by industry publicists plugging their product. And sometimes publicists even initiate the reactions, for the use of claques to exaggerate audience interest is not limited to the world of opera.

Just as we still know little about what components of a program are taken into account in viewer choices, so do we know little about what components generate involvement. Presumably, except for genres with built-in audience involvement mechanisms, viewers relate principally to characters. However, characters are themselves multifaceted and are also played by actors, but the West Enders taught me that some viewers relate both to character and to actor and then also evaluate the actor's ability to play the character (Gans, 1962, pp. 191–192).

Nor do we know how people relate to characters. Two decades ago, Horton and Wohl (1956) developed the concept of a parasocial interaction between audience and character, suggesting by that term that people develop vicarious social and personal relationships with a character. Although television programmers try to create characters with whom people can "identify," and thus apply the Horton and Wohl concept, little empirical work has been done on it, and as far as I know no one has investigated the kinds of parasocial relationships most viewers develop—if indeed, they develop any. The notion of intensity is relevant here as well, for viewers may relate in ways that differ in intensity. For some viewers, characters could be substitute or idealized selves, or mirror-images of selves. For others, characters could be surrogate parents, siblings, and spouses or lovers. For yet others they may be close or distant friends, or acquaintances. For many if not most, they are probably like nearby but not adjacent neighbors, people one meets once a week but who are forgotten between meetings.

It would also be useful to know how involvements with programs or relations to characters vary among different kinds of viewers, not only by amount of viewing but also by a variety of demographic, psychological, and sociological variables. For example, if it is true that the currently top-rated "Happy Days" and "LaVerne and Shirley" are popular because unlike most television series they are peopled by working-class characters, we still need to know how working-class viewers relate to them, and at what level of intensity. Or do they watch these characters, without relating to them, because they are glad to see characters resembling those they meet in their own sector of American society? Or because they are proud that some working-class characters have finally made it on prime-time television? Or because these characters, notably "The Fonz," frequently outwit middle-class characters, thus giving working-class viewers a chance to see in fiction what rarely takes place in fact? But how then do the same working-class viewers feel about Archie and Edith Bunker, who are closer to the traditional working-class character as lovable fool? Are they pleased to see even working-class fools on television, or do they use these characters—and the inaccuracies about real working-class life that abound in "All In the Family"—to feel pride that they are better than the foolish Bunkers? Or are all of these questions irrelevant, as Paul Klein (1971) has argued, because viewers watch programs peopled with working-class characters, and others, because they happen to be the least objectionable programs at that particular hour on that particular day?

Assuming that Klein is wrong, one can continue with further questions: How do upwardly-mobile working-class viewers feel about these characters—for example, about Gloria and Mike, the upwardly-mobile second generation on "All in the Family?" Also, what do middle-class viewers make of the working-class characters whom they must, given the size of the ratings, also be watching, and how do they feel about working-class characters who outwit middle-class characters? Similarly, how do black viewers relate to the programs with black characters, and what do they feel about them? And how do they feel about the fact that almost all of the characters they watch are white? Conversely, how do white viewers react to the black characters that have appeared on television in the last 10 years—and often on programs that get high ratings? Indeed, we do not even know why programs like "Good Times" and other have been popular. Do whites choose such programs because, once again, they are populated by non-middle-class characters? Or do they watch black characters because they have so few opportunities to meet blacks in real life? Or do they watch so that they do not have to feel guilty about their lack of contact with blacks? And how, if at all, do these programs affect white attitudes towards blacks they know and blacks in general? Or do all of these questions assume a higher level of involvement with what appears on television than actually exists?

Fourth, even without raising the difficult question of whether or not viewers have "needs" that are gratified or not gratified by television fare, one must ask to what extent the existing fare satisfies their conscious "wants." If viewers in fact have conscious wants on which they would act if they could still remains to be seen, but if they do, one must then ask to what extent these wants are dealt with by the handful of choices to which viewers are limited at any given time period. This is particularly relevant for those parts of the audience who are not primary targets for advertisers—that is, viewers outside the 18- to 49- year-old middle income population. For example, what do poor people want to watch if they chould choose other than the existing options— or old people, or children, and adolescents of various socioeconomic levels? Would poor people prefer programs with characters more like themselves, or does watching the more affluent provide diversion? Or does it create envy, or role models for potential or illusory upward mobility? These questions are obviously too simple, for they ignore diversities within the socioeconomic or age categories; for example, some old people seem to prefer to watch young characters and others want at least some oldsters—although we do not yet know what determines these variations in preference.

These questions also raise the more general question of how much social and other distance people want between themselves and television fare; or, to put it another way, how much they want "reality" or "fantasy." This question touches on the essence of entertainment, although even if the empirical question about people's wants can be answered it does not solve the programmers' problems, because one viewer's reality may be another's fantasy. The question also has policy implications and relevance for the audience role of the mass communications researcher, for whether viewers who want fantasy should get it is an issue that concerns many. Researchers sometimes argue that entertainment should hew close to reality and journalists have recently made the same argument about the so-called "docu-drama" that adds fictional situations and dialogue to programs based on real events and public figures. Researchers and journalists have a vested interest in the study of "reality," however. Even so, those who raise doubts about the use of fantasy raise important empirical and policy questions about the effects on viewers and for society of the emphasis on fantasy.

VIEWER INVOLVEMENT: RESEARCH AND
POLICY IMPLICATIONS

In this chapter, however, I want to concentrate on the question of viewer involvement, because it has so many implications for audience research. For example, if viewer involvement is low, then content analyses that make inferences from content about audience effects are inappropriate. Conversely, for viewers with high involvement, such content analyses may also be inappropriate, because their involvement stems from a highly specific interaction between personally relevant content and viewer that a content analysis, having to generalize, cannot take into account most of the time. Similarly, if involvement is low, it becomes doubtful that television fare can have many effects, at least on strongly held attitudes and culturally, structurally, or psychologically deeply rooted behavior. Still, as I noted earlier, television fare must also be analyzed, through audience research, for the presence or absence of personal relevance. Even if television programs that lack personal relevance turn out to have effects, these will be different from the effects of programs with personal relevance. The introduction of the personal relevancy of programs forces researchers to pay more attention to the social, economic, cultural, and other structures in which television viewers live, for these help to determine the personal relevance of programs; useful models for such research can be found in the now classic but neglected studies by Riley and Riley (1951) and Freidson (1953, 1953b).

Finally, low involvement also has implications for uses and gratifications research, for how much use can people make of television fare in which they are not involved, and what significant gratifications can they derive? And if viewer involvement is low, then questions about the societal functions—or even about the social significance—of television fare run the danger of taking that fare too seriously. To be sure, television programming is public culture and should be studied as such; public culture may, by its very presence, produce societal functions (and dysfunctions) even when no one takes it seriously. Because the major institutions of society obtain little direct feedback from people they often use the public culture as an indicator of people's wants and preferences and act on these indicators; thus American campaign politics has been transformed by the television coverage of election campaigns, even though audience studies of campaign television suggest that viewer involvement in the campaign, and in campaign television is not, with a few notable exceptions, very high. Nevertheless, a public culture that is taken seriously is quite different from one that is not taken seriously, and cultural analyses, or functional analyses of television, must keep the level of viewer involvement in mind.

The question of viewer involvement also has policy implications. For one thing, if public culture is not taken seriously by the audience, then policy aimed at shoring up its functions and eliminating its dysfunctions may be of low priority. For example, some observers have argued that the commonality of the television experience has an integrative function for society and is therefore valuable for a society torn by conflict and dissensus. However, if that experience does not actually integrate, or integrates only at a trivial level—for example, by providing conversational raw material for workplace, backyard and party chit-chat—then those who believe society requires more integration must resort to other, more effective policies.

Also, if viewer involvement is low, presumably many viewers have no significant investment in the current fare and would be amenable to more program variation, giving the suppliers of television programming far more leeway than they now believe they have in program innovation. In fact, program innovation of a kind is already at a peak, as new programs replace unsuccessful ones on short notice. Given the extreme competition for ratings between the networks, however, the now virtually instantaneous replacement of low-rated programs is a policy of desperation, which discourages risk-taking and thus encourages imitation rather than innovation.

Moreover, even when innovation does not take place under conditions of desperation, there are limits to leeway, and it would be illusory to think that the present audience, which is predominantly working and middle class, would choose high culture programs or the upper-middle culture-fare available on public television. Characters with upper-middle-class life styles generally do not do well on commercial television and the vaguely upper-middle-class characters developed by Norman Lear, such as "Maude," may be popular precisely because other characters make fun of their snobbish ways. As long as viewers can still choose the least objectionable program, they are likely to shun upper-middle-class fare most of the time (Gans, 1974, chap. 2) When choice is absent, of course, supplier leeway is maximized. Thus, English viewers loyally watched the BBC's mainly upper-middle-class television during the 1950's, when they had no other choice, but they quickly switched to the independent channel when it became available, subsequently forcing the BBC to alter its programming to compete with the independent channel. Smaller and poorer countries still have "choiceless television." Although one can argue that under certain conditions American viewers should be preempted from the possiblity of making choices—for example, to increase viewing of news programs on a major national issue—restricting people's choices deliberately is neither feasible nor, as a general principle, desirable.

Last but hardly least, the low involvement hypothesis is also, for me, a policy goal. If viewers became highly involved with television entertainment and took it seriously, then television might have more effects than it does, and it might reduce the involvement people have with more important institutions, such as family, work, neighborhood, community, and nation. Thus the lack of strong effects is, for me, a "prosocial" finding. The notion that people should be strongly involved with culture originated with 19th-century Romanticism—although even before that century, the Church and the State used culture for their own purposes. Romanticism prescribed involvement solely for high culture, but I doubt either that high culture always creates high involvement in its audience, or that it should do so.

None of these observations are meant to reject content analysis, effects research, or uses and gratifications studies. Rather, I suggest that, as I have noted elsewhere (Gans, 1978), media researchers must "decenter" the media, keeping in mind that their own high involvement with the media is an occupational attribute not shared by the audience and that television is mainly a tool for leisure. As a result it is far more affected by the dominant economic, political, social, and other institutions in society than our research indicates. To put it simply, society affects television more than television affects society. For example, media researchers have worried about television's impact on crime and violence but not enough about the impact of crime and violence on society and therefore on the viewers and on television. But if the recurring revival of television violence is actually an effect, albeit indirect, of the existence of crime and violence in America, then television research, whether about content, effects, uses, or gratifications, would have to ask many new and different questions about and of the audience.

Fieldwork and Audience Research

Many of the questions I have raised—and will raise—about the audience cannot be answered completely by the currently dominant research methods. If we are ever going to understand fully how viewers use television, what meanings they impute to or project on program content, how they relate to what they see, and what roles television plays in their lives, viewers will have to be studied in the process of watching, and shortly afterward, before the ephemeral effects of most programming wear off. This means finding a way of using fieldwork methods, particularly of sociology and anthropology, in audience research in order to be with people as they are choosing programs, watching them, and talking about them—and to know these people well enough in terms of their other roles and activities and their positions in community and society to be able to understand how television fits into their lives.

Fieldwork methods are, to be sure, difficult, time-consuming, and from the point of view of the currently dominant methodological paradigms, unscientific. Field studies of television viewing present yet another problem: They virtually require researchers to live with the people they study, with an almost clinical level of contact. Most field studies, notably of communities, have concentrated on the public spheres of life, which are easier to enter and in which the researcher intrudes less on people's privacy. Comparatively little work has been done, on the other hand, in the more private spheres of life.

Undoubtedly some people would let researchers visit them at home while they are watching, at least for a short period. However, people are often inhibited by or "perform" for fieldworkers at the outset of a field study. Although these effects disappear after a few days or weeks, they would not disappear during a short-term study. Some fieldwork on television viewing could be done as part of a community study, for fieldworkers doing such studies often interview people at home at night and in a quasi-social situation, but obviously this approach would be exorbitantly time-consuming. Some work can be done by watching with people in public settings—for example, at bars or in the public rooms of community centers, hospitals, and old-age homes—but then the samples would be atypical. Ingenious fieldworkers who can double as boarders, housekeeper, or babysitters could probably obtain the relevant data most easily, but they would be limited to studying one family at a time.

In short, the logistic and other problems of doing fieldwork on television viewing are great. Nevertheless, they have to be overcome if we are to deepen our understanding of how people watch television and what differences it makes to them.

THE SUPPLIERS OF TELEVISION
AND THE AUDIENCE

Mass communications researchers have only recently begun to pay empirical attention to the suppliers or producers of television fare. So far, most empirical research has been conducted among the suppliers of local news, probably because they are accessible to researchers, whereas studies of the suppliers of national news and entertainment fare must be conducted in New York and Los Angeles. The current interest in empirical research was preceeded by armchair studies, conducted by conservative and Marxist mass society theorists, who, working within a stimulus–response framework, viewed the suppliers of commercial entertainment as either catering to the vulgarity or exploiting the weakness of the "masses," depending on the ideology of the theorists (Bauer & Bauer, 1960; Gans, 1974, chap. 1). The work of these theorists still has a limited usefulness insofar as it raises, if only by omission, the question of whether the suppliers of commercial entertainment impose their product on the audience or whether the audience also imposes its preferences on the suppliers through one or another form of feedback.

To my mind the most useful approach to studying the suppliers has been developed by Paul Hirsch (e.g., 1972, 1977), who compares them to other suppliers of fashion goods and then studies the organizational imperatives— or limits to leeway—under which the industry, specific firms, and individuals within them operate. Because of my interest in the television audience, I shall limit my discussion to specifying the suppliers' relationships to the audiences and the processes by which suppliers and audiences have impact on each other, building here on empirical work already done by Hirsch (1972), Cantor (1971, 1972), and others.

The suppliers of television work within a well-institutionalized general feedback system, based largely on ratings and to a lesser extent on program pretests. Both provide only limited feedback information, however. The ratings supply data on audience sizes at a given time and permit inferences about what viewers did not watch as well as guesses about what they might have liked simply because they watched. Program pretesting methods obtain more direct and detailed feedback, but most reliably about familiar genres, characters, and situations—and some of the biggest successes on television flunked the pretests, for example, "Peyton Place" and, more recently, "All in the Family." In fact, much media audience research seems geared to certifying audience support for the familiar and thus fails to meet the need of content suppliers who want an answer to an admittedly difficult question: What will the audience accept next time? (I have never forgotten the response of a news magazine editor when I asked him why he eschewed audience research. He pointed out that the researchers once told him that the most popular cover illustration was of a pretty girl on a red background, but then he asked me "How often can I use a pretty girl and a red background on my covers?")

In other word, suppliers, who for commercial reasons must seek the largest possible audience, can obtain only limited guidance from audience research, at least as presently conducted in the industry. The question then is, how do they manage to provide the audience with acceptable fare? One possible answer is that they need not do so. For one thing, if audiences are involved only minimally with what they see, the suppliers must provide only one least objectionable program at any given time, and given that they deal with a diverse audience, what is most objectionable to one sector of the audience may be least so to another (Klein, 1971). Moreover, they need only supply a limited number of programs, because until the entire audience has access to cable television, it is forced to choose from a small number of alternatives at any given time period. Another possible answer is that even within these limits the suppliers are not providing acceptable fare, as suggested by the high and constantly increasing number of annual program failures.

The question can be rephrased as follows: Insofar as the suppliers provide acceptable television programs, how do they know in advance what the audience will accept? One hypothesis is that they have informal feedback ties to the audience; another, that they are themselves much like the audience and can therefore develop programs that please them and will also please the audience. A third hypothesis is that as professionals they resort to a set of professional guidelines and conventions based on past experience; a fourth is that the mass media recruit people who have an intuitive sense of what the audience will accept.

Based on my research among national television journalists and the available data on the suppliers of television entertainment, all of these hypotheses have some merit (Baldwin & Lewis, 1972; Cantor, 1971, 1972). Scriptwriters and producers often test their ideas on nearby audiences before production, on family members, and on friends; in some instances they get story ideas from what they observe and hear from family and friends. Meanwhile they must also test ideas on executives and sometimes advertisers, who are not only audience members but also veto groups. During production the reaction of technicians and secretaries is often monitored to provide clues to later audience reaction.

By all standard demographic variables, the suppliers of television programs are quite different from most audience members and should therefore not be able to please viewers by pleasing themselves. Moreover, some scriptwriters and producers are either frustrated "serious" novelists or dramatists—or are working on the Great American Novel after hours. Many others have "higher" personal tastes than the majority of the audience, which is not surprising, given their educational level and upper-middle-class status. Nevertheless, they have either grown a second, "professional" self, which is closer in taste to the audience, and learn to write for that self; or as professionals, they have developed an intuitive sense of what the audience will accept, even if they personally reject the audience's tastes and preferences. Popular novelists writing about Hollywood have often noted the status and taste difference between suppliers and their audiences, implying that it breeds a cynical rejection of the latter that is built into what the former produce. Whether or not this is accurate is an interesting empirical question, and it would be useful to discover if in television, as in other fashion industries, the status and taste differences between suppliers and audiences have any impact on the product and on audience acceptance. (Concurrently, it might also be worth studying whether or not the assumption that folk artists were and high culture artists are socially and culturally closer to their audience is actually true; my hypothesis being that it is not.) Moreover, some of the differences between creative people and audiences are probably reduced by television executives, advertisers, and others who are paid to worry about audience interest and suggest alterations in scripts to bring programs closer to median audience tastes. As such, they may function as representatives of at least part of the audience.

Last but hardly least, scriptwriters and producers also draw on professional guidelines, mostly unwritten and based on generalizations drawn from experience, which have two major components. One is an image of the audience, its demographic characteristics, and its preferences, which can be used to judge new ideas for their acceptance (Gans, 1957, 1979, chap. 7).

Research on the nature of that image and on the data sources on which it is based is needed, particularly because it may not be entirely accurate about the actual audience. The second set of guidelines comes from the genres and formulas of plots, situations, and characters within which most television fare is cast. Many of these are of venerable vintage, which means that they have been extensively "tested" on past audiences; some go back to the days of folk culture.

The creation of new programs invokes all of these sources of audience feedback, although in most cases program innovation involves only the addition of novelties to, or new combinations of, old genres and formulas. The main dilemma of suppliers appears to be in determining how much innovation would attract an audience bored with a present program and how much innovation would make the new program unfamiliar and therefore possibly objectionable. How suppliers solve this dilemma and what inferences they make about the audience during the innovation process deserves study.

Yet even more interesting are the instances in which suppliers take major innovating steps and under what conditions, in the industry, within the firm, and in the audience, such steps succeed or fail. The successes, for example, of "All in the Family," "Maude," and the "Mary Tyler Moore Show"—which represented, by industry standards, major innovations—are of course more visible than the failures, even though the latter are more numerous and perhaps empirically even more interesting in assessing the suppliers' image of the audience. Innovations also set precedents and encourage imitations, but the processes by which they encourage or discourage further innovation or provide the suppliers with clues as to the shape future innovation should take, need to be studied. My hunch is that despite all the feedback systems and professional guidelines, the suppliers of television are as surprised as laypersons when an innovation succeeds. This suggests how fragile and unpredictable the feedback ties between the suppliers and viewers actually are.

Whatever the role of direct and indirect feedback, however, it would be oversimple to explain television fare as the outcome of or a reaction to feedback. For one thing, the suppliers are not interested in feedback per se, for television fare is shaped by the advertisers' and networks' desire, or need, to attract the largest possible audience among the buying public, which is often equated with 18- to 49-year old women of median or higher family income. But even that public is diverse, so that suppliers can never satisfy everybody even in that audience.

More important, television fare cannot be designed by reaction to feedback alone—scriptwriters and others must, after all, create programs. Critics of popular culture have therefore argued that the suppliers impose content on the audience, although impose is too strong a word, because viewers can ignore programs they dislike. Still, television fare is always a combination of audience feedback and supplier input, and in theory one can distinguish between the two. This distinction can probably not be applied empirically, but one can study how scriptwriters get ideas for and fashion programs and subsequently try to elicit from them to what extent the components of a program have taken either feedback or the suppliers' image of the audience into account, meanwhile also tracing the sources of ideas that are not based on feedback.

Supplier input, as I called it, undoubtedly has both conscious and unconscious sources, like all other creative activity in popular or high culture. Because television programming is highly competitive and cost minimization is a major aim, there is a conscious effort to develop programs—and program components—that can be produced inexpensively. Thus, plots often turn in directions that cost the least to put on the air. But scriptwriters and producers also have "ideas" that they want to develop and in addition at least some have "messages" they want to insert into entertainment fare. Some years ago I met with a group of Hollywood television writers, who spoke of their attempts to insert moral or ideological themes and commentary on current events and issues into their scripts—as well as of their struggles with producers to keep them in the scripts. Cantor (1972) and Baldwin and Lewis (1972) have reported similarly about the conflicts between writers and producers over the amounts and forms of violence to be used in television scripts.

Still, many of the details of plot, character, and situation come into television content unconsciously, as writers draw on their environment and experience as well as on what they have read—and seen on television—in exercising their imagination. Presumably, television's prototypical brash teenage character is inspired at least to some extent by the children of middle-aged, upper-middle-class writers who have probably raised their offspring in the same child-centered way as upper-middle-class professionals elsewhere in America. Detective stories often draw on current events, particularly those relevant to people in a media industry, and an abnormal number, it seems to me, now deal with crooked or victimized film producers and Hugh-Hefner-like magazine publishers. This should not be surprising, for both reality and fantasy are socially constructed, and all other things being equal, the builders find their construction material close at hand.

This in turn raises questions about what institutions and sectors of American society are incorporated in television entertainment fare, even in the exaggerated forms of Hollywood fiction, and what parts are left out—and what impact, if any, this has on audiences, particularly those who reside in the left-out sectors. One can raise similar questions about what philosophical, moral, and political ideas enter into—and are left out of—the fare and with what consequences, if any, for the audience.

Eventually, industry studies must proceed from the macrostructural questions that Hirsch (e.g., 1977) particularly has raised, to these and other microcultural phenomena. Such studies would attempt to identify all of the inputs into different program types and the sources from which these inputs came. Once such research is done, it may be possible to consider more systematically to what extent television fare is made up of audience feedback or, rather, the suppliers' reactions to it and to what extent the suppliers themselves bring their own personal and social experience to bear on the scripts that viewers finally see on the small screen.

THE LARGER SOCIETY AND
THE TELEVISION AUDIENCE

Television does not exist in a vacuum but within a larger society, which affects both suppliers and viewers. The "larger society" is a vague term, and sometimes it is possible to identify specific institutions that affect television; sometimes, on the other hand, the general term will have to suffice until more research is available on the macrosocietal pressures impinging on television.

For example, both television fare and the viewers are affected by the state of the national economy. Although the income of television stations and networks does not seem to vary directly with ups and downs of the business cycle, inflation affects the cost of programming as it does everything else; thus the state of the economy presumably has a direct effect on the cost and therefore on the content of television fare. More narrowly, intermedia relations, such as the competition for advertising, will not affect only funds available for programming but to some extent the programs themselves. Bogart (1972, chap. 7) and others have shown how the print media have changed in response to the loss of advertising to television. Similarly, the movies have resorted to considerable programmatic innovation in order to cope with the departure of much of their audience to television. Conversely, no one has looked at the changes in television fare since the medium attracted so much of the magazine advertising and the movie audience.

The polity also has impact on both suppliers and viewers. The concern over the last 10 years with television violence reflects political pressures from parents who are upset about what their children are exposed to, although we still do not know why politicians responded to these parents in the late 1960s and not earlier. Politicians do not always respond to pressure, however. Sometimes they look for issues that will readily engender publicity and support for them, and television violence may fall into this category.

The concern with television violence was also generated by the Kennedy assassinations and the ghettto uprisings of the 1960s and is kept alive by the increases in the rates of violent crimes of the 1970s. Unique and highly visible events may, however, also be surrogates for other social forces, and I am struck by some similarities between the current concern with television violence and the earlier concern with movies. Both the Payne Fund studies of the movies in the mid 1930s and the recent massive research effort on television violence came about 20 years after the respective media had begun to be mass media. They also came during periods of economic downturn. Sklar's historical analysis of the movies argues that the Payne Fund studies were instituted because of fears of the more affluent about what the movies were doing to the less affluent and specifically whether or not films were threats to the established social and cultural order (Sklar, 1975, pp. 135–139). Such fears are often expressed by more affluent sectors of society during periods of economic and other difficulties. Whether or not a parallel can be drawn between the Payne Fund studies and the recent spate of television violence research remains to be seen.

Perhaps the most interesting set of questions arises, however, about the impact of the larger society on the audience. If television has uses and provides gratifications, they are served by the media but do not necessarily originate with them. After all, if there are needs that are gratified by television, the needs themselves come from elsewhere—either from universal characteristics of the human species; or from universal aspects of American culture, society, and personality; or from more specific sources. And if different populations have different needs, one must ask what factors in their backgrounds and experience create these needs.

Moreover, one can ask whether or not television has uses and provides gratifications that are not satisfied by other institutions or which were once satisfied by them. Do viewers want television comedy because work or family life never did, or do not now, provide opportunities for laughter? Must television provide suspense because for many people work in bureaucratized situations is now, or has always been, short of suspense? The more general question here is whether or not television complements viewers' other experiences, counteracts such experiences, or fills an experiential void.

If a major function of television entertainment is to provide diversion, it is relevant to ask what viewers seek diversion *from*. Perhaps everyone requires a certain amount of humor or suspense, but perhaps some people need more than others, and perhaps they need it because their other experiences—and the other institutions with which they come into contact—create a need or want for diversion from them. This means we must look at television viewing patterns and preferences in relation to other social conditions. People who are totally fed up or unable to cope with "reality" may become television addicts—or addicts to drugs and the like—but what about more typical viewers?

Both film and television producers have put forth some industry folklore on this topic. They suggest, for example, that during economic downturns, people are most diverted by "happy" pictures, especially about the rich; and that they like war films only after a war has been won (Landry, 1977). Whether these are accurate judgments about the audience or industry rationalizations is unknown, but they raise the important empirical question whether or not and how events in the larger society affect audience preferences, or at least choices. Does the state of the economy alter viewer choices? Are the recent spate of disaster films, and science fiction movies about outer space generated by macrosociological factors? Does sexual puritanism generate a demand for pornography, as Steven Marcus (1966) has suggested in his study of 19th-century Victorian England? Such questions can and should be raised contemporaneously, relating current television fare— including television violence—to conditions in the larger society. They can also be raised historically, to study the longer term patterns of television and movie content—and changes in them—in relation to events and structural changes in the larger society. Will Wright (1975) has argued that changes in Western heroes can be related to changes in the American economy, although his analysis is not convincing.

More simply, one can ask what demographic changes in the society do to and for television. If television fare is created for 18 to 49-year-olds, will a reduction in their number as the population ages force a change in this practice and in programs? Can we then expect older heroes and heroines— and young villains? Rising levels of education have obviously affected television fare in the last 20 years. Plots, characters, and dialogue have all become more complicated and sophisticated. What about changes in socioeconomic levels? Does a higher median income alter television fare? And what has the long-term increase in white collar workers, professionals, and technicians among viewers done to programs? More immediately, what changes in fare, if any, have resulted from the economic difficulties of the last several years? And from the fact that a larger number of viewers than ever before are unemployed or in fear of losing their jobs and frightened of inflation. Or perhaps changes in television programs develop independently of what goes on in the larger society; in that case, changes in "fantasy" bear little or no relation to changes in "reality."

Research on the effects of the larger society on television has one inherent problem; it can easily produce correlations between society and television, which are often imaginative, but cannot answer the causal question of how the former affects the latter. Causal analysis requires the tracing of social processes and their consequences, both on the suppliers and viewers of television, notably the former. If it is possible to identify the sources of supplier input, some of which presumably represent reactions to what is going on in the larger society, it may also be possible to see more clearly how the larger society leaves its mark on television fare.

THE TELEVISION RESEARCHERS
AND THE AUDIENCE

Finally, I want to raise some questions about our own work: what it does to and for audiences, and for what reasons. Television researchers are not very important participants in the social processes that create television fare and affect viewers, but they play enough of a role to justify some investigation.

Mass communications research began, one could argue, as the armchair speculations of the mass society theorists, with conservatives, such as Ortega y Gasset, worrying that the mass media would bring about the revolt of the masses, toppling the cultural elites from their high perch in the power structures of various European nations. Later, socialists such as Dwight MacDonald (1957) and Herbert Marcuse (1964) perceived the suppliers of entertainment as "lords of kitsch" who exploited the masses and tempted them away from socialism. Sklar (1975, pp. 135–139) has viewed the researchers who participated in the Payne Fund studies of the movies as unwitting agents of social and cultural control. They provided evidence and scientific legitimization for upper- and upper-middle-class "reformers" fearful that the films might help the so-called dangerous classes to overthrow the cultural and perhaps also the political order.

Conversely, modern academic television research grew out of the radio research that Paul Lazarsfeld and others carried out for the networks before they set up their own research departments. These researchers were thus helping to provide feedback to the suppliers, although they also used the applied studies for which they received funding to explore issues in basic research.

Today, the roles of academic researchers are much more complex, and the services they supply to others are much less visible, even to the researchers themselves. For one thing, researchers are now much freer to initiate their own studies and to satisfy either their personal curiosities or those of their discipline and colleagues. To be sure, studies that require grants are also influenced by the curiosities and interests of the foundations, the government, and the television networks, insofar as they sponsor or influence academic research. What we or granting agencies become curious about or fail to become curious about—and why—has not been much studied, however. Likewise, little attention has been paid to the metaphysical assumptions of effects research, such as its emphasis on "bad" effects. Until the recent interest in prosocial effects, researchers—much like the literary critics of the media— did not interest themselves in the good effects of the media, and even work on prosocial effects is more concerned with reforming the media than with identifying good effects in the present fare. Nor has anyone, as far as I know, thought about studying possible bad effects of high culture. The evaluation of effects is, nevertheless, a judgment of the evaluator, and perhaps the fact that researchers, like media critics, are denizens of the print culture and more at home with high than popular culture has affected their explicit and implicit evaluations of effects and their decision to pay most attention to effects they consider bad.

Many individual researchers are not aware of the values that underlie their curiosity, particularly those who use television mainly as a research site to explore theoretical issues of their behavioral science disciplines. But other individuals, and most if not all granting agencies, are concerned with television and either want to serve or affect the audience. But even researchers who use television only as a site for basic research may produce studies that have implications for the audience.

Without distinguishing between conscious and unconscious motives or implications, my hunch is that often research seeks, in one way or another, to justify "upgrading" television programming—or, to put it another way, to provide fare that is congruent with upper-middle-class tastes. Likewise, the increasing amount of research on television's reportage of national elections often seeks to upgrade the quality of political information, and perhaps even to change election campaigns themselves, in order to put more emphasis on the issues and less on the personalities of the candidates. Patterson's and McClure's (1976) research on campaign news and political advertising makes this goal virtually explicit. Insofar as an issue orientation to politics correlates with upper-middle-class status, the implicit goals of entertainment and election research are not very different.

The same type of question can be asked about the social research on television violence of the last 10 years. At one level, mass communication researchers were responding to the availability of research funds after many years of drought, but in the process they were also unwittingly co-opted into various economic and political agendas. For one thing, they became researchers for the politicians and interest groups seeking to alter network programming. At the same time they were also involved, if only indirectly, in the networks' attempt to maintain the status quo—reflected, for example by the networks' ability to exclude some researchers from the Surgeon General's Committee (Cater & Strickland, 1975). In addition, the television researchers gave an unintended assist to the movie industry and to other mass media that supply violent fare but that were not researched. More generally, they became involuntary participants in the general debate over violence in American society and by implication became allies of those political and economic interest groups and forces that sought causality in the mass media and as a result ignored the other forces and conditions in American society that help to produce a high rate of violence. Although the Violence Commission set up after the ghetto uprisings and the assassination of Robert Kennedy grappled with questions about the role of poverty, racial segregation, and the structure of the criminal justice system in encouraging violence, the television researchers could not do so, by the very nature of their assignment, although many pointed out in their reports that television was not the only factor in the increased rates of violence over the last decade.

I assume that the granting agencies, if not the researchers, hoped that the studies would justify a reduction of television violence, with the further hope that this in turn would reduce real violence. If there is, however, no significant causal relationship between television and real violence—as I believe to be the case—then the researchers were being used to support a spurious solution to a real social problem. Moreover, at the time it seemed like an easily achievable solution, and one that would not require having to deal with inequities in the economy and the criminal justice system. The researchers themselves may not have shared any of these values and perhaps were not even aware of them. They could do the research because they were personally opposed to television violence. In addition, they may also have wanted to eliminate television violence as a first step in upgrading television fare generally.

I do not mean here to suggest conspiracies. To the contrary, I think the basic impulse behind much of the television research on violence and other topics is more prosaic and much less political. Some researchers, I think, would like to upgrade television fare for themselves, using research here in place of market power, because researchers—and academics generally—are too small a population to be profitable to television's advertisers. But there is another aim, which is equally latent and lies behind a good deal of research on topics other than mass communications: To reduce the differences between intellectuals and nonintellectuals and between experts and laypersons—or, to put it more simply, to remake the American public to be more like us.

These are only hunches, to be sure, and I put them on paper mainly to raise the question of the role of television research in television itself; to ask what television researchers want to do and for the audience, and for the larger society, intentionally or otherwise. The next questions, then, are: What can television research actually do to and for the audience and what should it do, assuming it ever obtained the power to do anything significant.

REFERENCES

Baldwin, T., & Lewis, C. Violence in television: The industry looks at itself. In G. Comstock & E. Rubinstein (Eds.), *Television and social behavior* (Vol 1). Washington, D.C.: U.S. Government Printing Office, 1972.

Bauer, R., & Bauer, A. American mass society and mass media. *Journal of Social Issues*, 1960, *16*, 3-66.

Blumler, J., & Katz, E. (Eds). *The uses of mass communication*. Beverly Hills: Sage, 1974.

Blumler, J., & McQuail, D. *Television in politics*. Chicago: University of Chicago Press, 1969.

Bogart, L. *The age of television* (3rd ed.) New York: Ungar, 1972.

Cater, D., & Strickland, S. *TV violence and the child: The evolution and fate of the Surgeon General's Report*. New York: Russell Sage Foundation, 1975.

Cantor, M. *The Hollywood producer.* New York: Basic Books, 1971.

Cantor, M. The role of the producer in choosing children's television content. In G. Comstock & E. Rubinstein (Eds.). *Television and social behavior.* Washington, D.C.: U.S. Government Printing Office, 1972.

Freidson, E. Adult discount: An aspect of children's changing taste. *Child Development,* 1953, *24,* 39–49. (a)

Freidson, E. Communications research and the concept of the mass. *American Sociological Review,* 1953, *18,* 313–317. (b)

Gans, H. The creator–audience relationship in the mass media. In B. Rosenberg & D. White (Eds.), *Mass culture.* Glencoe: Free Press, 1957.

Gans, H. *The urban villagers.* New York: Free Press, 1962.

Gans, H. *Popular culture and high culture.* New York: Basic Books, 1974.

Gans, H. Social research on broadcasting: Some additional proposals. *Journal of Communications,* 1978, *28,* 100–105.

Gans, H. *Deciding what's news: A study of CBS Evening News, NBC Nightly News, Newsweek and Time.* New York: Pantheon Books, 1979.

Himmelweit, H., Oppenheim, A., & Vince, P. *Television and the child.* London: Oxford University Press, 1958.

Hirsch, P. Processing fads and fashions. *American Journal of Sociology,* 1972, *77,* 639-652.

Hirsch, P. Occupational, organizational and institutional models in communication research. In P. Hirsch, P. Miller, & F. Kline (Eds.), *Strategies for communications research.* Beverly Hills: Sage, 1977.

Horton, D., & Wohl, R. Mass communication and parasocial interaction. *Psychiatry,* 1956, *19,* 215-229.

Katz, E., & Lazarsfeld, P. *Personal influence.* Glencoe: Free Press, 1955.

Katz, E., Blumler, J., & Gurevitch, M. Utilization of mass communication by the individual. In J. Blumler & E. Katz (Eds.), *The uses of mass communications.* Beverly Hills: Sage, 1974.

Katz, E. *Social research on broadcasting; proposals for further development.* London: British Broadcasting Corporation, 1977.

Klein, P. The men who run TV aren't that stupid . . . they know us better than you think. *New York, 4,* January 25, 1971, 20–29.

Krugman, H. The impact of television advertising: Learning without involvement. *Public Opinion Quarterly,* 1965, *29,* 349-356.

Landry, R. Viet Nam, once No. 1 avoidance theme for U.S. films, emerging as "adjustment problem" cycle. *Variety,* October 26, 1977, p. 4.

Levy, M. The audience experience with news. *Journalism Monographs,* No. 55, April 1978.

MacDonald, D. A theory of mass culture. In B. Rosenberg & D. White (Eds.), *Mass Culture.* New York: Free Press, 1957.

Marcus, S. *The other Victorians.* New York: Basic Books, 1966.

Marcuse, H. *One dimensional man.* Boston: Beacon Press, 1964.

Patterson, T., & McClure, R. *The unseeing eye.* New York: Putnam's, 1976.

Riley, M., & Riley, J. A sociological approach to communications research. *Public Opinion Quarterly,* 1951, *15,* 445-460.

Sklar, R. *Movie made America.* New York: Random House, 1975.

Tannenbaum, P., & Zillmann, D. Emotional arousal in the facilitation of aggression through communication. In *Advances in Experimental Social Psychology* (Vol. 8). New York: Academic Press, 1975.

Wright, W. *Sixguns and society: A structural study of the Western.* Berkeley: University of California Press, 1975.

5

An Organizational Perspective on Television (Aided and Abetted by Models from Economics, Marketing, and the Humanities)

Paul M. Hirsch
Graduate School of Business
University of Chicago

On an analytical continuum, with art and commerce posed as polar opposites, American television producers and executives have long told us this medium is organized as a business. Indeed, American television is less "distracted" than any other mass medium by loyalties to such noneconomic goals as editorial policy and standards, generations of family ownership, or idiosyncratic decisions based on personal taste. It is an economic institution, first and foremost, responsive to market forces, and concerned only incidentally with questions about its broader cultural role or possible effects on a nation of viewers.

In this chapter I begin by placing the television industry in an organizational and economic context and by examining its products, markets, and strategies from the standpoint(s) of the organizations involved. Because much of what the industry has learned from its experience and research also pertains to questions and issues framed by government agencies and social scientists, I try to suggest areas in which findings and interests overlap and can inform each other. This survey of economic, organizational, and marketing perspectives on the industry also briefly discusses some of their implications for social scientists, cultural policy, and exploratory research designs.[1] Although the United States presents a far more commercial media system than other nations (as in most other sectors of American culture),

---

[1]For an elaboration on some themes that can only be touched on in this essay, see Hirsch, 1977a, 1977b, and 1978.

much of what follows should have applications for analyses of mass media in other nations as well. Especially where ratings are collected and acted on, many of the same pressures to maximize audience occur, though often to a less fearsome extent. And comparisons of the economic structure of American mass media, and the occasionally anomalous content that results from their primary commitment to profit and private enterprise, with that of nations whose media systems do differ in kind also may be instructive.

ECONOMIC AND ORGANIZATIONAL ASPECTS OF THE TELEVISION INDUSTRY

For practical purposes I limit my definition to the three major commercial networks, later adding the noncommercial Public Broadcasting System. Although the major networks own only seven television stations each, they reach a nationwide audience through approximately 700 affiliated stations that broadcast their programs simultaneously under the terms of negotiated contracts. So long as it remains more profitable for these "local" stations to affiliate with a major network, commercial television will continue to mean the presentation of programs by CBS, NBC, and ABC. Each network thus seeks to gather together a national audience for national advertisers, for whom it offers substantial economies of scale while providing affiliates (and viewers) free programming and paying them a percentage of the revenues obtained from transmitting national ads to their audience.[2] Although these arrangements are potentially fragile (local stations could refuse to renew their contracts), network affiliation is lucrative and unlikely to be discontinued for some time to come. (In the early 1950's, when affiliation with radio networks became an economic liability, radio stations did discontinue; the same networks, however, had already shifted most of their offerings and affiliation arrangements to "radio with pictures," as television was first called.) Consequently, my unit of analysis is the networks that procure and distribute much daytime and nearly all prime-time programming, rather than the stations that serve primarily as conduits for network programs.

Wherever three firms account for over 60% of an industry's sales, economists expect it will behave according to their models of an oligopoly. Television networks account for well over 60% of the entire industry's revenues and for more than 90% of all prime-time programming. Some of the organizational correlates, or probable consequences of this concentration at the program distribution end of the television industry, also characterized other mass media industries when they exhibited a similar market structure—

---

[2]These arrangements are well described by Noll, Peck, and McGowan (1973) in greater detail.

motion pictures and radio during the 1930s and 1940s. In fact it is likely the classical model of mass media and mass communication requires and unwittingly assumes an oligopolistic structure with few channels, a national market, and content designed for any and all members of society. Yet in response to television, American radio, movies, and magazines now exhibit greater diversity in content and appeal to more specific audience segments rather than to an undifferentiated populace. They also are far more competitive industries and less dominated by a restrictive number of suppliers and distribution channels. What used to be called the nature and "effects" of (inherent in) these media was really more simply a description of the cultural consequences of a particular and transient set of economic, organizational and distributional arrangements.[3]

In what is now often referred to as their "golden age" (since 1900 no mass medium has remained dominant for more than 30 years), these industries, like television today, relied on a small number of "reliable" production companies to supply the entertainment they distributed. Entry into the field of either production or distribution was severely limited; nearly all output was aimed at a single "mass" audience and adhered rigidly to a small number of familiar formats, genres, and plot outlines, often differentiated only in terms of which actors were featured. Relationships among the organizations and individuals in different sectors of these industries (then) and television (now) are characterized frequently by public disputes and personal rivalries, many of which are publicized, romanticized, and orchestrated by press releases that contribute further to the glamor and mystique of the entertainment business. In fact, much of the raw material for organizational analyses of disputes between networks and affiliates, or contract terms with actors, writers, production companies, producers, and network presidents, is already reported regularly in such journals as *Variety* and *Broadcasting*. Although they are less glamorous than the creators who produce the entertainment they disseminate, distributor organizations like networks, syndicators, and magazines appear to be the more powerful but less often studied of the two.[4]

For television specifically, Dominick and Pierce (1976) reported a narrowing in the range of program types offered in prime time between 1953 and 1974, with action–adventure offerings showing an increase at the expense of "reality"-oriented (documentaries, interview, public affairs) and variety show formats. In this respect, television networks have grown to resemble such other oligopolies as the automobile and movie industries, with "quality

---

[3]Paul Goodman (1963) suggested much the same point in a remarkable series of articles about television in the *New Republic*.

[4]An interesting exception to this statement is Lourenco and Glidewell's (1975) study of conflicts between a television network and one of its owned-and-operated local stations.

television," like small cars and art films, excluded from the domestic product line and traditionally imported from other countries.[5] One may also find parallels with, among others, the New York Stock Exchange, whose members' rates until recently failed to discriminate large from small buyers and who argued unsuccessfully against ending the practice in language very close to the television networks' stated rationale for opposing the growth of unrestricted cable television—for example, by "siphoning off" enough affluent buyers (viewers), the new marketing system (the competing medium) would decrease the quality of services (programs) for the less affluent (Welles, 1975).

In terms of program innovations, the tendency has been for the network with the lowest rating (formerly ABC) to differentiate its products from the others by introducing new concepts (Monday night football, higher levels of violence, "tough" documentaries); successful innovations, like price increases, have then been "matched," or imitated within a year by the remaining networks. Occasional efforts have been made by each network to integrate vertically (backward) by taking over entertainment production from outside firms, though in each case the federal government has threatened or brought antitrust suits. Network production, participation, or co-ownership of series has declined, though similar strategies are common to many (not necessarily oligopolistic) industries—for example, where automobile manufacturers purchase supplier firms, newspaper publishers purchase timberland or newsprint suppliers, or department store chains purchase part ownership of manufacturing firms producing goods sold under the retail chain's private trademarks.

High profits for the successful network(s) and barriers to the entry of new ones are attributable more to present channel scarcity than to activities on the part of existing networks (with the possible exception of their active opposition to cable television). Indeed, we have noted an interesting tendency for *each* dominant mass medium, historically, to be centralized and characterized by a few dominant firms—five or seven major movie studios (which also owned most movie theaters), four radio networks, two major wire services (plus smaller competitors and features syndicates). There also is some reason to expect that a successful cable television industry would itself seek economies of scale and higher profits by replicating a similarly oligopolistic industry structure dominated by pay cable networks. (Parenthetically, we also note that until recently the United States boasted *more* television networks and channels per city than most other nations, with all competing for the same audience, however, as distinct from the British model.)

---

[5]Television executives have been more responsive to their audiences than was Henry Ford, whose share of the automobile market dropped upon GM's introduction of a choice of colors for different cars. Ford's classic reaction at the time reportedly was, "They can have any color they want, so long as it's black."

A further characteristic of ologopolies is an interdependence and mutual awareness of how the market—in this case the entire U.S. population—is to be divided up. Here, the absence of additional (VHF) television channels has played a crucial role in the outcome. For rather than "segment" the market into component consumer (viewer) groups—much as detergent or toothpaste manufacturers target different brands for different user groups—the three networks compete for the largest possible share of an identical audience, preferably viewers between 18 and 49 years of age but otherwise largely undifferentiated. When this market structure of only three firms is combined with some salient features and characteristics of the television audience, we can begin to account better for the industry's choice of which programs to present in each television season.

TELEVISION VIEWING: A NEW TYPOLOGY

A most important characteristic of the market for national television, and basic to all advertising and programming strategies, is this fact: Within each day and time slot, *the size of the total television audience remains essentially the same*. That is, each network seeks to increase its *share* of the existing audience rather than to attract additional viewers whose sets are not turned on. The competition in business terms is for market share in a stable market, with little expectation that new offerings in any period will attract substantial numbers of people not already watching (regularly or occasionally) in that time period. A primary goal is to hold on to those presently tuned in while enticing viewers of the competing networks to switch. Such viewers are more likely to change channels than to turn off the set if the new program on a network currently viewed does not appeal to them. One interesting inference to be drawn from this information is a typology of viewing patterns, wherein (1) some "watch television," per se and are indifferent to which particular programs happen to come on; (2) others, although committed more to "watching something on television" than to viewing a particular program, nevertheless select among those offered at whatever times their sets are on; and (3) those whose loyalty is primarily to *specific programs*, who turn the set on for only those programs and otherwise leave the set off.

The basic analytic distinction here is between "watching television" versus "watching programs." *Social science models of television's effects on individual viewers, and on culture and society, do not consider such distinctions or their distribution among the viewer population.* I think these are important to our models, however, and discuss them in greater detail shortly. Several additional implications of the analysis so far include the following suggestions: (1) At any time, each network can expect a base of x million viewers to be tuned in by chance. This constitutes a "floor," to which

(2) programmers seek to add more viewers. (3) Even though the number in (1) may well be greater than in (2), the small but "crucial" difference in ratings points between "disappointing" and "high" ratings often depends on the (presumably) more conscious program selections of these additional viewers. Consequently (4) much TV programming—including, for example, scenes of "gratuitous" violence—is intended to attract people from this *subset* of viewers, many of whose sets are already tuned to one network, to stay with it or switch from another channel. In other words, (5) a TV program's success may well be determined by relatively small statistical minorities, who represent the increment over and above those millions who would have watched whatever came on any one of the three networks anyway, by chance. And (6) changes in programming, unless very substantial, will not affect aggregate viewing levels—which have not declined, for example, since fewer action–adventure programs were scheduled for the 1978 and 1979 television seasons.

Two recent changes in American commercial television seem to support these propositions. First is the apparent absence of a response in audience behavior to changes in programming wrought by the FCC's prime-time access rule. This required that local stations rather than networks program the first half hour of prime time (7:30–8:00 EST) and precipitated an "invasion" of syndicated game shows—a large change in content from the more glamorous and stylized genres that had been supplied by networks during this time slot. Yet I do not think there was any notable decline in the number of sets on at this hour, after the changes in program formats were instituted. Second is the advent of "family hour," and the relatedly high ratings achieved by ABC during the last three television seasons (1975–78). Once again, change in "which" programs were offered seem to have had little or no impact on the *aggregate* number of viewers choosing to watch television from 8 to 9 p.m. (EST). Although the extent to which program genres changed fundamentally is less than in the instance of the prime-time access rule, the ABC network's somewhat innovative response to the "family hour" concept provides an interesting example of how network shares can shift in a stable aggregate market. I suspect its provision of highly successful programs built around fancifully bionic men, women, and comic book heroines also represents the development of (or return to) a basically nonviolent action–adventure format.[6]

In seeking to learn from and account for the overall success of ABC's programming under Fred Silverman's direction, one has to be less impressed

---

[6]My impressionistic coding of these shows' degree of realism and level and intensity of violent sequences places them in the same low range of antisocial activities as characterized the "Lone Ranger," whose episodes are still running in syndication. In this, my perception apparently differs from that of coders employing the Gerbner index, who placed the "Six Million Dollar Man" among the 10 most violent programs on television this season (Gerbner & Gross, 1976).

by the content of specific programs than by the decisions made concerning the *scheduling* and promotion of programs, as well as the attention accorded the *forms* and *genres* within which programs are based. Much of the credit for the great success of ABC's "miniseries" and of "Roots," for example, has been attributed to the experimentation in scheduling that each entailed.[7] Industry experience also suggests such high ratings for specific programs yield additional benefits for a network in the form of an "inheritance effect": viewers make more of a habit of watching other offerings on the same network after they have switched to it for any single program ("Roots," the Olympics, political conventions, or election coverage). High ratings for a program need not mean that each viewer deliberately chose it over the alternatives, however. If ABC's "normal" ratings for the time period during which "Roots" appeared are subtracted from the ratings record set by this series, the "increment" might well be substantially attributable to the combination of black viewers switching to the network en masse, plus a "residual" of persons who consciously selected the episodes for viewing. Should it matter, in social science models of television's effects, whether a program was deliberately chosen or watched accidentally? Should media effects vary in terms of the variety of ways viewers may become exposed to the programs they see? Or should we expect individuals' reactions and interpretations to be independent of how they came to be exposed to any particular program? I suspect that characteristically neither the network(s) nor advertising agencies have conducted any studies of the distribution of viewers' interpretations of "Roots," or stated reasons for watching it, and it is hazardous to speculate too seriously on its social or cultural meaning(s).[8] However, one of the most striking organizational indicators it provides regarding American society in the 1970s is that every local affiliate of ABC, in both the North and the Deep South, carried or "cleared time" for all episodes.

TELEVISION AS A CONSUMER GOOD

Television networks provide and purvey an unending stream of audiovisual symbols, presented in the form of regularly scheduled series of "specials" (Williams, 1975). The product with which networks attract consumers is intangible. But, as in the case of many other cultural and consumer goods, the aggregate demand level is predictable. Much as the annual sales volume of

---

[7]"Roots" was scheduled to run two hours nightly for six nights in a row; a "miniseries" consists of only 3 to about 8 episodes of each program, rather than the more conventional provision of 26 episodes per series per year. It is most conducive to serializing dramatic stories and was innovated, I believe, by the BBC.

[8]One of the most interesting explorations of this question to appear so far is by Peter Wood (1977).

aspirin, deodorants, automobiles, and, to a slightly lesser extent, of books, records, and movies can be accurately projected, we can anticipate with some confidence that next year Nielsen will continue to report television sets in the U.S. remained on for an average of between six and seven hours daily—no matter what programs are offered. Less certain, however, is what volume or share of these aggregate markets each competing firm (for example, Sterling Drugs, Bristol-Myers, Ford, GM, Random House, Capitol Records, and MCA) will obtain. Most of these large, publicly held corporations manufacture or produce a large variety of different products or different brands of the same basic product.[9] None wishes to be overly dependent for its success on the sales performance of any one item. And all determine their performance, as an organization, by averaging across all products and brands to determine their net profit.

Television networks operate much like large multibrand firms in these other consumer goods industries, although their marginally differentiated and advertised "brands" consist of programs, and their product (entertainment, visual images) is purchased in a slightly more complicated manner. Viewers "purchase" programs with their time, which networks in turn sell to advertisers wishing to reach them. Each program's ratings provide the network information as to its market share for each day or evening time slot. Although the success of single programs will obviously concern its producers and the programming department (just as brand managers are acutely sensitive to how well General Foods' Maxim and Maxwell House brands of coffee are selling individually), the *organization*'s success—that is, its advertising rates for the coming year—will be substantially determined by averaging *across* its ratings for all programs in the current season.

An important difference in perspective between a television network and many viewers and social researchers is that although the executive sums across all his network's (or record company's or movie studio's) productions, much of the audience and critics focus their interests and attention, as consumers, on particular programs (or individual records, movies, performers, or genres). However, a growing body of market research provides empirical support for McLuhan's assertion that "the medium is the message." That is, because people vary in the amount of time they are able and willing to watch (expend on) television, only a minority of episodes of a favorite program are likely to be seen by any one individual; much television viewing will be accidental or random; it is difficult to construct *any* distinctive viewer (demographic, psychographic) profile for most television network

---

[9]Sterling Drug's product line includes many well-known brands, such as Bayer Aspirin; Bristol-Myers makes Bufferin, Excedrin, Ban, Ipana; Random House is a subsidiary of RCA; Capitol Records releases several hundred records a year and has no other lines of business; MCA is an entertainment conglomerate involved in record manufacturing, movie, and television program production.

programs—beyond their mandatory appeal to 18- to 49-year olds;[10] viewers of a detective show at t_1 are no more likely to watch another detective program at t_2 than others who earlier viewed a different type of program at t_1; and, statistically, the best predictor of what a given viewer will watch tomorrow is whichever channel his set is tuned to today. These findings, by Goodhardt, Ehrenberg, and Collins (1975)[11] buttress Leo Bogart's (1972) observation and the long-standing contention by Paul Klein (of NBC) and others that people are more attracted to *television* as a product category than to specific programs (brands)—much as when people shop for aspirin many (though certainly not all) are indifferent to which brand of aspirin they will purchase. To a degree, these findings also buttress Klein's faith in the art of scheduling (the time and day of broadcast) as a prime determinant of a program's success, as well as Gans's hypothesis (chapter 4, this volume) that the personal involvement of the audience in the program it views is minimal.

At the same time, these researchers' findings neither explain all of the variance (R^2 statistics are not provided, and Kirsch and Banks, 1962, earlier reported that the network or channel "factor" leaves much unexplained variance), nor have they sought to explain why television viewers behave as they do or how they interpret what they see. Nevertheless, their findings provide impressive support for McQuail's (1970) conclusion that a majority are indifferent to *what* they view on television (in terms of its quality more so than broad moral themes). They also define parameters within which it may be most fruitful to explore audience perceptions, uses and gratifications, and to relate them to broader models of cultural meaning and industry practice.

MARKET RESEARCH FINDINGS AND ISSUES FOR SOCIAL SCIENTISTS

How and why people watch television and what symbolic content and messages viewers encode are questions of interest to a broad spectrum of social researchers and cultural analysts. Different aspects of each are posed and investigated in a variety of organizational and disciplinary contexts, ranging from English and anthropology departments and academic social research institutes, to television-industry-based research units and creative and media departments in advertising agencies, to congressmen and government agencies. Typically, each of these parties asks different questions

---

[10]As one network researcher I interviewed explained it: "Any program whose demographic profile was distinctive would mean it is attracting some segments at the expense of others we don't want to lose; it would be cancelled."

[11]I am indebted to Hilde Himmelweit, Elihu Katz, and John Robinson for calling this important book to my attention.

TABLE 5.1
Organizational and Disciplinary Contexts of Research and Inquiries About Television

Research group or discipline	Main goals of inquiry	Typical methods employed	Examples
Ad agencies	1. To improve reach and frequency 2. To increase advertising effectiveness	Focus group interviews Statistical modeling Panel surveys	Gensch & Ranganathan (1974) Bogart (1967); Friedman (1971)
Television industry	1. To monitor reach and frequency 2. To test program concepts and pilots 3. To increase program ratings	Sample surveys Statistical modeling Galvanic skin tests	BBC (1976); Goodhardt, Ehrenberg, & Collins (1975); Klein (1971)
The humanities	1. Cultural analysis 2. Explication of common experience 3. To relate symbolic content to linguistic and structuralist analyses of culture	Textual analysis Literary criticism Ethnographic studies Oral History	Cater & Adler (1975; 1976); Newcomb (1974); White (1973)
The social sciences	1. To test and develop theories of media usage and effects on individual *attitudes* or social institutions 2. To use mass media content and patterns of use as a convenient research site or adjunct to other, non-media-related intellectual problems	Sample surveys Statistical modeling Field studies Content analysis	Patterson & McClure (1976); Blumler & Katz (1974); Bagdikian (1971)

and employs concepts and methods likely to be unfamiliar to the others. Some of these are outlined in Table 5.1.

Each of these research groups and disciplines has contributed valuable ideas and information on the general topic of television and social behavior. The division of labor accords pretty well with differences in the main goals of inquiry listed in the table. As a general rule, market research results (such as those of Goodhardt et al., 1975) contribute most to our understanding of *how* individuals and households behave toward and utilize the medium-that is, of main patterns of audience behavior. The promise of social science studies lies in explaining *why* people act toward (or with) it as they do. Here, whereas improved theories, models, and data may be called for, we are hardly starting from scratch and can rely on the impressive foundations already built in psychology and sociology. The humanities also have a strong contribution to make toward our learning more concerning just *what* there is in television and its programs for people and cultures to experience, perceive, and possibly react to. The all-too-often simple and reductionistic categories usually employed by social scientists performing content analyses, to take an example, have changed little in over 30 years; additional categories might profitably follow from examining recent work by folklorists, ethnographers, linguists, literary critics, and historians-such as (1) the papers commissioned for the Aspen Institute's collection (Cater & Adler, 1975, 1976) on television as a social and cultural force, (2) Horace Newcomb's (1974, 1978) treatments of the basic formulaic elements of most television genres, or (3) parallels between Elizabeth Eisenstein's (1968) research on the impact of standardized typesetting and a greater quantity of printed works on medieval culture, and the more current "social science" concerns about television's standardization of American culture and the redundancy of a small number of themes brought on by the sheer quantity of material it presents.

My assessment of the strengths of each area is, of course, oversimplified. Television and advertising organizations, for example, although grouped under market research, are not necessarily interested in all of the same questions. Each television network's primary concern is to increase its audience size in accordance with the demographics desired by advertisers, but it has little responsibility or concern for the advertiser's additional problem of whether the commercials shown in the time periods purchased are attractive, effective, "wear out" fast, or help to induce repeat buying. Similarly, areas overlap and there are topics in which all three share a common interest.

In seeking to learn more about how people watch television and to develop theories about how commercials help to maintain or increase brand sales, market research has developed several models of clear interest to social scientists. Basically, social scientists utilize a "marketing" model of their own when seeking to assess the impact of issues like televised violence on viewers or the "image" of minorities conveyed by the medium. Typical models employed for these purposes, however, stress imitation, modeling, and attitude conversion, whereas several marketing theories now strongly assert that television does not "work" through these processes at all. Rather, Krugman (1965) and Ehrenberg (1974) among others stress awareness levels of a brand or product category (rather than attitudes toward it) and argue that viewers initially purchase many brands of products simply on the basis of awareness levels and familiarity. *The real test of whether or not they have been influenced is if they repeat-purchase the same brand the next time they buy something in that product category.* Further, these models suggest that attitudes become favorable only *after* the product or brand has been tried and found useful—that is, that merely purchasing it once (on a trial basis) does not indicate a favorable disposition toward it. (Contrasts between the "learning hierarchy" and these marketing models are reviewed nicely by Ray, 1973. The latter also resemble Rogers' (1962) model of innovation diffusion.)

One implication of these models from advertising is that if "violent behavior" shown on television is conceived as a product being merchandised, two variables of special importance become whether or not population awareness levels (of particular forms) are raised by television exposure and whether or not the behavior shown is then tried *twice or more* by any given viewer. (Of course, I am not suggsting that to "try" violence once is of no consequence.) Imagine the product category to be "conflict resolution," with "violence" only one of the brands available to the consumer, others being "negotiation," "withdrawal," and so on. If violence is tried but found wanting and not repeated, several advertising models would view the "commercial" as having failed. Although I certainly am no expert on this topic, I wonder if something might not be gained from a more careful examination and possibly from greater utilization of distinctions such as these (which in turn may be derived from theories of cognitive dissonance and self-perception). In this area, as with commercial products, there would be a substantial difference between behavior that is repeated regularly and behavior tried once (perhaps in a lab) and then discontinued.

A related distinction that I have borrowed from advertising models pertains to the extent a message is simple, focused, and repeated regularly and over long time periods. Population awareness levels for new and heavily television-advertised brands often exceed 60% within a year of their introduction. (Most 1976–77 viewers, for example, were surely aware of "Short 'n Sassy" shampoo and Farrah Fawcett-Majors, two of the year's new product introductions.) Yet a vast majority probably cannot name three members of Jimmy Carter's cabinet—all of whose publicity was short-term, albeit focused and repetitive prior to their respective confirmations. An appropriate question, regarding television's "effects" in the arena of conveying information, is not only whether or not the information is salient but also whether or not, like advertising, it is simple, focused, and repeated over time. Most political news exhibits none of these characteristics, so if a large majority cannot identify Ray Marshall the reasons for this may say less about the mass media's inability to convey information than about the manner in which the information was (not) conveyed. *The overridingly pessimistic conclusions stated in much social science literature on national television as a medium for information diffusion may be exaggerated, because they are based disproportionately on cases in which the types of information involved failed to meet these conditions* (and were probably of low salience as well). Awareness of information is a much simpler and different dependent variable than attitude or behavior change. For advertisers and political candidates, a media campaign may be a great success if "only" 4% of its viewers switch over to the advertised brand or cross over from the competing candidate. Few such campaigns are intended to impress or activate all viewers equally. Although the uses and gratifications perspective in social science research assumes this for other areas as well, these expectations for, and experiences with, advertised products and marketing models have implications for other theories of mass media effects.

A final issue raised by market research studies is how findings like those of Goodhardt, Ehrenberg, and Collins (1975) about *how* viewers watch television can and should be interpreted. Because a large segment of viewers appears to prefer "television" per se to specific programs and program genres, does it follow that they are basically indifferent to television, that people have low involvement with this medium and/or its content, that it is epiphenomenal—a way to pass leisure time and (therefore) of little cultural consequence? Here is a possible example of separate goals for market and social science research. For where it may be sufficient for network programmers to know they have a "floor" of viewers per program and can obtain more by smart scheduling, the social science question remains "Why?" or "How come?" Little *explanation* of these descriptive findings is sought or offered by (or especially relevant for) most market research professionals, but an analytical theory to account for it remains a central province of social science research. Several complementary and alternative hypotheses, to add to the indifference–low involvement interpretation of the findings, are offered by studies in cultural anthropology, such as Lloyd Warner's (1959) discussions of the common mythic themes underlying news stories, entertainment genres, and mass-media-produced symbolic content in general; additionally, recent developments in sociolinguistics and structuralism may be examined for applications to mass communication. Several examples follow of what these fields offer that might be rewarding to explore further and build on.

Considering the amount of time people spend with the media and the high levels of exposure this assures, mass-media-disseminated information and perspectives constitute a form of cultural oxygen (pollution?) breathed in by all. Network television, currently the most prominent force in creating and disseminating our national popular culture, provides a set of common and consistent markers and normative reference points for all members of the population. It provides an "official" framed version of expected sets and types of events, relationships, explanations, outcomes, and solutions. Where this is approached at the "surface" level, polls show Americans consistently list as the "most pressing issues" those which the mass media have covered and singled out as significant. Concerning studies of television entertainment, it appears many content analysts (most notably Gerbner and Gross, 1976) have anticipated the general finding that most viewers are relatively "disloyal" to specific programs and genres. If so much audience behavior results in "random" viewing, then the practice of collecting "census" like data, *across genres*, on racial and sexual stereotypes or the number of violent exchanges, now has more of an empirical referent, and receives some support in terms of how viewers actually see programs. A critical question remaining to be answered, however, is whether the appropriate unit of analysis is the entire program, or discrete program segments. This is particularly true for "heavy viewers" and for the segment suggested earlier that watches television *per se*; however, it may be less appropriate for the segment of more selective, program-loyal viewers who plan certain selections in advance and avoid most others. Because much public discussion of television's cultural significance is based around specific programs and studies of particular shows continue to be of interest (e.g., the question Do viewers of "All in the Family" become more or less bigoted? is an example of this type of concern) both types of inquiries into television's surface messages remain informative and valuable.

Pursuing the finding that so much of the television audience appears oblivious about which programs it consumes, we can also hypothesize that *people appear indifferent because the basic formulaic elements, plots, themes, and messages of so many programs are identical.* For example, if viewers are attracted by an underlying myth common to nearly all television drama—that good triumphs over evil—then there may be no clear cut reason for many to prefer or discriminate one (say, "Hawaii Five-O") over another rendition of the same thematic elements ("Starsky and Hutch," "The Waltons").[12] Following along these lines (perhaps borrowing notions like "deep structure"

---

[12]The actual finding is that individuals do have clear television program preferences, but are often too busy with other things to watch regularly during their time periods. They also watch other, less favored programs in other time periods as well, so that "favorite shows" become a poor statistical predictor of actual overall viewing patterns. Additionally, there are few measurable demographic or psychographic differences here *between* individuals to distinguish viewers with one set of preferences from their counterparts; this seems to yield a random distribution of aesthetic preferences (as between particular programs) across population subgroups whose scores are matched on standard demographic and psychographic items. Were each network seeking a different audience, these findings might be different. For example, viewers of many programs on the Public Broadcasting Network are drawn disproportionately from higher SES households; viewers of nonnetwork entertainment programs (Lawrence Welk) are also more distinctive demographically.

and seeking to decipher Levi-Strauss along the way), we may postulate that people are far from indifferent about the ritual of the hero overcoming each episode's particular forms of adversity. Rather, the low involvement and indifference may end with the question of whether or not the marginal differentiation between leading actors, directors' styles, and (less often) writing style is sufficient to motivate fine discrimination among programs. In a very different sense from McLuhan's, the "medium" of underlying myths, universal values, and rituals becomes the "message," and many of the seemingly bored, lazy, indifferent viewers come to look more active, aware, and sensible. If so, one would predict a widespread (negative) viewer response if situation comedies were to end on sour notes and the "wrong" persons (say, Columbo or Kojak) jailed (or worse) at the end of every detective show.[13]

This analytical framework is taken as a given by many humanistic analysts of popular culture and is seen occasionally in a minority of content analyses conducted by social scientists, mainly emphasizing "surface" features like each character's race and sex. It directs greater attention to the underlying themes and mythic elements of popular culture (Cawelti, 1976; Newcomb, 1974), and the dramatic framing and literary devices employed in and across genres, taking in "news" and advertising as well as fiction (Goffman, 1974). One of its policy implications is that if viewers are attracted to a small number of mythic paradigms that cut across genres and underlie the surface plot action, then the coding categories widely used in content analysis are incomplete and conceivably misdirected. This cultural perspective, which warrants more serious consideration, harks back to (1) Breed's (1958) suggestion that mass media provide less learning than a standardized patterned ritual toward which we look and expect things at specified times; (2) Warner's (1959) conclusion that mass art provides continuity of myth and legend, thereby contributing to the integration and persistence of culture; and (3) Durkheim's theory of such collective representations as unifying complex industrial societies. Anthropology provides a further point of departure from which these ideas might well be profitable explored in greater depth. Of particular interest is Bennett's (1976) suggestion (also in line with symbolic interaction theory) that there need be no contradiction between conceiving of television as conveying the basic elements of a shared common culture on the one hand while also noting that individual members of society need not aggree with or be uniformly aware of its every prescriptive component.[14]

---

[13]Two interesting examples of programs that "violated" these types of viewer expectations may be "Rhoda" and "Mash," The first, supposed to be a situation comedy, lost ratings (viewers) after the story lines became less comedic and both more sour and serious. Producers of "Mash" received an unusually large number of irate letters when they violated viewer expectations of the program's genre by having a popular officer, based in a Korean combat zone, killed by enemy fire during the episode in which he was supposed to have been discharged. These viewers asserted that such "realism" had no place on a comedy about a medical unit based in a war zone.

[14]I am grateful to Stephen Withey for bringing these related concepts to my attention.

A basic issue for the social sciences has long been, "How do we measure this sort of thing?" Another question that might be posed in answer to the first is, "Why haven't we at least kept these questions more alive while awaiting some more precise measures for them?" Not surprisingly, there is some risk in hewing too closely to those topics most amenable to methodological convenience (and often conceptually inflated for the same reason). For example, in a recent review of his research on attitude change and consumer behavior, Krugman (1977) offered the rather startling conclusion that findings of low recall of messages among television viewers mask far higher levels of message and pattern recognition than the recall percentages imply. Use of recall particularly "obscures non-commercial cultural effects of the medium [p. 7]." Krugman, along with some prominent literary critics, believes the main impact of television as a medium (whether organized into a structure of few channels or many) lies in its presentation of continuous visual images. These are nonverbal and, he notes (1977), would be processed by the right side of the brain. In contrast,

> It is interesting that our tradition of research in public opinion is so heavily in-vested in reporting public reaction to the news, initially newspaper news. The news is very factual stuff, left brain stuff. The continuous, and very prominent, reporting of public reaction to news probably over-represents the extent of that reaction, while the cultural impact of "right-brain" television, though presumed to be enormous, is difficult to demonstrate. We need new and different research techniques. To develop those tools, and in addition to studying learning without involvement, or behavioral change without prior attitude change, you may have to study memory without recall [p. 7].

If Krugman is correct, a standard criterion for measuring mass media effects has been providing a distorted picture of its relation to the audience. What would this suggest for social science research?

SOCIAL SCIENCE RESEARCH ON TELEVISION AND MASS COMMUNICATION

One suggestion, which Krugman's comments reinforce, is that this whole area is "wide open" in terms of the potential contribution of research utilizing a wide variety of concepts and designs. I do not think this is widely perceived by social scientists, many of whom wrongly believe that our present models of how media-provided myths, genres, programs, and advertising diffuse, gratify, or influence audience members are explanatory; or (among sociologists) that opinion leadership and low recall of "important" messages (like "Who is Ray Marshall?") render these questions moot; or that commercial networks and advertising agencies already have collected the data needed and have the answers. This review of the field has found each of these widespread beliefs unsupported by the evidence. That there are findings to be found should be more effectively communicated to the social science community.

Television and its relation to society offers a number of unusual advantages as a research topic. One is that practically every member of society can speak articulately about it from personal experience and interest. Commercial agencies come across very few individuals who are uncomfortable discussing or keeping diaries of their television-related attitudes and experience. Yet very few projects have invited people to define their perceptions of what elements make for good and bad programs,[15] how programs should be classified,[16] what happens on them, or whether they watch programs or television. Indirect and unobtrusive measures (short of spying in living rooms) should not be that difficult to develop. An attraction for increasing the number of studies in these areas might be that whatever is learned could well be a finding, and as small-scale studies accumulate we might soon discover we have been learning a substantial amount indeed.

Quantitative modeling and simulations have been employed fruitfully but so far almost exclusively by advertising agencies for the relatively narrow purpose of estimating individuals' probabilities of being exposed to an advertisement under conditions $a, b, c, \ldots, n$. As McPhee (1966) showed over a decade ago, however, mathematical and heuristic models for types of cultural content, innovations, and their likelihood of attracting an audience (of size $q, r, s, \ldots z$) are equally feasible and intellectually fascinating. There is plenty of room for more. For example, variations in the size of the viewer segments described as "television" versus "program" watchers might yield substantially different predictions of which types of programs at what time period will be seen by each segment and why. Similarly, if the work force went to flexible hours, or a four-day week, what sets of changes in the availability of audiences might be expected, and how might this in turn alter television programming?[17]

So far, we have talked almost exclusively about the relation of mass media to mass audiences. However, one of the most promising areas for organizational research, small-scale surveys, and simulated situations is the topic of how elites and organizations that are affected by mass media interact with and influence them as well. We know, for example, that political leaders tend to treat stories reported in newspapers and on television as highly significant—simply because they received mass media coverage. Marc

---

[15]Herbert Gans suggests this idea in Popular Culture and High Culture (1975), and Himmelweit et al. recently have conducted an important study on "The Audience as Critic" (1980).

[16]Elihu Katz suggested this at a recent meeting of the Committee on Television and Social Behavior.

[17]I am grateful to F. Gerald Kline for bringing this question to my attention. There already is evidence of a smaller audience for daytime soap operas, as the proportion of women in the labor force continues to rise.

Franklin and other law professors believe that although television coverage of the House Judiciary Committee hearing on Watergate showed no effects on public opinion (Robinson, 1974), the fact of their televised coverage greatly increased the likelihood that *Congress* would have impeached then-President Nixon. What is their model of its importance where, for example, a poll may show widespread public ignorance or indifference to so many news items? Why do politicians perceive television as a more powerful medium than newspapers? Under what circumstances do political leaders and their various constituencies, respectively, believe one to have more credibility than the other? Studies of these kinds of questions are inexpensive to mount; they can address city council and state representatives as well as Washington-based congressmen, for example. At the organizational level, political scientists can examine the effect on political campaigns of having to allocate most of their budgets to television advertising; organization researchers can inexpensively study issues involving media industries at several levels via the trade press and personal interviews. An encouraging sign is the increasing number of studies beginning to appear on the organization of print and broadcast journalism (e.g., Epstein, 1973; Sigal, 1973).

Perhaps the most interesting theoretical payoffs will come from increasing collaboration across disciplines and research areas. I find exploring and monitoring developments in the areas of economic and marketing research and the humanities among the most promising. Others, such as in administrative law (the impact and basis of FCC decisions), business, linguistics, and anthropology also will continue to remain important to keep track of. In Comstock et al.'s (1975) recent and valuable compendia of all social science studies related to television and behavior, however, most or all of these fields were (by definition) excluded. Collaborations across these lines are likely to pay off, however, and lines of communication should be kept open.

I am an optimist. During my recent explorations of this subject, I have found both colleagues in other disciplines and industry representatives accessible and open to a degree that surprised me. Curiosity about and interest in television and social behavior are widespread. If enough social scientists become aware of the potential of this topic for fruitful research and agree with most of the outside world that these questions are worth answering and researchable, then we can expect our knowlede of mass communication processes to increase at an encouraging pace during the next five to ten years.

ACKNOWLEDGMENTS

I am grateful to all of the members of the SSRC Committee on Television and Social Behavior for their comments and suggestions on the original version of this paper. The research on which this paper is based was generously supported by a Rockefeller

Foundation Humanities Fellowship and by the John and Mary R. Markle Foundation.

REFERENCES

Bagdikian, B. *The information machines.* New York: Harper, 1971.

Bennett, J. W. Anticipation, adaptation, and the concept of culture in anthropology. *Science,* 1976, *192,* 23.

Blumler, J., & Katz, E. (Eds.). *The uses of mass communication.* Beverly Hills: Sage, 1974.

Bogart, L. *Strategy in advertising.* New York: Harcourt, Brace, 1967.

Bogart, L. *The age of television (3rd ed.).* New York: Ungar, 1972.

Breed, W. Mass communication and sociocultural integration. *Social Forces,* 1958, *37,* 109–116.

British Broadcasting Corporation. *Annual review of audience research findings,* No. 2. London: British Broadcasting Corporation, 1976.

Cater, D., & Adler, R. (Eds.). *Television as a social force: New approaches to TV criticism.* New York: Praeger, 1975.

Cater, D., & Adler, R. *Television as a cultural force.* New York: Praeger, 1976.

Cawelti, J. *Adventure, mystery, and romance.* Chicago: University of Chicago Press, 1976.

Comstock, G. A., Lindsey, G., & Fisher, M. *Television and human behavior.* Santa Monica: Rand, 1975.

Dominick, J., & Pierce, M. Trends in network prime-time programming 1953–74. *Journal of Communication,* 1976, *26,* 70–80.

Ehrenberg, A. A. C. Repetitive advertising and the consumer. *Journal of Advertising Research,* 1974, *14,* 25–34.

Eisenstein, E. Some conjectures about the impact of printing on Western society and thought: A preliminary report. *Journal of Modern History,* 1968, *40,* 1–53.

Epstein, E. J. *News from nowhere.* New York: Random House, 1973.

Friedman, L. Calculating TV reach and frequency. *Journal of Advertising Research,* 1971, *11,* 21–26.

Gans, H. *Popular culture and high culture.* New York: Basic Books, 1975.

Gensch, D., & Ranganathan, B. Evaluation of television program content for the purpose of promotional segmentation. *Journal of Marketing Research,* 1974, *11,* 390–398.

Gerbner, G., & Gross, L. Living with television: The violence profile. *Journal of Communication,* 1976, *26,* 172–199.

Goffman, E. *Frame analysis.* New York: Harper, 1974.

Goodhardt, G. J., Ehrenberg, A. S. C., & Collins M. A. *The television audience.* Lexington, Mass: Lexington Books, 1975.

Goodman, P. Television: The continuing disaster. *The New Republic,* January 26, 1963, pp. 24–26.

Himmelweit, H., Swift, B., & Jaegar, M. E. The audience as critic: A conceptual analysis of television entertainment. In P. Tannenbaum (Ed.), *The entertainment functions of television.* Hillsdale, N.J.: Lawrence Erlbaum Associates, 1980.

Hirsch, P. Public policy toward television: Mass media and education in American society. *School Review,* 1977, *85,* 481–512.

Hirsch, P. Social science approaches to popular culture: A review and critique. *Journal of Popular Culture,* 1977, *11,* 401–413.

Hirsch, P. Production and distribution roles among cultural organizations: On the division of labor across intellectual disciplines. *Social Research,* 1978, *45,* 315–330.

Kirsch, A., & Banks, S. Program types defined by factor analysis. *Journal of Advertising Research,* 1962, *2,* 29–31.

Klein, P. The men who ran TV aren't that stupid ... *New York,* January 25, 1971, pp. 20, 21, 29.

Krugman, H. The impact of television advertising: Learning without involvement. *Public Opinion Quarterly,* 1965, *28,* 349-356.

Krugman, H. As quoted in "New Studies of Brain Functioning Show Need for Marketing Strategy Revisions: Krugman." *Marketing News,* March 25, 1977, p. 7.

Lourenco, S., & Glidewell, J. C. A dialectical analysis of organizational conflict. *Administrative Science Quarterly,* 1975,*20,* 489-508.

McPhee, W. When culture becomes a business. In S. Berger, M. Zelditch, & B. Anderson (Eds.), *Sociological theories in progress.* Boston: Houton-Mifflin, 1966. [Also in P. Hirsch et al. (Eds.), *Strategies for communication research.* Beverly Hills: Sage, 1977.]

McQuail, D. The audience for television plays. In Jeremy Turnstall (Ed.), *Media Sociology.* London: Constable, 1970.

Newcomb, H. *TV: The most popular art.* New York: Anchor, 1974.

Newcomb, H. Assessing the violence profile studies of Gerbner and Gross: A Humanistic critique and suggestion. *Communication Research,* 1978, *5,* 264-282.

Noll, R., Peck, M., & Mc Gowan, J. *Economic aspects of television regulation.* Washington D.C.: Brookings, 1973.

Patterson, T., & McClure, R. *The unseeing eye.* New York: Putnam's, 1976.

Ray, M. Marketing communication and the hierarchy of effects. In Peter Clarke (Ed.), *New models for mass communication research.* Beverly Hills: Sage, 1973.

Robinson, J. Public opinion during the Watergate crisis. *Communication Research,* 1974, *1,* 391-405.

Rogers, E. *Diffusion of innovations.* Glencoe: The Free Press, 1962.

Sigal, L. V. *Reporters and officials.* Lexington, Mass.: Lexington Books, 1973.

Warner, W. L. *The living and the dead.* New Haven: Yale University Press, 1959.

Welles, C. *The last days of the club.* New York: Dutton, 1975.

White, H. Structuralism and popular culture. *Journal of Popular Culture,* 1973, *7,* 759-75.

Williams, R. *Television: Technology and cultural form.* New York: Schocken, 1975.

Wood, P. H. Roots of victory, roots of defeat. *The New Republic,* March 12, 1977, pp. 27-28.

6 After the Surgeon General's Report: Another Look Backward

Leo Bogart
Newspaper Advertising Bureau

Customs and culture change continuously. It is no easy matter to identify the critical points at which change accelerates, to connect simultaneous changes in human values and in social institutions, and to explain why change took just the path it did and not some other way.

Cultural change presupposes *ex*change, a web of human contact and communication. The evolving communications technology of recent centuries has transformed culture by vastly increasing the flow of standardized symbols. This flow of symbols is under centralized management. What we call the mass media are not merely technical means of disseminating communications; they are also human organizations that self-consciously exercise vast power over the information that society has at its disposal, over common fantasies and accepted standards of judgment and conduct. Inevitably there must be disagreements as to how this power should be socially controlled.

This chapter traces the recent history of controversy, conflict, and policy change surrounding one component of television content–violence. Its conclusion is that without any loss of audience, content can undergo substantial modification as a result of pressures that arise from outside the normal institutional processes of commercial broadcasting. This implies that there is considerable leeway in the critical decisions regarding the programming menu from which the viewing public selects its fare. Freedom of viewing choice can be exercised only within the limits set by those who provide the choices, and their own judgments can always be rationalized with the argument that they are merely instruments of popular taste. That argument is hard to sustain in the face of the changes that television programming has undergone in recent years. These changes represent, in part, a response to forces set in motion by social research.

The research in question is embodied essentially in the 1972 report of the U.S. Surgeon General's Scientific Advisory Committee on Television and Social Behavior (Bogart, 1972–73; Cater & Strickland, 1975; Comstock & Rubinstein, 1972, 1972b; Comstock, Rubinstein & Murray, 1972; Murray, Rubinstein, & Comstock, 1972; Rubinstein, Comstock, & Murray, 1972). This committee itself was set up in response to Congressional demands for definitive evidence regarding the effects of televised violence, a subject that has been debated and researched almost since the beginnings of commercial television and that has antecedents in earlier debates over violence in film and in comic books.

Amid the substantial literature of the subject, the surgeon general's committee's effort stands out because of its official sponsorship, substantial ($1.5 million) budget, and methodological diversity. Its publication may, therefore, be considered as a watershed in the continuing discussion of television content and its influence on public values and behavior.

THE TERMS OF DEBATE

The debate over the effects of American media violence has always had strong overtones of moral feeling, but it has had research at its center. Although from time to time individual studies have been the focus of discussion, none has presumed to be comprehensive and all-conclusive, and few if any have been beyond criticism. Thus the construction of a debating case has required the comparison and synthesis of many different isolated pieces of research of varying quality, techniques, and theoretical foundations.

The catalogue of complaints against television violence is well known and can be summarized as follows:

(1) The presentation of violent or aggressive acts is more emotionally arousing through the audiovisual media (motion pictures and television) than it is in either radio or print.

(2) Arousal may heighten awareness, involvement, or pleasure, but it also has the potential of being emotionally disturbing and consequently of evoking aggressive feelings.

(3) Media content that is perceived as literally real, or as resembling reality, is more arousing than content that is accepted as fictional.

(4) Children are more impressionable and more susceptible to arousal than are adults.

(5) People who are intellectually and emotionally disadvantaged are less capable than others of gauging the reality of media content and are more readily aroused.

(6) A single highly charged media episode can be perceived as intensely realistic by an emotionally troubled individual, with consequences that are threatening either to himself or to others.

(7) Even when it is presented as fiction or fantasy, a single media episode can provide a role model or suggest techniques of conduct to a troubled individual, with antisocial consequences.

(8) A single episode of media content may be sufficiently arousing to have emotionally disquieting aftereffects on many of those who have been exposed to it.

(9) Repeated exposure to media content that shows violence routinely used as a technique of handling interpersonal relations and enforcing social controls can lead to a more general acceptance of violence exercised by private individuals and by public authorities.

(10) Repeated exposure to media presentations of fictional violence might dull sensibilities to the horror of real violence.

The evidence in support of these propositions is of varying degrees of persuasiveness, and in at least some cases the statements go well beyond the evidence. Still, all appear to be congruent with common sense and none of them have been scientifically contradicted.

In defense of existing practices, a different set of propositions has been formulated along the following lines:

(1) The evidence regarding the causal connection between exposure to fictional violence and subsequent behavior is still inconclusive; there is no Supreme Court of research that has determined once and for all that media violence has antisocial effects.

(2) Violence in the media is extremely hard to define. The term covers a vast variety of social interactions, ranging from the trivial to the horrendous. When a mouse in an animated cartoon hoists a cat by his own petard, the critics count it as a "violent" act, just as they might count an axe murder!

(3) Violence never can be judged in the abstract; it must be considered in the context of its motivation and consequences, and in respect to the characters engaged in it as actors and victims. Violence may, as in *Romeo and Juliet*, be inherent in a story line.

(4) Anticipation or dread of violence, not violence itself, is an essential ingredient of drama; it is a time-tested device to arouse the interest of the audience.

(5) Decisions on media content are determined by audience interests and not by any preoccupation with violence for its own sake.

(6) The mass media accurately reflect the society and its cultural tastes. Any mass medium whose management tried to impose its own values in contradiction or defiance of these tastes would quickly be forced out under the economic pressures of the commercial media system.

(7) Standards of public taste and of morality have changed rapidly in the past dozen years. The critics of media content are characteristically out of touch with the prevailing mores.

Besides these points, an additional argument has been used in the past but has faded in the absence of research substantiation:

(8) In a troubled and violent world, action and excitement in media content serve a positive, cathartic function and thus may reduce the level of violence that is manifested in real life.

These two sets of reasonable-sounding propositions accord with sharply contrasting philosophies regarding decision-making in the mass media and social regulation of their content. The contrast between them raises the question of what constitutes proof (as distinct from expert judgment) in social science and how it differs from proof in the legal system.

Public concern over media content characteristically becomes of concern to media managements when it leads to public hearings before legislative or regulatory bodies that have actual or latent power to challenge the media's financial interests. The legal definition of "proof" beyond a reasonable doubt is skillfully used by broadcasters defending their franchises. Yet the legal definition, which demands an unequivocal "yes" or "no" answer to every question, is totally irrelevant to the procedures of social science research in which the shadow, large or small, of a doubt is never absent and leads constantly to the reinterpretation of existing evidence and the search for new evidence.

PREMISES AND PARTIES IN THE DEBATE

The debate over socially questionable media content long antecedes the invention of television and may be as old as the mass media themselves. Here, for example, is a contemporary critique (*Livesey's Moral Reformer* 1833, cited in James, 1976) of the early nineteenth century popular press, which makes exactly the same points that television's critics make today:

The extensive circulation of newspapers is a sure criterion of the *mental* activity of the people of this country, but by no means of the advancement of moral principles and virtuous habits. This is certain from the circumstance, that the most licentious papers usually command the largest sale. The "Life in London," an abominable print, has an amazing circulation, whilst others of a similar character are sought after in proportion as they publish anything that is vile and destructive to virtue and religion. The press is degraded by adventurers, who constantly prostitute their talents for gain. Knowing the depraved taste of our immoral population, they suit their articles to their readers, and are thus openly, and with an unsparing hand, sowing and watering the seeds of moral deformity [p. 258].

The popular editors of the time might have replied to such imprecations by questioning the moralist's credentials, and the debate is still characterized by disagreement over the question of what constitutes expert judgment.

The participants in the controversy over television violence take certain premises for granted:

(1) Among the mass media, television represents a uniquely powerful cultural force by virtue of its universality and the large amounts of leisure time it occupies for most people.

(2) Over half of all viewing time is concentrated on the broadcasts that originate with the three commercial networks.

(3) The intense competition among the networks is dedicated primarily to the achievement of larger audiences and larger audience shares than their competitors, because the absolute overall size of the viewing public at any given season and time of day is virtually invariable.

(4) Programming schedules are carefully planned in order to retain casual viewers as far as possible from one time period to the next, and losses are cut short by quickly eliminating weak programs that do not perform to expectations.

(5) Research techniques derived from the social sciences are relied on strongly both to measure the size and composition of audiences for sale to advertisers but also to evaluate program concepts, personalities, and "pilots" before any decision is made to put them on the air.

Even without any active controversies, programming decisions are subject to a number of institutional cross-pressures. The significant parties include:

(1) Network programming managements, who must keep in mind the whole array of possible alternatives, the audience flow through the course of each evening, and possible competitive responses to every move they make.

(2) The managements of affiliate stations, including the powerful groups like Westinghouse Broadcasting and Storer. They not only have a keen concern about ratings levels but are most sensitive to community and advertising pressure at the grass roots, where the mores differ from those of Radio City and Beverly Hills.

(3) The program producers. They are an important source of new ideas, but paradoxically they are also a force for the homogenization of content. Because they deal with all three networks and (through syndication) with independent stations as well as with the motion picture industry and the rest of "show business," they spin a universal web of negotiation and gossip.

(4) Advertisers and agencies. Only a handful of each represent real power, but they are the broadcasters' only source of income, and their pronouncements cannot be ignored. They would prefer to think of television programming as a stable system within which they can jockey for the most advantageous "cost-per-thousand" for their commercials.

(5) Government officials, including Congressional committees and staffs and the Federal Communications Commission and its staff. Merely by their existence and by the periodic paperwork they exact from stations, they are a symbol of the public's sovereign interest in what goes out on the air, although they rarely move to censure violations of this interest.

(6) Public action organizations, lobbying to achieve specific modifications in broadcasting practices. Violence and advertising addressed to children have been two perennial subjects of their concern.

All these institutions interact to produce the continuous evolution of television content.

TELEVISION CONTENT AND TELEVISION RESEARCH

In 1978, U.S. television was an $8.7 billion industry with net profits (before taxes) of over $1.5 billion. Anything that changes the balance of television programming affects its ability to generate audiences and advertising revenues. Thus it must threaten all those whose power and wealth derive from things as they are.

The effect of having a commonly held advertising sales objective (as well as an advertising market that takes audience size as its primary yardstick) has been to produce generally similar programming philosophies, formulas, and schedules from all three networks. This has been facilitated by the frequent movement of managerial and performing talent from one network to another, by the common use of the independent film production firms that produce a substantial proportion of on-air material, and by the practice of rapidly imitating innovations that seem to be striking the right note with the viewing public.

In its totality, television programming in the United States has retained a certain stability of content derived from formulas that worked well in the "golden days" of radio and Hollywood and that had their antecedents in nineteenth century popular theater. But this overall continuity masks substantial shifts in the mix of programming—shifts that rarely manifest themselves dramatically when we compare the lineup for one season with the next (see Table 6.1).

TABLE 6.1
Total Network Programming Composition by Program Type

Program type	1972–73	1973–74	1974–75	1975–76	1976–77	1977–78
Movies	23.0	25.2	22.8	15.1	15.2	15.2
Western Drama	4.8	1.6	1.6	—	1.5	1.5
Action/Adventure/Mystery	31.7	34.6	39.3	40.9	34.1	30.3
General & Family Drama	7.9	7.9	17.3	18.1	12.1	18.2
Situation Comedy	16.7	18.9	11.8	16.7	19.7	21.2
Music/Comedy/Variety	11.1	7.9	3.2	6.1	10.6	6.1
News	1.6	0.8	0.8	—	1.5	1.5
Sports	3.2	3.1	3.2	3.0	3.0	3.0
Miscellaneous programming	—	—	—	—	2.3	3.0

Note. Yearly figures are in percentage.

Although the distribution of air time by program types is readily traceable, changes in the symbolic content of programs are difficult to describe with scientific precision. Such changes have been tracked in trend studies of the portrayal of women and minorities, but research has seldom dealt with the subtler value statements that television producers and writers make.

It may well be that the analytical techniques most appropriate to the elucidation of television's changing symbolic content are those of literary and art criticism rather than those of social research. Yet research has played an important part in television's history and has been used as a weapon in the argument over its content. Thus the perceptions of social research by the general public, by legislators, by social activists, and by spokesmen for commercial interests are all illuminated by the case history of how it was used in the discussion of television violence (Rubinstein, 1976).

Television is of course a mainstay of the American marketing system, and the television business is itself operated by marketing principles. Research is the key element in the marketing concept, which begins with the definition of public demands and shapes products to suit those demands. In the "democracy of the marketplace," suppliers compete to introduce and improve products that satisfy the special needs of every consumer segment.

The preoccupation with audience measurement may therefore be explained not merely as a necessity to meet the requirements of advertisers who buy time on the basis of so and so many "gross rating points"; it is also justified as the public's guarantee that its desires and wishes will be met. In response to the objection that a uniform policy of catering to the majority reduces programming to the level of the least common denominator, the broadcasters can point proudly to their news and public service programming and to the many cultural events they air at a loss because of their sense of responsibility to the highbrow minority.

To evaluate this theory of "cultural democracy" raises the question of how the public's desires are shaped in the first place. (For a sympathetic exposition of the "cultural democracy" thesis, see Gans, 1974.)

If they simply existed in a static state, there would be no difficulty in understanding the role of television and other mass media operating by commercial methods. But public tastes, needs, and interests are in fact always changing, and creative inventions in media content are in themselves one of the powerful forces for change. A plausible theory of how this happens might go along these lines:

Television time becomes attractive to viewers to the degree that it is filled with constantly varying (albeit comfortingly familiar) content. No one would gladly suffer endless reruns of the same program. This means that there is an insatiable appetite for talent, which is heightened by the sharp competition for time on the air.

Because there is always more material (or potential material) at hand than can ever be used, decisions must constantly be made to activate certain programming ideas and to select specific programs. The key network executives responsible for these decisions have (or develop) an extraordinary sensitivity to public taste and are exceptionally shrewd in their ability to anticipate public response to particular kinds of competitive programming lineups. (It is not accidental that a high proportion of these key programming executives received their original experience as researchers, especially as ratings analysts.) The decisions are heavily guided by program research, but they are also responsive to shifts in the public mood, as these are cued by political and economic events and by successful developments in other realms of popular culture.

As new programming comes on the air it is enveloped in a haze of promotion and publicity. New celebrities are created and new styles are celebrated. When these promotions generate excitement, there is vigorous follow-through and appropriate imitation. When they fail, they are quickly disposed of.

Given the public's wide latitude of tolerance and given the virtually closed system within which the three networks operate, any program that is put on the air is automatically guaranteed a substantial audience. As viewers become familiar with it, they also come to accept it or even to like it. Surveys of the audience play back affirmative attitudes. At this point the public has been given "what it wants" and democracy has been served.

How does this process differ from the workings of the commercial market for other forms of popular culture—magazines or motion pictures? There are at least four critical differences: (1) The public pays for on-air broadcast programming in an extremely indirect manner, through purchases of advertised goods and services rather than through direct purchase of publications, where conscious value decisions must be made. (With pay television, this distinction of course disappears.) (2) Unlike the other media, broadcasters use as their principal resource the free air; this gives the public, at least theoretically, a proprietary interest in what is done with this resource; social ideals and social inhibitions may interfere with the pure operations of the market mechanism. (3) The high degree of audience concentration in the time filled by the three networks creates a commensurately high concentration of decision-making power. Newspaper and magazine audiences are by contrast highly fragmented. (4) Research is usually used routinely not only as a basis for making programming decisions and for modifying program content but also to supply inexorable judgments of success and failure. (Magazine editors also watch their single-copy sales from issue to issue and commission postmortems to determine the readership scores for different articles and features, but they do *not* routinely research content before deciding whether or not to run it or utilize research to tell them how to have stories rewritten after they decide to publish them.)

The way in which television programs are developed has been described by Richard M. Powell, chairman of the Writers' Guild of America:

> The network programmers let it be known what they're looking for.... The writer or writer-producer comes in with a format which is quickly reduced to the lowest common denominator by the network people—something like a group of men with large feet stamping out a grass fire. What is left generally is a series about two cops with a warm human caring relationship toward one another, leaving behind them—as they work their warm and caring way through the TV season—a mountain of dead bodies. In other words, the format is a vacuum—which must be filled by violence.

The last phrase is an overstatement, but the interaction between network programmers and the actual producers rings true: Television content *is* hammered out to specifications. Violence, or the elimination of violence, may be among the features specified.

TELEVISION AND FILM

To understand the changing character of television entertainment content, a digression is necessary. Violence on television cannot be viewed separate from the change in the content of Hollywood films since a series of court rulings struck down local censorships in the early 1960's.

Of all the mass media, motion pictures were most directly hit by the growth of television. Paid attendance went from 33 times a year for the average person in 1948 to 5 times a year in 1977. Moreover, a major change took place in the composition of the film-going public as the downtown movie palaces closed down and television took over the family audience. Young people of courting age and blacks came to represent increasingly large proportions of the attendance, and films were increasingly designed to their interests.

A new ratings system was initiated by the Motion Picture Association in November 1968 (see Table 6.2). In the period since the rating of films began, a smaller proportion of those produced and released have been rated G (suitable for general audiences) and a higher proportion have been rated R.

It would certainly be unfair to attribute the accompanying changes in film content (increasing mayhem and bloodletting, nudity, explicit sex, profanity and obscene language, etc.) to the adoption of the ratings system. These changes corresponded to far-reaching structural transformations in American society and population. They reflected shifts both in the general mores and in public morale during the age of the nuclear stalemate and the period of the Vietnam war (Winick, 1977). But it also seems likely that the presence of the ratings itself became a force that encouraged the more casual

TABLE 6.2
MPA Ratings 1968–1977

	11/68–10/69	11/72–10/73	11/76–10/77
G	32	15	12
PG	39	34	39
R	23	47	41
X	6	4	8
N =	(441)	(550)	(405)

Note: Yearly figures are in percentage.

use of scenes, language, and imagery that would have been censored by the old Hays MPA Code Authority.

As long as there was but a single standard of acceptability by which all films were judged, there was no incentive to introduce scenes or dialogue that might be cut by the censor. When the rules were changed, such content was not only acceptable under a heading other than G for the family audience but actually was considered a positive attraction for young filmgoers who wanted precisely what was banned for those younger still. Thus it became commonplace for films constructed to the traditional old Hollywood formulas to incorporate bedroom sequences or street language gratuitously, as though for the express purpose of restricting the rating and making them appear more daring than they really were.[1]

Hollywood naturally sought to exploit, for its theatrical audiences, areas of content which were barred from television. This trend went along with technological innovations—the wide screen (first introduced with Cinerama in 1952), flawed and unsuccessful attempts to develop a three-dimensional projection system, and a nearly universal use of color (at a time when television still did not have it).

All these developments raised film budgets at a time when producers were retrenching their production schedules. With fewer films being made for fewer theaters, the doublefeature virtually disappeared and the super spectacular high-budget production became more and more the norm. Films were backed by intensive television advertising promotions that often exceeded their production costs. With expensive, colorful, large-screen, lengthy entertainments, Hollywood mined sectors of the market that television could not reach. Similarly, its subject matter and its portrayals of human relationships deviated farther and farther from the restrictions of the old Hays Office. Crime went not merely unexpiated; it was routinely glorified.

---

[1]In addition to this trend toward more explicit sexual content and profanity in films rated by the Motion Picture Association, there has been a substantial increase in the number of self-designated "X-rated" pornographic films shown without the benefit of an official rating.

The illusional skills perfected by the film industry were applied to project realistic scenes of sadistic brutality and mutilation. In the domain of sex, little was left to the imagination, and rape became an increasingly popular motif.

It was inevitable that television programming should feel the effects of these great changes in the medium with which it is so closely linked. The relationship of television to films is not one-way. What happens in uncensored Hollywood directly relates to what happens on television. The financial and production structures of television programming and feature film-making are now thoroughly integrated. As already mentioned, both draw on a common pool of talent. Both use the same studios and facilities. Hollywood is the center of production for most prime time network programs and major syndicated shows. A great deal of television fare, including some with the highest ratings, now consists of feature films that have already had a theatrical release. In addition, there has been a growing use of other films designed especially for *initial* showings on television and subsequent showings in movie theaters. All this has made it inevitable that the changing style and substance of feature film-making should be reflected on television. Even the theme of male prostitution in *Midnight Cowboy* could be adapted for television with a few minor cuts. But the need for such cuts became less compelling as late evening television was increasingly regarded as a refuge for "mature adult" viewers.

Yet another factor to be mentioned is the increasing distribution and commercial success of low-budget pornographic films, sometimes produced under criminal auspices. So-called "X-rated" films are available on the pay channels of many hotel room television sets. The growth of pay cable television creates an expanded opportunity for them to be made available to home viewers as well. The manager of one cable system reports, "We asked viewers what they wanted, and 90% said hard-core movies" (*Wall Street Journal,* May 17, 1977). Whether or not such attactions would significantly affect the size of the audience for conventional television programming, it seemed reasonable to suppose that this new competition would be met by further adjustments in style and content on the part of the networks.[2]

"MEASURING" VIOLENCE

The Milton Eisenhower Commission on the Causes and Prevention of Violence paid particular attention to the effects of violence in the media (Baker & Ball, 1969). Witnesses before the Commission proposed a systematic and continuing measurement of violent content in television

---

[2]In July 1978 the Supreme Court ruled in the WBAI case that the FCC could regulate on-air obscenity. The full effects of this decision were not immediately apparent.

programming. This led to the development of a "violence index" by George Gerbner and his associates at the Annenberg School of Communications. Gerbner's studies represented an important contribution to the report of the Surgeon General's Scientific Advisory Committee on Television and Social Behavior.

The summary report of the surgeon general's committee was edited with extraordinary care to avoid any overstatement of inferences from the data. Nonetheless it stated clearly that there was a causal connection between television violence and subsequent antisocial behavior, at least for some children. Moreover, a poll of the authors who contributed research papers to the report found that most of them had independently concluded that television violence was harmful to children (Paisley, 1972).

This conclusion was entirely congruent with a substantial body of literature that had accumulated on this subject over a generation (Bogart, 1972; Liebert, Neale, & Davidson, 1973). The academic literature continued to build up further evidence that pointed in the same direction. (A comprehensive review by F. Scott Andison [1977] summarizes the data by saying that "television, as it is shown today, probably does stimulate a higher amount of aggression in individuals within society" [p.323].)[3]

In 1976 the National Science Foundation allocated $1.5 million for grants on "policy-related research on the effects of television." However, many research professionals in the field appeared to believe that the necessary evidence was already in. A three-day workshop held under the auspices of the Ford and Markle foundations in November 1975 for the purpose of defining priorities for future research on television did not place the subject of violence high on the list.

At the behest of the Senate Communications Subcommittee, the National Institute of Mental Health commissioned further studies by Gerbner and his associates. These studies found that the level of televised violence did not diminish in the three years following the publication of the surgeon general's report. NIMH also commissioned the Social Science Research Council's Committee on Television and Social Behavior to prepare a paper on the feasibility of independently devising a violence index. This paper examined the criteria for defining and measuring violence, with particular stress on the motivation and context of violent acts. Although the Committee's conclusions were thickly enveloped in all the necessary qualifications, the paper may be interpreted as essentially an endorsement of Gerbner's methods and data.

---

[3]See also Comstock, Lindsey, and Fisher (1975) and Comstock (1977). In his review of the literature, Comstock points to the importance of situational factors in determining the type of response to violence. He concludes that behavior is more important than verbal advocacy and that similarities between the audience and the figures in the portrayal, the consequences of violence, and the opportunities to imitate fictional violent behavior all enter into the mix.

Gerbner's violence index was widely criticized by television spokesmen as "mechanistic" on the grounds that it lumped together violence in different types of programming regardless of its character or context. (When a CBS research report [April 21, 1977] took the Gerbner index to task, Gerbner cited the SSRC paper in his rebuttal.)

The National Citizen's Committee for Broadcasting, under a $25,000 grant from the American Medical Association, did two parallel analyses. One was based on the Gerbner definition. The other was based on a definition acceptable to the television industry, measuring only such clean-cut incidents of violence as murders, killings, and beatings. The results showed a strong agreement in the violence ratings produced by these two methods. The conclusion was that the use of the Gerbner definition is "sound, accurate, and fair when measured against other standards" (National Citizens Committee for Broadcasting, 1976).

In a further corroboration of Gerbner's conclusions, Bradley Greenberg and Charles K. Atkin analyzed 92 episodes broadcast during peak children's viewing hours in 1975 and 1976 and found almost twice as much physical aggression in Saturday morning cartoons as in prime time programs and as much antisocial behavior during family viewing hours as in the period after 9 p.m. (Greenberg, 1977).

Nonetheless the networks persisted in their efforts to find an alternative to the Gerbner index, presumably one that would not only be more realistic in their terms but also one that would reflect violence.

Alfred R. Schneider, vice president of ABC, told an Association of National Advertisers' Workshop in March 1977 that his company was

> working on and hopefully will shortly be able to introduce a method of examining and reviewing the portrayal of violence on both a qualitative (and) quantitative basis.... We are doing this because we believe that *the primary issue is whether violent sequences in entertainment programs affect behavior, rather than perceptions or attitudes or taste levels* [my italics].

The notion of behavioral consequences divorced from changes in perception, attitude, and taste is in itself worth pondering. The distinction may defy explanation on theoretical grounds, but it is certainly pragmatically interesting in that behavior represents a much more difficult subject of research.[4]

---

[4]As an example of the importance of "perceptions," one might ponder the findings of a study by Gerbner, Gross, Jackson-Beeck, Jeffries-Fox, and Signorielli (1977). They compared the responses of heavy TV-watchers with those of light viewers to questions about the likelihood that they themselves would be involved in violence. The heavy viewers "significantly over-estimate the extent of violence and danger in the world." Those under 30 were "even more imbued with the television view of life."

The definition of violence, it could be concluded, was inseparable from the definition of its effects. The terms of debate, as laid down by Schneider, rule out any concern with the influence of television content on the hearts and minds of the viewers, on their notions of what is good and bad. Yet few voices have been raised to suggest that "perceptions," "attitudes," "taste levels" are precisely the subjects of real concern.

QUESTIONING THE EVIDENCE

The two television network research executives on the surgeon general's committee, both had subscribed to the Committee's report and in hearings before the Senate Communications Subcommittee publicly acknowledged that there was some causal connection between television violence and aggression.

But almost from the outset, other broadcasting executives disassociated themselves from the consensus position of the surgeon general's committee. The contention that the evidence to date was incomplete and inconclusive was facilitated by the widespread misreporting of the committee's findings in the press, a good deal of it apparently the handiwork of headline-writers.[5]

Testifying before the House Communications Subcommittee, spokesmen for the three television networks "refused to agree that there might be a causal link between violence on television and aggressive actions by young viewers because, they said, 'clinical evidence is lacking to prove such a link' " Broadcasting, March 7, 1977.)

Bob Howard, president of the NBC Television Network, said, "If we feel something is worth doing, we're going to do it. NBC has its own code....

---

[5]An analysis of press coverage (Tankard & Showalter, 1977) of the surgeon general's committee report compared a random sample of 20 Texas newspapers, a sample of 10 prestige newspapers throughout the United States, and all magazines indexed by the Reader's Guide for 1972. There were 19 articles in the national newspaper sample, 21 in the Texas newspaper sample, and 16 in the magazine sample. Of the 20 Texas newspapers, 4 never published anything on the report. Most of the newspaper articles suggested that the relationship between viewing violence and aggressiveness depends on a third set of variables, but in few cases did the headlines reflect this. The magazine articles were more likely to indicate that viewing of television violence increases aggressiveness, and a substantial number of articles published in Texas newspapers gave the impression that viewing television violence has no effect on aggressiveness. The widely cited New York Times headline read, "TV Violence Held Unharmful to Youth." The headline in the Fort Worth Star Telegram said, "Study Finds Violence Not So Bad." The Milwaukee Journal said, "Panel Rules TV Violence is Safe for Most Children." By contrast, the Washington Post's headline read, "Tendencies Increase: Study Links TV, Child Aggression." And the St. Louis Post Dispatch said, "TV Said to Add to Aggression." The authors characterize press treatment as "confused and indefinite.... Few newspapers ran follow-up news coverage that might have clarified the ambiguity of the findings" (Tankard & Showalter, 1977, p. 298). Only 23% of the papers studied put the story on the front page, and few used editorials or commentary.

What people have been talking about are programs that are not true to life. In 'Police Story,' you can feel the agony of a police widow."

Tom Swafford, CBS vice president for program practices, said, "Violence in CBS shows is in context, is germane to the plot, advances character development, and is never gratuitous."

Some time later, as head of the Code Authority of the National Association of Broadcasters, Swafford told a meeting of the International Radio and Television Society that the role of his organization was "to defuse the criticism." Taking issue with David Pearl, chief of behavioral sciences for NIMH, Swafford said that Pearl could only use phrases like "'there might be,' 'could have an effect,' and 'may cause' in discussing television violence because there was no absolute evidence to back up Pearl's assertions [p. 46]." *(Media Decisions,* March 1977).

Frederick S. Pierce, president of ABC television, noted that violence represented a smaller percentage of his network's output. According to *Advertising Age* (February 28, 1977), he "accused the American Medical Association of making unfounded and unproved allegations based on the unscientific though well-intentioned research from BI Associates and the National Citizens' Committee for Broadcasting." He invited the AMA to join his network "in a *meaningful scientific analysis* [my italics] of the relationship of programming to viewer behavior."

NETWORK-SPONSORED RESEARCH

While the Surgeon General's report was still under way, CBS had underwritten a series of experimental studies directed by Stanley Milgram and R. Lance Schotland that tested the hypothesis that exposure to specific antisocial acts shown in television programming would result in specific antisocial imitative actions on the part of viewers (Milgram & Shotland, 1973). The evidence did not support the hypothesis. In the authors' words, "If television is on trial, the judgment of this investigation must be the Scottish verdict: Not proven [p. 68]." But the argument that antisocial acts seen on television are sometimes imitated rests on deviant cases that occur rarely in the normal course of events, and the Milgram-Schotland studies merely demonstrated the difficulty of re-creating complex natural phenomena in the laboratory.

Another industry-sponsored study was a replication by William Wells (1973) of an earlier experiment by Seymour M. Feshbach and Robert D. Singer (1971), whose conclusions supported the "catharsis" theory that exposure to violence would reduce aggression. Wells did find that children deprived of their normal diet of violent programming became more verbally aggressive, but he concluded that it was the loss of their favorite shows that irritated them.

In the aftermath of the surgeon general's report, the television networks stepped up substantially their own direct support for research on the subject. Although CBS and NBC had long maintained active and excellently staffed programs of social research, ABC initiated its activity only after the subject was pressed by the Senate Subcommittee. (The timing of this new interest coincided with ABC's remarkable rise in ratings and revenues; the Gerbner index had shown it to lead consistently in its use of violent programming content.)

NBC's Social Research Division, under Ronald Milavsky (unpublished), initiated a major field study of television violence in 1969 and subsequently invested over $1.5 million on it. The research was conducted between 1970 and 1973 in two Midwestern metropolitan areas that are comparable in size and other criteria, except that one of them had a much higher homicide and aggravated assault rate than the other. One part of the study covered approximately 1200 elementary school boys and 1200 girls and used six waves of self-administered questionnaires. Data were also obtained from the boys' parents. Children were evaluated by others in their classes to identify their levels of aggression, verbal and physical. The other part of the study covered 800 teenage boys who answered five waves of questionnaires.

Milavsky and his associates began with the premise that there would be an initial high relationship between viewing of television violence and aggressiveness but that this was not a demonstration of causality. They wanted to see how much of the difference in aggressive behavior between the initial questionnaire and subsequent contacts could be accounted for by earlier exposure to violent television once the original levels of aggressive behavior were controlled. (That is to say, would children who were watching a lot of violent television in 1970 be *comparatively* more aggressive in 1973, given the fact that they were more aggressive to start with?) They found no discernible effect of this kind, nor any discernible effect of the cumulative exposure to television violence over the duration of the study, even when exposure to violence was looked at in relation to total television viewing.

For that matter, they also found no evidence to support either (1) the "catharsis" theory, (2) the theory that television violence especially affects children predisposed by a high level of aggressive behavior at home, or (3) the "desensitization" hypothesis that television violence especially affects those children who like real television violence (as shown on the news). (Quite apart from its media-related aspects, the research produced a number of results that illuminated the general subject of aggressive behavior.)

Comment on the NBC findings can properly be made only after their publication, but it appeared inevitable that questions would be raised about the salience of the principal hypothesis. If a 10-year-old boy (the youngest age contacted on Wave 1) who watches a lot of violent television is already more aggressive than his counterpart who doesn't, is it reasonable to expect that he would become progressively and proportionately more aggressive by the age of 13? Most of those who are concerned about the effects of television violence would probably argue that they are experienced most acutely at an earlier age.

ABC released two studies that suggested that children already predisposed to aggression could be susceptible to the influence of televised violence. Apart from that, they concluded that normal and abnormal children alike were in no way influenced in the direction of aggressive behavior by what they saw on television.

ABC's first study consisted of a series of laboratory experiments by Lieberman Research (no date) among 10,000 children 8 to 13, using an "Electronic Pounding Platform" as a substitute for the old-fashioned Bobo doll. The experiments found that "pre–post increases in inclination toward aggression were found to be associated with exposure to violent elements in television programs."

The research found

> sharp differences in impact depending upon the type of violence portrayed, with killings and woundings producing the most, and chase scenes and verbal violence the least... The more realistic, the greater the effect. The more children could identify with the characters, the greater the effect.... Although most children indicated they were not scared by any of the programs and indicated a preference for programs that scared them, being scared is associated with greater aggressive tendencies.

Sudden and unexpected violence was especially disturbing; a large screen created more excitement than a TV screen; commercial interruptions inhibited arousal.

Another series of eleven studies was conducted by Melvin A. Heller and Samuel Polsky (no date) among emotionally impaired children, children from broken homes, and others in prison for crimes of violence. They found that "exposure to aggressive television content did not lead to heightened aggressive behavior" but that "programs with more aggressive content produce more aggressive fantasies" and "susceptible children tended to use aggressive television materials to bind their own aggressive drives." "Although no causative relationship was found to criminality as such, there are indications TV can affect the style or technique of crimes." (In fact, of a group of 100 youthful criminals, "22% confessed to having imitated or tried out criminal techniques they had first seen demonstrated on television.") In an interview (1977) following the release of his report, Heller expressed the opinion that "some violence is good for children to see when in the context of

programs such as 'Roots,' for example, that provide strong social messages."
(Broadcasting, February 28, 1977.)

A review of the Heller and Polsky research by Robert Liebert and his collaborators (1973) takes issue with the conclusion that televised violence can have an effect only on those already predisposed by basic personality structure and early childhood experiences. "The persistent message is that the television medium should be let off the hook as a cause of societal problems of any kind." Reexamining the data, they find that

> the majority of statistically significant effects reported in this book tend to support further the conclusion of the Surgeon General's report that certain common televised examples contribute to both aggressive attitudes and overt aggressive behavior among some children and youths.

Another ambitious research project conducted under broadcast industry auspices was a study among 1565 London boys 13 to 16. The research, funded by a $290,000 CBS grant, was conducted by William B. Belson (in press), a pioneer student of television's social effects.[6] Belson's findings were based on the boys' own reports of their behavior. He concluded that "the evidence is strongly supportive of the hypothesis that long-term exposure to television violence increases the degree to which boys engage in serious violence" and the degree to which they use bad language, swear, and are violent in sport or play. He found serious violence increased by long-term exposure to content that featured close personal relationships together with verbal or physical violence, to programs in which violence is thrown in gratuitously, to programs featuring realistic fictional violence, to violence "in a good cause," and to Westerns.

Belson found little evidence of effects from violence in sports (as in wrestling or boxing), in cartoons, science fiction, or slapstick comedy. Nor did he find evidence that television violence makes viewers callous, makes them accept violence as a way to solve problems, or makes them consider violence to be inevitable or attractive. Belson (in press) found no effects on respect for authority or on preoccupation with the results of violence. He states: "The violence most increased by television is of the unplanned, spontaneous, unskilled kind."

Belson matched his more violent and less violent subjects on a variety of variables in order to control for the specific effects of exposure to television

---

[6]The boys were questioned on their exposure to 68 television programs during the period 1959–1971. Each program had been rated for the level of violence in its content. The sample was separated into a high-scoring half in terms of exposure to television violence and a control group (the lower-scoring half). The two groups were carefully matched in terms of 227 possible predictors of violent behavior that could operate independently (Belson, in press).

violence. Still, his findings were dismissed as "correlational" by CBS's research vice-president, David Blank.

Although the report of the NBC staff-directed study has not yet been published, it seems fair to conclude that the research done by or for the networks has not contradicted or substantially qualified the conclusions of the surgeon general's report.

THE PUBLIC RELATIONS OF TV VIOLENCE

The broadcasters' insistence that the case against violence was not proven became all the more forceful as a number of public action organizations took a more aggressive adversary position. Medical journal articles attacking televised violence as "a risk factor threatening the health and welfare of American children" received heavy publicity support from the American Medical Association (Rothenberg, 1975; Somers, 1978).

The editors of the *New England Journal of Medicine* wrote to the one hundred leading television advertisers asserting that "the pervasive portrayal of violence on TV has a destructive influence in our society" and asking if they intended to sponsor programs clearly identified as violent? They termed the results "encouraging," but their meetings with network officials were less so. The broadcasters "stressed the point that they consider the parent primarily responsible for a child's television viewing habits" (Feingold & Johnson, 1977).

Action for Children's Television (ACT) continued its vigorous lobbying for restrictions on the content of both programming and advertising. The Communications Office of the United Church of Christ delivered a strong attack on television's "commercialized" violence and sex.

The U.S. Conference of Mayors sponsored a study of 73 hours of primetime "action" shows in which weapons appeared at the rate of 8.87 per hour, 68% in confrontations between the user and someone else. However, only 36 on-screen deaths resulted, thus attaching "an unrealistic and potentially dangerous lack of severity to the consequences of weapon use" (Wilson & Higgins, 1977).

The National Parent–Teachers Association organized a series of meetings around the United States to coordinate parents' opposition. The Association's "public hearings" brought forth a variety of emotional witnesses. All this commotion kept the subject of television violence alive as a public issue to which legislators and officials had to remain responsive. For the most part, the activity emerged as an expression of long-standing institutional concerns or commitments, but it was also stimulated periodically by news items that focused attention on the subject in defiance of Milgram and Shotland (1973).

For example, in October of 1973, a film called "Fuzz" on the ABC network showed juvenile delinquents in Boston burning derelicts to death for fun. Several days later, six youths poured gasoline over a 24-year-old woman and burned her to death (1973).

Daredevil Evel Knievel's televised motorcycle jump was alleged to have resulted in a number of accidents when children emulated it (Singer & Singer, 1974).

NBC and KRON-TV in San Francisco were sued for $11 million after four teenagers sexually attacked a 9-year-old girl, allegedly after one of them "got the idea" from a scene in a television program, "Born Innocent." The case was subsequently dismissed after the trial judge's ruling that the plaintiff would have to demonstrate that the network had intentionally incited the act.

In the fall of 1977, the television violence issue was made a central feature of the defense brought in the case of a 15-year-old Florida boy, Ronald Zamora, accused (and ultimately convicted) of killing a neighbor. His attorney sought to prove that his client's ability to make moral judgments had been destroyed by his long exposure to a diet of televised violence. The ingenuity (or peculiarity) of this defense brought extensive attention to the subject at the start of the new (1977–78) broadcast season, in which (according to the networks) violent programming content was being substantially reduced. This reduction, accompanied by self-congratulatory statements, occurred at a time when the networks were still releasing critiques of the Gerbner Violence Index and still sponsoring research intended to prove that violence had no effects on child behavior.

A January 1977 Gallup poll found that 70% of the public saw a correlation between TV violence and the rising crime rate. A Harris poll in August found 71% who thought there was too much violence on television.

At the same time the broadcasters continued to protest against the validity of the evidence that had been raised against them, they professed their intentions to lower the levels of violence. The Code Authority of the National Association of Broadcasters announced that it would publish guidelines for this purpose. ABC promised a 50–60% cutback in the detective–action category.

The institution of "family time" for an hour in the early evening following the network news was a direct response (in September 1975) to the surgeon general's report and to the resulting public clamor. The idea was to restrict violent and sexually explicit programming before 9 p.m., EST, at a time when young children were watching. However, much of the viewing of television by children occurs after 9 p.m. The net effect of restricting violence during the family hour may have been simply to encourage a freer use of it outside the designated period. (This hypothesis cannot be verified, because changes in the balance and content of programming were already under way for other reasons.)

Variety heralded the family hour with the headline, "TV Networks Raise Gore Curtain; Open Season on Violence; Slime at Nine," and labeled 9 p.m. EST "the starting gun."

The practice of the family hour was challenged in the courts and encountered an adverse decision that the networks appealed. In any case, it was apparent that during family time the networks has lost some share of the audience to independent stations that continued to feature old movies and syndicated programming that held to the old violent formulas.

There were, unfortunately, different interpretations of what effects the broadcasters' new policies had had on programming content. On January 19, 1976, *Television/Radio Age* reported a decline in the amount of prime time violence during the 1975–1976 season "not because of public outcry, but because viewers are now rejecting these shows instead of just complaining about them." By contrast, George Gerbner reported in testimony before a Congressional committee that 1976 was the highest year for violence in the eight years (to date) of his annual profile. All three networks showed considerable increases in all categories. Gerbner was challenged by a CBS study using a different definition of violence. It showed that for 13 weeks' monitoring of primetime shows, the amount of violence on CBS declined slightly from the 1975 and 1976 broadcast years—from 32.5 to 31 incidents. (ABC remained steady at 49.3. NBC rose from 37.9 to 55.9.)

By the following year, however, the new look in programming had taken effect. In 1977, violence approached the record low of 1973, among the eleven years measured in the Gerbner studies. Seventy-six percent of all network programs contained some incidents, compared with 80-90% in the preceding two years. But more important, there were only 6.7 violent episodes an hour, compared with 9.5 in 1976. (ABC was rated the "least violent" network, NBC the "most violent.") (Gerbner, *et al,* 1978).

General Foods' Media Services Director, Archa Knowlton, reported that unacceptably violent programming would be 19% of the total, down from 36% in 1976 and 43% in 1975. The Television Bureau of Advertising offered reassurance to media buyers alarmed by the reductions in violence that "advertisers will not miss any of their demographics that they've been buying in the past."

In fending off the demand for change in programming content, the network spokesmen raised a new issue—censorship. This specter was especially frightening in the post-Nixon era when sensitivities had been heightened to the vulnerability of publicly licensed broadcasters to governmental intervention. Interference with their programming practices was therefore, by implication, a genuine threat to democracy.

"There are too many critics, do-gooders and would-be censors," said a "top-web exec" quoted by *Television/Radio Age* (January 19, 1976). "Nobody ever got hurt knocking TV. Most of the criticism we receive is orchestrated by a few organizations. It could hardly be considered a true cross-section of American sentiment." James Duffy, president of ABC-TV, warned that advertisers should not be swayed by outside pressure groups, which he characterized as made up of people who had not seen the programs.

ABC's programming chief, Fred Silverman (later president of NBC), told an American Association of Advertising Agencies meeting in 1977 that "programming to please pressure groups will emasculate television." Silverman said, "Every time you remove a commercial from a controversial show, we all take a step backward to a mass medium that is less free, less diverse, less creative, and less democratic." According to *Advertising Age,* he "compared pressure groups, lists of advertisers associated with violent or sexually suggestive programming, to the blacklists of the McCarthy period." "Our responsibility is to the public, not to pressure groups, and our chief responsibility is to see that the public has the final say over what goes out over the public's airwaves, not the pressure groups' airwaves."

For once, the networks found allies for their position among the independent program producers, especially those who had won commercial benefits by adapting old formulas to new standards of acceptability.

Producer Norman Lear warned that if the networks "bend to certain interest groups, it might force all television back to the "blandness" of the late 50's and early 60's. Speaking to the Broadcast Advertising Club of Chicago, Lear accused the American Medical Association of taking a stand against television violence "to draw attention from its own violence."

Yet another defense for the broadcasters came about as a result of their insistence that any unwanted consequences of exposure to television content arose from problems at the receiving rather than the transmitting end. Some government officials found this shifting of responsibility quite convenient. Thus Senator John Tower of Texas acknowledged that "the amount of violence on television is reprehensible." But he went on to say, "I would not for one moment entertain the idea that the state should take it upon itself to deprive us of this subject matter. The responsibility for the exercise of wisdom in this matter lies with the parents." (He was apparently unconcerned with the practicality of inducing 100 million parents to "exercise wisdom" as an alternative to having three network managements do so.)

An October 1977 report of the House Communications Subcommittee similarly denied that there was a proven and precise cause and effect relationship between televised violence and aggressive behavior. According to *The New York Times,* the report "concluded that while the networks were chiefly responsible for the present level of televised violence, some of the blame must also be assumed by affiliated stations, program producers, advertisers, *and the viewing public* [my italics]." The report asserted that parental supervision was "probably the most effective way to curb the negative effects of excessive viewing of television violence by children." Six dissenting members of the Committee, in a minority report of their own, accused the majority of having reached "the meaningless conclusion that we are all to blame for the excessive levels of violence on television—thereby implicitly stating that no one is really to be held accountable."

DOES HE WHO PAYS THE PIPER CALL THE TUNE?

"The crusade against violence on television has assumed the proportions of a major assault," said *Television/Radio Age* (February 28, 1977), "and while the television industry has not quite raised the white flag, some strategic retreats and regroupings of forces are in evidence."

In the face of an impassive stance on the part of the broadcasters, Richard E. Palmer, president of the American Medical Association, called on advertisers to reduce their sponsorship of violent programming content. Indeed, to quite a few observers the only effective pressure to which broadcasters would be responsive appeared to be the economic leverage exercised by advertisers. Although in the early days of television, advertisers and agencies themselves controlled many programs, this is rarely true today because the high costs of advertising have virtually eliminated the old practice of exclusive program sponsorship. But the most powerful advertisers, being most in the public eye, are also the most nervous about using pressure on the media they use.

A statement by the Association of National Advertisers denied that advertisers were in any sense responsible for the changing character of program content. By contrast, the J. Walter Thompson advertising agency made a strong effort, through a series of presentations, speeches, and research projects, to decry the effects of violence and to question the value of advertising placed in a violent context. A public opinion survey conducted by the agency found that one-third of the public referred to "bloody scenes" on television, although "there is little if any blood actually shown in entertainment programming." Joel Baumwoll, the agency's research director, concluded that "the findings indicate that the public's perception of gratuitous violent acts exceeds the actual existence of such acts portrayed on the screen." The survey found that 63% of the public felt that television's violence is "very harmful" or "extremely harmful" to children, and 40% felt it is "very" or "extremely harmful" to the general public. Forty-two percent say it is either "very" or "extremely widespread."

The National Citizens Committee for Broadcasting early in 1977 singled out the 10 leading sponsors of violent television shows. The list of advertisers was headed by Sears Roebuck and included General Motors. Arthur M. Wood, president of Sears, quarreled with the criteria used in the assessment but ordered a revision of the company's advertising guidelines. In 1977 Sears removed 104 commercials from programs that were "not suitable viewing for the entire family." Nonetheless, Sears's headquarters and stores were the target of demonstrations by the National Federation of Decency (headquartered in Tupelo, Mississippi), which said, "Sears has brought violence, illicit sex, and gutter-level profanity into our homes."

Robert Lund, general manager of the Chevrolet Division of General Motors, also issued new policy guidelines against excessive television violence. His statement read, "We've got to go beyond ratings, beyond market share, beyond pragmatism. Advertising today has the power to help shape the quality of life... the creation of powerful advertising is not an end in itself."

"A blood and guts environment is a terrible place to put a commercial for Jello," said General Foods' Archa Knowlton. The *Wall Street Journal* (April 5, 1977) observed that "few advertisers have gone so far as to withdraw commercials from entire series, even those rated most violent by the Citizen Committee....Instead they say they avoid particularly 'objectionable episodes.'"

Moving from its previous hands-off position, the Association of National Advertisers' Board of Directors recommended that advertisers keep close watch on television programming content. The ANA statement recommended that advertisers keep a week-by-week watch on television shows carrying their commercials. The statement noted that advertisers had to accept programming responsibility because they "are generally blamed when viewers dislike something they see."

Similarly, the American Association of Advertising Agencies rejected government regulation and censorship and added cautiously,

The role television plays in shaping the attitudes of the American public—particularly its young people—requires that we carefully examine our support of programs which portray violence to the point that they may be promoting violence. *Precise research on this relationship is lacking* [my italics], but the complete absence of positive values should render complete proof of negative value unnecessary. Common sense tells us that a repeated diet of physical brutality could stimulate individuals to similar violence or general public insensitivity to violence. Commercial sense—which we as agencies are paid to provide—warns us that the context in which our clients' messages are viewed is a vital consideration. We believe that since advertising supports TV, it is obligatory for agencies—as a service to their clients—to restudy advertising commitments to programs which feature violence.

As *Advertising Age* (March 28, 1977) commented, this expression of concern did not represent "a specific stand" on the subject.

At a *Media Decisions* forum (1977), Howard Lelchuk, senior vice president of the N. W. Ayer agency, proclaimed that "the axiom 'violence begets violence' is invalid; or at least has not yet been proven." Lelchuk concluded that "there is a demand for violence and more mature themes.... Should [advertisers and agencies] decide what should or should not be on the air?

They can—if they are willing to gamble their company's profits by increasing the cost of doing business by buying non-violent programming which might be lower rated."

An investment analyst interviewed by the *Wall Street Journal* suggested that the expressions of concern over violence on the part of advertisers were actually part of a negotiating game and were merely being used as a pretext in order to cut down television advertising rates. A spokesman for McDonald's noted that half the prime time programming in 1976–77 was "action stuff." And "companies that make products aimed mainly at men—razor blades, say, or trucks—find themselves in a more delicate position. Action shows deliver just the kind of audience they're interested in: young males."

Media Decisions columnist Ed Papazian (1977) wrote:

> If we take the 18–34 group and within it take only the highest income level, the penalty for switching completely to non-violent shows could rise to 25%.... Advertisers selling razor blades, cars, shampoos, hair dressing, grooming aids and scores of other products and services... had better take a long hard look at the media implications.... It's almost like asking a brand manager, who is spending five million dollars now to get the same impression weight he bought for three million only four or five years ago, to ante up $750,000 out of his media budget to support the anti-violence cause. Put it this way and I doubt the idea would get any serious support.

In any case, Papazian writes: "Despite a succession of well-funded government studies into the effects of violence on children, and many private investigations, *there is little hard evidence on the subject* [my italics.] (*Media Decisions*, May 1977, pp. 12–16.)

FROM VIOLENCE TO SEX, OR WORSE

By the end of 1977, the issue of TV violence was no longer under active discussion in broadcasting circles. The two years just past had shown extraordinary increases in revenues for both the networks and stations and sharply accelerated competition for ratings. A series of reshufflings in ratings preeminence had led to convulsions in management and personnel. An unprecedented number of new programs had been yanked off the air with new ones substituted. Regular prime time programming was routinely being interrupted by sports events and mini specials that created a less predictable pattern of viewing. The "action shows" that had drawn criticism for their violent character were a far less visible component of the evening television program mix. The attention of critics was drawn instead to a new development: the steady, and studied, infusion of sexually suggestive themes, references, and language into television programming.

No scientific measurements exist in this area comparable to Gerbner's Violence Index, which permits trends to be tracked from one broadcast season to the next. The National Federation of Decency did produce a report on television sex, in emulation of Gerbner, but without his systematic procedures. Its members viewed 865 hours of prime time network programs in 1977 and found 2433 scenes that were "sexually suggestive" (89% involving extramarital sex).

Casual observation indicates an increasing departure from the stiff standards of decorum that have traditionally prevailed in network broadcasting. This has been true of daytime serials, whose fictionalized emotional entanglements have for some time expanded beyond family bounds. Since the establishment of the family hour, illegitimacy, abortion, incest, homosexuality, rape, and extramarital affairs have all become subjects for prime time programming after 9 p.m., with varying degrees of dramatic integrity or cynicism.

As the clamor over television violence became attenuated out of sheer boredom, a new controversy arose over television sex. In the 1977–78 broadcast season this controversy quickly focused on a new ABC prime time comedy series, "Soap." Described by ABC as "a sophisticated adult farce," "Soap" featured transvestism, promiscuous marital infidelity, impotence, and other humorous themes (Stutzman, 1978). Before the first episode was aired, ABC received over 22,000 letters of protest. Some affiliates refused to carry the program, and all of the originally scheduled advertisers withdrew their commercials.

The network knew the limits of permissiveness and ordered some rewriting of the original scripts. The homosexual who was planning a sex-change operation fell in love with a girl. A directive from the producer read, "Corrine's affair with a Jesuit priest, her subsequent pregnancy as a result and later exorcism are all unacceptable."

The volume of complaints diminished after the program was on the air, and advertisers (not necessarily the original ones) came back. "Soap"'s ratings, although not spectacular, were satisfactory. Viewer surveys in Richmond, Virginia, and Lexington, Kentucky, found only about one in four who thought the program was "offensive," and found that churchgoers were no more often turned off than other viewers.

As in the case of television violence, the issue of censorship was quickly raised. The program's coproducer, Paul Junger Witt, said that if ABC gave in, it would be "an irreversible step toward selling out freedom of expression." Edwin T. Vane (1978), ABC's programming vice president, declared that, "A basic industry principle was at stake, namely, who is going to control what goes on the air—special interest groups who lobby, or the American public?" (*New York Times*, February 23, 1978.) Aryeh Neier, executive director of the American Civil Liberties Union, argued that "it is reprehensible that groups should try to pressure television, through its advertisers, not to deal in controversial material."

NBC Research Vice President William Rubens agreed that "TV is not and does not want to be a trend setter of the sexual revolution. In fact, it lags behind the prevailing attitudes." Broadcast industry executives agreed that the trend toward sex-oriented programming would continue (a self-fulfilling prophecy, because those who did the predicting also made the decisions.)

There appeared to be little pertinent comment from social scientists at this point in the debate. The report of President Nixon's Commisssion on Pornography and Obscenity (angrily rejected by him) had found no empirical evidence to demonstrate that exposure to such material had concrete behavioral consequences. (The contrast between this conclusion and that of the surgeon general's committee on television violence became a topical matter of commentary,[7] though it was pointed out that fictionalized violence and sex occurred in sharply differing contexts.) Inevitably the lay discussion of whether or not changes in subject matter would lead to antisocial acts reduced itself to the cause and effect model of laboratory science. There appeared to be little willingness to accept the fuzzier notion of cultural modification, arising from the interaction between members of the society and the shifting array of symbols and models presented to them by the mass media.

TWO CONCLUDING NOTES

This is an interim report in a continuing story, but even at this early stage two observations may be derived from what has happened:

(1) The issue of violence in television content is only one of innumerable elements that could be examined in terms of the processes described at the start of this chapter. If the system works to give the public "what it wants," then arbitrarily imposed changes in the level of violence would in effect give the public what it does not want and should be met by an outcry, or at least by a reduction of viewing levels.[8] However, if the use of violence represents autonomous judgments on the part of decision-makers, to which the viewing public becomes accustomed, then a modification should be equally

---

[7]Richard Dienstbier (1977) compares the findings of the report on television violence and the report of the Pornography Commission and suggests that the conclusions of the two groups are inherently contradictory. He suggests that the social learning model is not equally applicable to sexuality and violence but that there is no substitute model available.

[8]During 1977 television viewing levels moved slightly downward for the first time in years, with a 3%-4% dip in "Homes Using Television." But levels tended to recover in the first half of 1978. There was no clear indication that the changes in viewing patterns were linked either to any reductions in the level of violence or to the new wave of "adult" content.

acceptable. By extension, what is true for violence would also hold for sexually suggestive or explicit material; for the representation of minority group members (only when they were shown exclusively in servile roles would the majority tolerate their presence); for definitions of the good life; for the use of tobacco, alcohol and drugs; and for the portrayals of businessmen, prison inmates, the elderly, welfare mothers, and academic overachievers. In short, the entire symbolic structure of television could be substantially altered by executive fiat, and if this were done gradually and intelligently, the public would go along. Persuasive evidence to this effect may be found in the similarity of viewing levels in different countries (Robinson, 1977). When allowances are made for variations in the length of the daily broadcasting schedule, the total amount of viewing time in television households remains quite comparable—in spite of considerable variations in the amount of choice and in programming philosophies and policies.

(2) The short-run response of many social scientists to the surgeon general's committee report was one of disappointment that its release was not immediately followed by social action. No remedial legislation was passed; no stern orders issued forth from the FCC. The broadcasters "stonewalled."

Yet only five years later it seems that the report did have an effect, because it became a landmark reference point in a debate that was older than television itself. In spite of their vigorous resistance and their refusal to acknowledge any harm in their past practices, the broadcasters were merely carrying on a holding action while their main forces abandoned the field. Television content had shifted away from violence, and rather quickly at that.

How had this come about? One must venture the guess that it came about primarily because of the changed position of key advertisers, who no longer wished to be associated with violent programs, even in a seller's market for commercial time.

What made the advertisers move? Again, one must guess that it was the pressure of powerful and respectable mass organizations like the AMA and PTA, these in turn activated by smaller special-purpose lobbying groups like ACT and the NCCB, which refused to let the issue rest. These groups drew sustenance from the report of the surgeon general's committee. They successfully argued that although the ultimate truth might defy discovery, the jury was in. Thus they demonstrated that social research can influence social policy, even in the face of rather heavy odds.

REFERENCES

Andison, F. S. TV violence and viewer aggression: A cumulation of study results, 1956–76. *Public Opinion Quarterly*, 1977, *41*(3), 314–331.

Baker, R. K., & Ball, S. J. (eds.), *Violence and the media*. Washington, D. C.: General Printing Office, 1969.

Belson, W. *Television violence and the adolescent boy,* London: Teakfield (in press).

Bogart, L. *The age of television* (3rd ed.). New York: Ungar, 1972.

Bogart, L. Warning: The surgeon general has determined that TV violence is moderately dangerous to your child's mental health. *Public Opinion Quarterly,* 1972-73, *36*(4), 491-521.

Cater, D. & Strickland, S. *TV violence and the child: The evolution and fate of the surgeon general's report.* New York: Sage, 1975.

Comstock, G. A., Lindsey, G., & Fisher, M. *Television and human behavior.* Santa Monica: Rand, 1975.

Comstock, G. A. Types of portrayal and aggressive behavior. *Journal of Communication,* 1977, *27* (3), 189ff.

Comstock, G. A., & Rubinstein, E. A. (Eds.), *Television and social behavior: Vol. I, media content and control.* Washington, D.C.: U.S. Government Printing Office, 1972. (a)

Comstock, G. A., & Rubinstein E. A. (Eds.), *Television and social behavior: Vol. III, television and adolescent aggressiveness.* Washington, D.C.: U.S. Government Printing Office, 1972. (b)

Comstock, G. A., Rubinstein, E. A., Murray, J. P. (Eds.). *Television and social behavior: Vol. V, television's effects—further explorations.* Washington, D.C.: U.S. Government Printing Office, 1972.

Dienstbier, P. A. Sex and violence: Can research have it both ways? *Journal of Communications,* 1977, *27*(3), 176.

Feingold, M., & Johnson, G. T. Television violence—reactions from physicians, advertisers and the networks. *New England Journal of Medicine,* 1977, *296*(8), 424-427.

Feshbach, S. M., & Singer, R. D. *Television and aggression: An experimental field study,* San Francisco: Jossey-Bass, 1971.

Gans, H. J. *Popular culture and high culture.* New York: Basic Books, 1974.

Gerbner, G., Gross, L., Eleey, M. F., Jackson-Beeck, M., Jeffries-Fox, S. & Signorielli, N. *Violence profile No. 8; trends in network television drama and viewer conceptions of social reality, 1967-76.* Philadelphia: University of Pennsylvania, 1977.

Gerbner, G., Gross, L., Jackson-Beeck, M., Jeffries-Fox, S., & Signorielli, N. Cultural indicators: Violence profile No. 9. *Journal of Communications,* 1978, *28*(3), 176-207.

Greenberg, B. S. *Pro-social and anti-social behaviors on commercial television in 1975-76.* East Lansing: Michigan State University, 1977.

Heller, M. S., & Polsky, S. *Overview: Five year review of research sponsored by the American Broadcasting Company, September 1970 through August 1975.* New York (no date).

James, L. *English popular literature: 1819-1851.* New York: Columbia University Press, 1976.

Lelchuk, H. *Sex and violence in the media: Whose responsibility?* (unpublished talk, December, 1977).

Lieberman Research, Inc. *Overview: Five year review of research* sponsored by the American Broadcasting Company, September 1970 through September 1975. New York (no date).

Liebert, R. M., Neale, J. M., & Davidson, E. S. *The early window: Effects of television on children and youth.* New York: Pergamon Press, 1973.

Milgram, S., Shotland, R. L. *Television and antisocial behavior: Field experiments.* New York: Academic Press, 1973.

Murray, J. P., Rubinstein, E. A., & Comstock, G. A. (Eds.), *Television and social behavior: Vol. II, television and social learning.* Washington, D.C.: U.S. Government Printing Office, 1972.

National Citizens Committee for Broadcasting, *Prime time violence profiles.* Washington, D.C.: 1976.

Paisley, M. B. *Social policy research and the realities of the system: Violence done to TV research.* Institute of Communication Research, Stanford University, 1972.

Robinson, J. P. *How Americans use time: A social-psychological analysis of everyday behavior.* New York: Praeger, 1977.

Rothenberg, M. Effect of television violence on children and youth. *Journal of the American Medical Association*, 1975, *234*(10), 1043–1046.

Rubinstein, E. Warning: The surgeon general's research program may be dangerous to preconceived notions. *Journal of Social Issues*, 1976, *32*(4), 18–34.

Rubinstein, E. A., Comstock, G. A., & Murray, J. P. (Eds.). *Television and social behavior: Vol. IV, television in day-to-day life.* Washington, D.C.: U.S. Government Printing Office, 1972.

Singer, J. L. & Singer, D. G. A member of the family, *Yale Alumni Magazine*, March 1974, pp. 10–15.

Somers, A. R. Violence, television and the health of American youth. *New England Journal of Medicine*, 1978, *294*(15), 811–817.

Stutzman, B. Television's "Soap" controversy. *Freedom of Information Center Report No. 386*, Columbia, Missouri, 1978.

The Surgeon General's Scientific Advisory Committee on Television and Social Behavior. *Television and growing up: The impact of televised violence.* Washington, D.C.: U.S. Government Printing Office, 1972.

Tankard, J. W., & Showalter, S. W. Press coverage of the 1972 report on television and social behavior. *Journalism Quarterly*, 1977, *54*(2), 293–306.

Wells, W. *Television and aggression: Replication of an experimental field study.* Unpublished manuscript, 1973.

Wilson, M., & Higgins, P. B. Television's actional arsenal: Weapons use in prime time. Handgun Control Staff, U.S. Conference of Mayors, 1977.

Winick, C. From deviant to normative: Changes in the social acceptability of sexually explicit material. In E. Sagarin (Ed.), *Deviance and social change.* Berkeley: Sage, 1977.

7

Social Influence and Television

Hilde T. Himmelweit
Social Psychology Department
London School of Economics
and Political Science

The title Social Influence *and* Television has been chosen deliberately in preference to a more traditional one: Television as Social Influence. The difference is important. Despite the emphasis of researchers on the viewers as active processors of the output of the media, both gratification and effects studies have been concerned primarily with the influence of television on the individual and on institutions. The counterpart, that of society's influence on the media (their structure, organization, and output), has received very little attention. Moscovici (1976) in his important book *Social Influence and Social Change* likens the emphasis that the social sciences place on responses to output to a *functional* model of social change. He suggests instead what he terms a *genetic* model. The former stresses adaptation (even if this implies transformation of the message to fit the individual's needs); the latter, interdependence. He argues that it is through interaction that there is continuous growth and change. Neither institutions nor individuals can stay put or fail to contribute to that change. The difference between the two models is not solely one of emphasis. Depending on which model one adopts, rather different research questions will be raised.

It is hardly necessary to point out that an institution like broadcasting (which Stuart Hall[1976] likens to the nervous system of a society) affects and is affected by the larger society, by other institutions, and by the viewers and listeners it serves. Yet an inspection of social science research in this field shows that the implications of this interdependence have been insufficiently spelt out. I shall try to correct this imbalance by presenting a conceptual model in which interdependence is highlighted and use it to draw attention to areas of research that might repay study.

My views about broadcasting have been much influenced by two recent experiences. As the only foreign member on the American Social Science Research Council Committee on Television and Social Behavior, I became aware of how much the effects of broadcasting are treated as generally applicable without examining the extent to which they might be limited to the particular broadcasting system or to the particular cultural values of a society. For example, the description of a broadcasting company as "arguably the single most important cultural organisation in the nation" would be unthinkable in the American context. Yet this is how the Annan Report (1977) opens its discussions of the BBC and its services. By contrast, the president of the National Association of Broadcasters in the United States (Barnouw, 1968), on the occasion of the twenty-fifth anniversary of American broadcasting, offered a rather different view of broadcasting:

> American radio is the product of American business! It is just as much the kind of product as the vacuum cleaner, the washing machine, the automobile and the aeroplane... if the legend still persists that a radio station is some kind of art centre... then the first official act of the second quarter century should be to list it along with local dairies, laundries, banks, restaurants and filling stations.

Not surprisingly, compared with American broadcasting, British broadcasting raises different expectations, forges different links to other institutions, and attracts a different type of staff. A cross-cultural approach is, therefore, basic if the role of the media is to be understood.

The second, and more important, experience in shaping my views came from two and one-half years' work as a member of the Annan Committee, set up by the British government to consider the future of broadcasting. I was the only social scientist on a committee of 16 drawn from different regions of the country, different occupations, and political persuasions. The work of that committee, which I briefly outline in the following, allowed me to see broadcasting from more vantage points than are usually available to social scientists.

THE ANNAN COMMITTEE

The Committee was the sixth committee of enquiry into broadcasting, generally set up 10 to 15 years apart, shortly before the expiry of the Charter and of the Act through which Parliament lays down broad guidelines for the conduct of the BBC and the Independent Broadcasting Service, respectively. The present committee was to provide the government with an independent assessment of the services and their future role. Legislation follows on the reports of such committees (even though its recommendations may, of course, be ignored). This has the advantage that the committee's deliberations are taken seriously. It also means that the committee is exposed to the full force of pressure groups. The disadvantage is that such a committee tends to get too concerned with what is economically and politically feasible in the short run. As a result, the reports of most of the committees of enquiry into broadcasting have tended to be conservative reports, looking at the next 10 or 15 years, when the field requires a much longer time perspective.

The Committee's charge was to consider "the future of the Broadcasting Services in the United Kingdom, the implications for present, and recommended additional services, of new techniques, and to propose what constitutional, organizational and financial arrangements and what conditions should apply to the conduct of these services [p. 3]." The work began in 1974, and the report was presented to Parliament in 1977 (Annan, 1977). A year later the goverment produced a White Paper (Broadcasting, 1978) outlining its legislative proposals. The Conservative government now in office will shortly publish different proposals.

The Committee invited submissions: 6000 individuals wrote letters complaining, praising, or making suggestions, and 750 memoranda were received from a variety of organizations. The Committee visited all television and many radio stations of both services, interviewing management, broadcasters, and the trade unions. Even though the visits were of necessity short, the fact that many of the same issues were raised in the different broadcasting stations made these visits rather more significant than might otherwise have been the case. In addition, many other interested parties, as well as the broadcasting organizations, gave written and oral evidence.

The life of such committees is generally too short to conduct research. Instead, review papers were commissioned on a variety of issues, including the profits made by the independent television companies, the relation of broadcasting to politics, and the effects of likely population changes by 1990, the end of the period of the Committee's remit. The International Institute of Communications provided accounts of the role and responsibility of the producer in four countries as well as case studies of the Dutch, Swedish, German, and Canadian broadcasting systems. These countries were chosen because they offered different solutions to some of the key issues that concerned the Committee: access, accountability, and editorial independence.

Issues Addressed By the Committee

Following are examples of the types of questions that the Committee, implicitly or explicitly, needed to answer before making recommendations. Even though some of the problems are specific to the British scene, I have included them because they might well represent issues of future concern in the United States. Others have strict counterparts and are likely to occupy the attention of the Carnegie Commission, which is concerned with the evaluation of Public Broadcasting.

The first set of issues concerns questions about the relation of broadcasting to other institutions in society and the second with the evaluation of the existing services together with an assessment of their potential. I have selected from among the concerns those that point to questions to which social scientists might address themselves, even though there is little indication either of interest or activity in that direction. I shall come back to this point later. Many of the issues are intimately linked to particular views about society, its needs, and the way in which society might develop within the period of the Committee's remit.

In line with the genetic model of social change, we need to distinguish between issues specific to broadcasting and those which, though they appear in the broadcasting context, reflect the general concerns of society at large. For example, during the zenith of broadcasting in the Sixties, questions of *accountability* were hardly raised at all. In the Seventies, they loomed large in the documents received. They were a reflection of a general distrust of authority that extended to many institutions, not just broadcasting.

If accountability was to be increased, what form should it take? How can a dividing line be drawn between accountability and control, the enemy of creative freedom?

Access was another issue. The right to communicate, as distinct from the right to have access to information, has received increasing emphasis in recent years. People, it is argued, resent the fact that broadcasting, unlike print, is not accessible to them. How can this demand be reconciled with another equally valid one—namely, that the majority of the audience is little interested in access and want the air time for entertainment programs. Where should access be allowed full play? At the national or the local level?

How should access be handled? Should we follow the example of some American states, where the broadcasting organization effectively becomes a carrier and is, therefore, not responsible for content? Or does the British penchant for compromise produce a better result—the "edited access programmes" in which the corporation's facilities are placed at the disposal of the organization or individuals who wish to present their viewpoint.

The essence of access programs is that they represent committed viewpoints. Most of them are about social or political issues. Parliament requires of broadcasting balance and impartiality, yet this requirement would be violated in access programs unless special provision were made for the "right to reply." If such programs are exempt from this requirement, why restrict such exemption to access programs? Why not extend it to other programs suitably labeled as committed programs but that are part of the BBC's or Independent Television's (ITV's) own output?

Given the interdependence of the media, should the Committee be concerned solely with the healthy development of broadcasting, the object of its remit, even if this might endanger the viability of the press or the film industry, or should it make some assessment about the relative importance of the different media and make some predictions about the economic consequences of given strategies?

Should the BBC's claims that *editorial independence* requires that its revenue comes not from government but directly from the viewers (by license fee) take precedence over the equally valid claim that the license fee is a regressive tax in a country where everyone views, because it hits hardest those with the lowest income? The old age pensioners, for instance, and others with small resources are generally also those most in need of television, because few alternate sources of stimulation and entertainment are within their reach.

To what extent should the need of a healthy democracy to have many editorial outlets take precedence over the *need for economy of scale*? Dividing the BBC, for instance (responsible for four radio and two television channels as well as for local radio), into BBC radio and BBC television would be a way of ensuring that a greater number of editorial voices would be heard.

And what about the *new technologies*? Once again, should the Committee be concerned primarily with ensuring the continued health or even expansion of television and radio—that is, make their future its prime concern—or should its first consideration be how best the new technologies can add to the enjoyment and instruction of the public? If it is the latter, its recommendations need to ensure that their development is given free rein, even if this affects, as it will, the present pre-eminence of television and radio. It is the same debate that took place in the United States between cable television owners, the FCC, and the three national networks. It is the old debate of protectionism against free enterprise.

What should be the criteria for determining whether or not a given operation constitutes a broadcast and so becomes the province of broadcasters? Teletext is a case in point. Teletext, a device developed in Britain, makes it possible for the viewer, with the help of a push button console, to select for projection onto the television screens any of one hundred continuously updated pages of print produced by the television companies. Like a broadcast it is distributed from a central source via the television screen. Once the attachment to the set has been purchased to make the print appear on the screen, the service is free. These are all characteristics of broadcasting. Or, alternatively, should the fact that Teletext provides printed material be the operative factor and the screen seen simply as a distribution agency, rather like the local news boy bringing the paper to one's door? If we take the latter view, how justified is it for editorial control of 100 continuously updated paragraphs to remain in the hands of the broadcasters? What link, if any, should there be with a press whose viability Teletext might threaten?

Governments in most countries provide guidelines to broadcasting organizations when they do not control them directly, and the reasons given are that such organizations use *a scarce national resource*. As more air waves become available, together with more channels (through transmissions by cable and satellite), how justifiable is such control? How far should broadcasting increasingly be treated like the press that seeks its own funds and writes its own ticket, provided that it does not offend against the laws of the land? Again, how much does a society like ours continue to want programs aimed at mass audiences? To what extent does that aim conflict with the equally worthy aim of "letting a hundred flowers bloom"?

Do the broadcasting services have exclusive rights over the channels assigned to them or only first call? When, as in the case of BBC2, the channel is not in constant use, should others be able to lease time, with the broadcasting service acting as common carrier without editorial responsibility for content?

A related problem is the extent to which *freedom of choice* should be restricted by making regulations that favor the showing of programs produced in the country itself; Britain, France, and, in particular, Canada, have such restrictions. In the case of Britain, the restrictions have much to do with the trade unions; in the case of Canada, with her fight against cultural swamping by the United States.

How far should the fact that the British public has not expressed any desire for a fourth channel influence the Committee? Or should it bear in mind that predictions about the reluctance of the public to embrace the new have always been wrong. Each time, purchases (often quite expensive ones), necessary to sample the new, have been greater and more rapid than expected. This was so when commercial television was first introduced, when BBC2, and later when color became available. These are some of the examples of policy decisions in which the broadcasting services are considered as a whole in relation to other institutions.

The second set of issues has to do with *evaluation of the services* themselves. What yardstick should be used? How can a distinction be made between those aspects of broadcasting basic to broadcasting and those due to the ethos or habit built up in the course of time—that is, to the professionalization of management and producers (Burns, 1977)?

Take the example of the news. By what criteria should one determine the adequacy of a news service? What is meant by unbiased? Is there some objective criterion of impartiality similar to Thurstone's assessment of the adequacy of a scale? Or should the criterion be the reception of the news by the audience, so that it is sufficient if the audience believes the news to be unbiased? Is a public served by having a series of news items presented with no, or insufficient, time for explanation, or does this mislead, that is, create a "bias against understanding" (Birt, 1975; Katz, 1977)? What would be the effect of lengthening the news or adopting the American pattern of having a TV personality such as Walter Cronkite interpret the news? How far is the distinction between news and current affairs, with the former providing facts and the latter comment, upheld in practice?

The Committee devoted a great deal of time to an analysis of news and current affairs programs, including the unintended side effects produced by the *need for balance*. Because conflicting sides need to appear on the same program, this often leads to an exaggeration of real differences and becomes a gladiatorial confrontation rather than a search for common ground. In Britain most television journalists first worked for the press; is this the right training for news presentation on television? What do we know about the relation of words to pictures?

And what about the social science findings concerning the effects of *television violence*? Accepting the weight of evidence that violence on television is more likely than not to have an effect, how should the broadcasters be asked to respond?

How far is it justified for the BBC to act as if it were a commercial service whose survival depends on the audience size or should it aim at interesting— possibly a smaller audience—in high quality entertainment and information programs by showing these at peak viewing times? To what extent should television act in a paternalistic way, aiming at the development of taste, and to what extent should it reflect public taste? Parliament enjoins the services to be responsive to the public; what does this mean in practice? These are some of the issues examined. Nearly all represent conflicting desirable objectives. To arrive at a decision required not only assignment of priorities but also assessment of the probabilities that a given change will bring about the desired effect.

The problems as well as the solutions are intimately connected with the way in which the institutions of a society are run, controlled, and interrelate. I come back to this issue later. Certain of the problems are philosophical and political, reflecting different views about society and about the role of government. Others are amenable to economic, sociological, or psychological analysis. Examples of the last are evaluation of public responsiveness to new ideas, of the way information is absorbed, attitudes organized and changed, tastes developed, and choices made.

With one exception, I do not intend to discuss the more than 160 recommendations or the response of the public and the government to them. I do describe later one of the key recommendations—what to do with the unused fourth channel—because it illustrates the extent to which particular recommendations about changes in broadcasting make sense only in their "local" context. They cannot easily be exported because they are built on a particular relationship with government and intertwined with the traditions of a country.

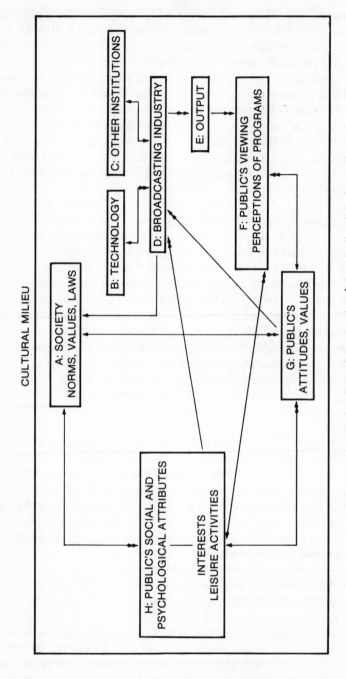

FIG. 7.1. Model of the Role of broadcasting. ↕ denotes direction of strong influences. ↑ denotes direction of some influence.

A conceptual model is presented in Figure 7.1 that brings out the interdependence of broadcasting with other elements in the society. This model offers a way of analyzing some of the questions raised and provides a schema for the analysis of different broadcasting systems and a means of examining the contribution of the social sciences to an understanding of the role of broadcasting. Examples drawn from different countries will be used to illustrate the model.

A CONCEPTUAL MODEL OF
THE INTERDEPENDENCE OF BROADCASTING
AND SOCIETY

A graphic representation of the model is given in Figure 7.1. The arrows indicate the direction and the number of arrows (one or two) the strength of the influence. Because broadcasting is essentially an "open system," the arrows nearly always point in two or more directions.

To reflect adequately the many strands of influence, the model should really be represented as a sphere. If that were done, the *cultural milieu* now indicated by the frame would be shown in the center rather like the bulb inside a globe that lights up the countries of the world The cultural milieu is like the air we breathe. It has its roots in the history, geography, and climate of a country and in its traditions and talents, actual and potential. To give one instance: Writing and acting for the stage have always been of a very high standard in Britain. To make good television drama, the industry did not have to create a tradition; it had only to adapt the existing one to the requirements of the small screen. Also, British acting understates; this made the transition from the theater to television, which demands understatement, that much easier by comparison, for instance, with France, with its more theatrical style of acting in the tradition of La Comédie Française.

The cultural milieu is what others would define as the national character or flavor of a country. It is expressed in the institutions and in the quality of life the society has developed and in the priorities it assigns to given values. It concerns the country's class system, its style of work and leisure, its openness to outside influences, its mode of resolving conflict and of welcoming or retarding change.

Box A represents *Society:* its values, norms, tastes, traditions of censorship and control, its laws and regulatory frameworks that define the freedoms and constraints under which broadcasting functions in a society.

It goes without saying that the extent to which the media are directly or indirectly controlled by government profoundly affects its output. There is a continuum of degrees of government control. At one end is the direct-state-controlled broadcasting system of the USSR or Iran and at the other end the free-for-all that at present operates in Italy. In Italy, following a recent decision of the Italian Supreme Court, any individual or organization can set up a radio or television station. Other countries operate a halfway system, with stations or services licensed by government. In the case of the United States this is done by the federal government, whereas in Germany it is a matter for each land. In the United States editorial independence is safeguarded through the First Amendment, in Britain it is achieved through special institutional arrangements. Parliament issues general guidelines (that the service should be of high standard and should inform and educate, as well as entertain) and then erects a buffer between government and broadcasting, delegating to Authorities, one for each service (whose members it appoints), the task of running broadcasting. Such Authorities have the dual role of ensuring that the broadcasters operate in the public interest and that the broadcasters' editorial and creative independence is safeguarded.

Society's views about broadcasting also affect the manner in which it is financed. When broadcasting is seen as a public service and not simply as an entertainment medium, it tends to be financed either directly by a government grant, as in Canada, or, exclusively or largely, by the public in the form of a license fee, as in most European countries. The difference in degree of independence from government between those services that depend on a grant and those that depend on license fees is less great than might at first appear, because governments authorize the size of the fee. The present inflationary times in Britain, for instance, require frequent changes in the size of the license fee. At times of greater financial stability, this potentially indirect means of government control would be less in evidence.

The list of direct and indirect effects of the regulatory provisions is a long one. It must also include an international perspective. For example, broadcasting, especially television, came to Europe after it had already been established in the United States. The European countries, in devising their system, were of necessity much influenced by their dislike of American television operating as a commodity industry aimed at maximizing audiences at minimum cost rather than as a public, all-round service designed to "open a window on the world."

Equally, within each country, the perceived consequences of earlier regulatory provisions affect the type of changes recommended or made. Thus in Britain, for instance, the Annan Committee considered that the BBC and the commercial system took competition too seriously, as if they were commercial enterprises whose profitability, prestige, or even existence depended on maximizing audiences. Yet each service has the privileges of a monopoly: ITV in being the only source of advertising on the air and the BBC in being the only public service. The BBC could justifiably argue that its remit was different from that of a commercial service. The Committee concluded that the "duopoly," in practice, had become a straitjacket so that, on occasion, competition for audience size and not the interests of the public appeared to be the primary goal of the broadcasters. Yet the IBA itself had pointed out that the relation of size of audience to enjoyment of a program by those who viewed it was low. Audience size was therefore an inadequate measure of public satisfaction. In deciding on the use of the fourth channel, the Committee felt that it should not be used to provide more of the same. Rather, regulated diversity should replace the existing regulated duopoly. The Committee used the term *regulated* to express its belief that to ensure a high quality, balanced service, a service needed to work within appropriate guidelines laid down by Parliament. There should be no free-for-all to attract maximum audiences. What was needed was a structure that would encourage innovation and inject new ideas and new talents and so provide the same shake-up, this time for both services, that in the 50s the institution of commercial broadcasting had provided for the BBC. The Committee therefore recommended that the fourth channel should be run as an Open Broadcasting Authority,[1] which, unlike the IBA, would not license companies but instead commission or buy programs from many outlets, including the BBC and ITV. Unlike ITV or the BBC, it was to be financed from a diversity of sources, including advertisers, sponsors, and grants.

In the model, Society is influenced by the broadcasters; the decision about the fourth channel is the result of such an influence—that is, of the output of the existing services and management's explanation of their policy. Because society's effects are stronger and more direct than those of broadcasting *on* society, the former is indicated by a stronger arrow than the latter.

So far I have discussed direct influences through regulations and finance. Society also exerts influence through the norms and values it takes for granted and that determine what it finds or what appears on the screen. Here, the interdependence on equal terms as between broadcasters, the public, and society is greater than in the case of the regulatory provisions.

Broadcasting is both influenced by, and influences, public opinion in artistic, social, and political matters. For example, in the 60s there were a number of programs that sympathetically portrayed the lives of homosexuals, bringing the public face to face with their problems and the extent to which, given the law, homosexuals had to use deception and were vulnerable to blackmail. Such programs, and those citing the evidence against capital punishment, contributed to the climate of public opinion by providing a platform for the airing of different views about these difficult issues.[2] Direct cause and effect cannot, of course, be assigned; it may simply be one more

---

[1]It is worth remembering that the essential institutional arrangement for broadcasting in Britain involves a two-tier structure. Despite reluctance to add to bureaucracy, a truly novel service—paradoxically just because it is novel—could not be run by any of the existing Authorities but requires that a new Authority be set up, however small.

[2]In the early 60's, capital punishment was abolished and the law repealed which made homosexual acts between consenting adults illegal.

instance of the extent to which certain issues are in the air and the media, government, and the public react to them.

There is a subtle chemistry between the climate of opinion, the selection of topics by the broadcasters, and the effects of these in turn on the climate of opinion. Without the growing distrust of authority in the 60s, the satirical late night show put on by the BBC might not have occurred or might have been far less popular. It was the popularity of the program "That Was the Week That Was," which satirized such sacred cows as the government, the opposition, the monarchy, and the church, that in turn led to a change in norms of what is permissible and what offends.

Box B represents *new technological developments* ranging from satellite, cable videocassettes, and recorders to teletext and viewdata. Included here are also the electronic news-gathering service (ENS)/ and the electronic Press. In the developed countries the supremacy of a few networks as in the United States or of a small number of channels with all the near monopoly position that these entail will soon be on the decline. As the amount of time people are likely to devote to viewing or listening is probably relatively inelastic, it follows that each development that increases the number of choices open to the public represents a threat to the existing services. The provision by the FCC, when allocating cable licenses, of a "cordon sanitaire" around each of the three networks indicates that influence works both ways. In the long run, however, the direction of influence from the technological developments to broadcasting is stronger than that of broadcasting in influencing the direction of such developments. Both depend on the extent to which the respective governments and the public value broadcasting.

The Annan Committee's decision to treat Teletext not as print but as broadcasting is a case in point. It decided that this technological development can be introduced to enlarge the province of broadcasting. Not all agreed, myself among them. We felt that Teletext was essentially a print medium and that control over its development and use should not be exclusively in the hands of broadcasters. We further argued that there was danger that if it remained in the hands of broadcasters it would be developed primarily as an adjunct to broadcasting and that it might threaten the viability of newspapers, especially evening newspapers, often bought for news about sports, stock shares, or entertainment.

That new technological developments should influence broadcasting needs little explanation. Less obvious is the *interdependence of broadcasting with other institutions* (Box C). To name but a few: In Britain education, for example directly influences some broadcasting output. In Britain the control over the content of educational programmes is in the hands of an outside body largely composed of educationalists. The one exception is the Open University whose material is shown on the BBC where it is the University and not the BBC which exercises editorial control. In China, television and radio have always been used for educational purposes; so has broadcasting in most developing countries. Here, too, the interdependence is clear. Hornick (1974), for instance, has shown convincingly how much the success of television programs depends on the appropriate infrastructure inside the schools.

There is also the interdependence between law, medicine, and broadcasting. Medical advisors prescribe advertising standards and have influenced the government to develop legislation that prohibits advertising of cigarettes. Soap operas about hospitals or programs on medical practices and advances presented on the screen influence the demand for particular medical services and create expectancies within the public about the relation of doctor to patient and about medical provisions. This is so in developed countries and particularly in developing countries in which educative programs designed to inform the public about the value of vaccination, for instance, may stimulate a demand for such service that outstrips the capabilities of the service to meet it. This is another example of the interdependence of broadcasting and other institutions, the effectiveness of the broadcast depending on the adequacy of the existing infrastructure.

The Law is another institution that similarly affects broadcasting in that its rules determine what can be broadcast, and it in turn is affected by the discussions on the air of the adequacies of the legal system.

Mention should be made here, particularly in the British context, of the important role played by the media in relation to industry. The Glasgow Media Research Team (1976) analyzed for a period of six months the news reporting of industrial affairs. Although there were some doubts about the group's interpretation of the reasons behind their findings, the content analysis clearly showed that the media largely concentrate on one aspect of industry—namely, on its industrial disputes. It also showed that the causes behind the disputes are rarely discussed and that management and workers are very differently represented. The former are shown sitting at their desks, calm, patient, and conciliatory; the latter, at the picket line shouting abuse. Above all, hardly any time was devoted to the reasons behind the strikes or to the strategies both sides had used to avert a strike. Management complained to the Annan Committee about the way they were presented in current affairs programs (generally traveling in the back of a large, chauffeur-driven limousine). The Trade Unions also complained, in their case about the way they are shown in the news on the picket line, and both sides were disturbed that so few other industrial matters make news. Here is an interesting example of an instiution that judges itself to be misrepresented on the media by what is *not* shown relative to what is shown.

The interdependence of advertising and broadcasting cannot be stressed too much. In the United States the power or influence of advertisers is overt and direct; it affects the choice of programs and their scheduling. Displeasure of advertisers or any concern they express about content tends to result in changes being made. The recent successful attempt, engineered in the United States by Nicholas Johnson (a former FCC commissioner), to use the advertisers to achieve a reduction of violence on television is evidence of the strength of their influence. By linking the names of products with the 10 most violent programs, Nicholas Johnson induced certain manufacturers with large advertising accounts to refuse to advertise close to programs that contained much violence. What social science research failed to effect, the advertisers achieved with ease—a neat demonstration of their relative influence.

In countries like Britain, in which the conduct of the commercial services is overseen by the Independent Broadcasting Authority, the influence of advertisers, though muted relative to that which they exercise in the States, is still considerable. It will remain so, especially on scheduling, in any country in which audience size determines advertising rates and hence the profitability of a service.

I have left to the last under this rubric the interdependence between broadcasting and the body politic. This varies from country to country but is everywhere sensitive and perceived by both as influential. Each side sees the other as manipulative and itself as needing to guard against manipulation. In some countries censorship and control are direct. In the case of Britain, this is so in one instance only: government through the use of *D-notices* can prevent or delay discussion of an issue in the press or on the air. More important are the subtler relationships set up not by edict but by convention. No one doubts the *agenda-setting function* of broadcasting or that through what it shows on the news—the order in which it is presented as well as the way the information is conveyed—broadcasting determines what events constitute "figure" or "ground." And although broadcasters stress that they draw a distinction between fact and analysis, a content analysis of British TV news shows that this distinction in practice is not very sharp and that the reporting of a news event often includes emotive and evaluative terms (Svennevig, 1979 personal communication based on content analyses of news items).

The interdependence is further highlighted in that the degree of trust between broadcasters and politicians determines who will be prepared to be interviewed and the types of questions, politicians agree to answer. Harold Wilson, the former Prime Minister of Britain, waged a vendetta against the press and broadcasting. A great deal of space in his autobiography (1971) is devoted to discussion of what he saw as the bias and interference of broadcasters. In Britain and elsewhere it is generally the treatment or supposed treatment of politicians on the air that touches off a general debate about the arrogance of the broadcasters and about the need for greater accountability. Politicians need the media and because of their need develop a love-hate relationship. The media in turn, especially in countries like France, which are politically very sensitive, feel constrained. But even more important, if the media determine the agenda of public concern, then as the examples of Northern Ireland and Vietnam have shown, this raises the question of the responsibility of the media to alert the public to an ongoing problem *before* it erupts into violence and the question of whether or not earlier presentation would have reduced or enhanced the likelihood of such violence.

The period prior to elections sharpens the conflict. Nearly all countries have laid down guidelines for the presentation of party political programs or views; in nearly every country the managers of political campaigns try to gain extra air time through making politicians' appearances newsworthy. The media, therefore, add to the sensationalization of a campaign. The Kennedy–Nixon and the Ford–Carter debates offer direct evidence of the relation of broadcasting to the body politic.

Indeed it is hard to think of any institution, including the Arts and the Leisure Industry, that is not affected by broadcasting or does not in turn influence the media's coverage of its activities and interests.

Box D represents the *television and radio industry.* In Britain, for instance, differences in programming result from the fact that the BBC combines both radio and television in one national service, whereas the commercial service separates radio and television. The commercial service is regional, even local, with each station licensed to a different company, a decentralization that is much greater in countries like the United States and Canada. By comparison with Britain these countries have a larger number of editorial outlets—at least in theory. In practice, networking and syndication make the difference far smaller than it might otherwise have been.

Diversification and the industry's style of management are important factors. In the case of the BBC it is a hierarchical organization and may have more in common in its management structure with the Civil Service than with that of the fashion industry (an analogy which Paul Hirsch [1975] draws convincingly about the organization of the media industry in the United States). Tom Burns (1977) in a recent study of the BBC suggests that professionalization has superseded the earlier ethos of public service. As a consequence much is being produced and evaluated with reference to a mythical public with little effort devoted to finding out how the public really reacts. Although the programs are made on behalf of and for the public, the Annan Committee found little evidence in Britain of determined effort (judged by expenditure of time and money) by the services to inform themselves of the public's response. Given the difficulties of the task, security is to be gained from perceiving the public as either incapable of expressing views or as having so many different views that it cannot provide a guide for action, with the result that colleagues become increasingly important sources of feedback. The socialization that occurs within the service provides the ethos and shapes style.

How far this is the case varies of course with the industry's source of finance, with the degree of competition, and with the profitability of a large audience. Where, as in the United States, competition is intense and audience size the important factor in determining profitability, the views of the public are sought. Under those circumstances the industry behaves like an advertising agency whose existence depends on demonstrating the effectiveness of their commercials. In Britain, France, or Sweden this is not as important. The views of the public tend to be sought when there have been accusations of bias (e.g., in Britain regarding the handling of Northern Ireland or the elections) or where there is concern about effects (e.g., violence on television). Formative research is rare in such countries either because there is no one-to-one relationship between audience size and profit or because the lack of competition puts the service in a privileged position. Under those circumstances the service is viewed primarily as an artistic and journalistic enterprise and formative research as interfering with the free exercise of creativity. In those countries even the effectiveness of educational programs aimed at achieving quite explicit objectives is rarely examined.

In analyzing the nature of the industry, analogies should be sought from other enterprises. We have already suggested that the BBC has certain, though not of course all, elements in common with the civil service, especially its structural arrangements, its concept of public service and, not least its implied suggestion that the BBC knows best. In other countries, notably in Canada and in the States, the television industry is more like the fashion industry, relying on hardy perennials as well as investing in a small number of high prestige or high risk goods. Paul Hirsch's anlaysis of the industry is very relevant here (chapter 5 this volume).

The nature of the industry, its structural arrangements, its mode of financing, its link to other institutions, all affect its concerns and its perception of its role vis-à-vis the public. They also affect the process of socialization through which a broadcaster is taught his craft. Training in most organizations is thought to occur by osmosis; training schemes are few and short and the real training is through an informal apprenticeship. Because the task of producing programs is difficult, given the size and heterogeneity of the audience, the industry develops all kinds of strategies to protect itself against failure. There is the slogan often used but little examined: "We would if we could, but we know we can't." There is the readiness of imitate, to take over another's success: the "Six Million Dollar Man" spawns the "Bionic Woman," the "Police Woman, "Charlies's Angels." Minor and expensive changes to existing formats are often made. Conventions built up a long time ago remain unchanged—for example, that series are made in units of 13 episodes or multiples thereof.

The treatment of series illustrates one difference between the British and the American broadcasting systems. In the case of most British series—a few soap operas apart—the number of episodes is laid down from the start, or a series finishes when the author feels he has nothing more to say; it is not continued ad infinitum. Huw Wheldon (1975) pointed out that "Till Death Us

Do Part"—a highly popular BBC series—was taken off the air because its author felt he had run out of ideas. Its American imitation "All in the Family", on the other hand, is the work of a team or a relay team of writers and may well continue until the public votes for its demise by switching to another channel.

Finance, as Katz and Wedell (1978), have so convincingly shown, determines output in another way. The United States is the greatest exporter of soap operas, Westerns and crime programs to developing and, indeed to the developed countries. In part this is so because they are well made but in the main because they are cheap, their costs having already been written off. For many developing countries their view of American life is almost exclusively derived from such programs.

Box E represents the *programs: their content, style, and scheduling.* It will be seen that this is the only box where there are no arrows pointing to influence from society or the public. It seemed more appropriate to show the influence of both directed at the industry that creates the output (Box D).

Boxes F and G concern *the public or the audience.* Box F is self-explanatory: The audience's perception and enjoyment of the programs feeds back into the television industry and also affects the viewing behavior, attitudes to the television industry and to television, and the taste of the public. Tastes are certainly created or at least fostered, by the media (Himmelweit & Swift, 1976). But they do more than that. As is indicated in Box G they influence outlook and as Gerbner and Gross (1976) have convincingly argued, provide the *symbolic reality* that the viewer, in the absence of information to the contrary from his own environment, uses to understand and react to the society in which he lives. The arrows here point in all directions. The public, like the television industry, represents an *open system* influenced by society as well as shaping that society. The public's background, its socialization through home, school, and work, affects its leisure habits, outlook, and taste, and these in turn are influenced by the amount the public views and by what it views.

It is hardly necessary to stress that the audience is an active processor of the output of the media, not a tabula rasa on which the media make their imprint. Box H denotes *social and psychological characteristics of the audience* that relate to leisure pursuits and interests and affect its attitudes and outlook as well as the gratification it seeks from the media.

SOCIAL SCIENCE AND TELEVISION

Inspection of the model shows that the social sciences have concentrated on the study of only a small segment. Nearly all efforts have been centered on Boxes F, G, and H—that is, they have been concerned with the response to output by the public and with accounting for variations in response in terms of the public's social and psychological attributes. Studies of the effects of broadcasting display an interesting historical progression related to the spread of the medium in society. When television was novel—that is, "figure" rather than "ground"—researchers were concerned primarily with delineating the *range* of its influence. Using the same subjects, they sought to examine *stimulation effects* and *displacement effects* (what activities were reduced, given up, or transformed as a consequence of the time taken up by viewing). The studies of children by Himmelweit, Oppenheim, & Vince, (1958); Schramm, Lyle, & Parker (1961), and Furu (1971) in Britain, the United States, and Japan, respectively, are illustrative of this phase. Each of the researchers realized that his or her study might well be the last study before television became so widespread that it would no longer be possible to compare children who view with a meaningful control group.[3] Given this "last ditch" situation, the researchers chose to paint a broad canvas rather than to focus more searchingly on one or two aspects. This made good sense at a time when television's influence was little understood and when initiative for such studies came because of concern about television's potentially harmful, rather than beneficial, influence.

It makes equally good sense now that television has become "ground," is part and parcel of the fabric of the society, that attention should be focused more narrowly and hence more searchingly on the medium's influence in one or other significant domains. Two, in particular, have been studied: the effect of televised violence and the effects of political broadcasting. One of the reasons for the concentration on one or the other effect is clearly expediency: the pressure of policy-makers and fund-giving bodies. Another is the exciting possibility for the social scientist that his results may not only be of interest to academics but may also play a significant role in policy-making. This is an intoxicating situation, but it is well for social scientists not to confuse fact-finding with social influence.

This confusion becomes apparent when we examine the 160 or so studies of the effects of televised fictional violence. The important finding that needed to be carefully documented is that there is nothing specific to violence but that it affects the viewer in much the same way as does any other message repeatedly given. Over time, given the "right" conditions, such messages make an inroad on cognition, evaluation, and behavior (Himmelweit & Swift, 1976). The range of studies further show that the strength, duration, and range of effects varies more with the viewer and his or her social context (educational ability, personality, age, and neighborhood) than with the type of violence portrayed. Yet here too some differences have been established: Stabbings are more frightening than gunshots (at least to young Americans and Britons) and

---

[3]Schramm had to look for a control group in Canada, there being none in the United States.

violence in ordinary surroundings, which aid identification, more so than violence that occurs in a historical or science fiction setting. Many useful findings have been obtained. But enough is enough. There is no way in which we can demonstrate with any precision *the increment* brought about by the experience of viewing televised violence. If indeed we could measure the impact of particular experiences that are self-selected, it would be tantamount to knowing a great deal more about causes of behavior than we do.

If, however, research is to influence policy-makers, then researchers need to concern themselves with a different segment of the model—namely, with the television industry and its constraints. Two types of study are needed: one that looks at the "attractiveness of violence (i.e., the extent to which it is the violence itself that gives a program its appeal) and another that looks at the power structure of the industry so as to determine where public opinion could be maximally effective. For example, in a study that I recently completed (Himmelweit, Swift, & Jaeger, 1980), the BBC viewing panel (N over 1000) rated, using 19 attributes, each of 20 programs, 11 of them action programs. All programs (the majority were series) were shown at prime viewing times. We found that the viewers distinguished between exciting and violent; the former was related to enjoyment, the latter rarely so.

Using regression analyses, we predicted the success of the programs from the ratings of the attributes. There was a great deal of agreement in rating the degree of violence in the 20 programs, but presence or absence of violence was related to enjoyment in only 4 out of the 11 programs. In the case of two British detective series, it marginally detracted from and in the case of two American series it marginally added to enjoyment. The emptier of other attributes, the flatter the characterization; the more repetitive the plot, the more violence enlivened; where there was no such dearth, violence was either irrelevant or detracted. Here, then, is evidence that reducing violence does not *eo ipso* mean that the programs would be less well liked; thus the industry needs to be less afraid than it is of the effect on enjoyment of cutting down on violence.

NEEDED RESEARCH

Once attention shifts from the study of television's negative to its potentially positive influence, there will be more opportunity for field experiments. One could vary the introduction to and subsequent discussion of programs to assess the relative role of each in changing outlook and/or behavior. The studies of prosocial behavior (Rushton, 1979) are a case in point. Alternatively, studies could be carried out, as was done in Britain, by taking an educative program—in this case sex education for primary school children—to map not only the influence of the program itself, but also the influence of the teachers' discussions and parents' responses. Complex interactive modes of anlaysis need to be used. Rogers (1975) carried out such a study in Britain and found that change in attitude of primary school children (about nudity or pain in childbirth) occurred but was *unrelated* to information gain. Attitude change and information gain were considerable and still present after retest three months later. The lack of correlation between attitude change and information gain poses interesting theoretical questions, because it is generally assumed, especially in studying the effects of political broadcasting, that information gain and attitude change covary. It may equally well be that attitude change is brought about by identification and by modeling without the viewer becoming more knowledgeable, at least with regard to the rather narrow range of information generally sampled in such studies.

Thus the emphasis must be on research that involves experimentation with different contexts. There is also a need to inspect more closely what in a program makes for influence. Of interest here are the respective roles in news programs of commentary (voice-over) and picture. Kline (1977) varied experimentally the relation of voice-over to picture; where they were dissonant, voice-over seemed to affect beliefs whereas the picture influenced evaluation. More such studies with better controls need to be done; they go to the heart of what television is all about—namely, the presentation of ideas, facts, and fantasy, through sound and pictures.

More formative research is needed, not so much to determine what viewers like, though this is important, as to test certain hypotheses about the social influence of messages and about information gain. For example, we need to learn more about the effects of creating particular mental sets by the title of a program, by the introduction to it on the screen, or, in the case of school television, by the teacher's preliminary remarks. Such studies can be done in the laboratory or in special sessions with small audiences. Producer collaboration would be needed to develop and present the appropriate material.

Some field experiments can often be mounted relatively easily to answer precise questions. For example, does repetition increase information gain or does the viewer or listener learn "to switch off" so effectively that the does not "hear" the message a second or third time? Lorry drivers spend their day listening to the radio and, in the course of such extensive listening, will, whether they like it or not, listen to several news bulletins. As a consequence, how much better informed are they about "near" and "distant" events compared with people of similar education who listen to the news only in their leisure time? If they are better informed, how long do they retain the information? Is it possible that with such surfeit of information, information is absorbed for a very short period, only to be discarded?

The link between the public's response and other institutions needs study. Field trials might be carried out to determine the extent to which the effectiveness of a message can be enhanced by appropriate community support. I have in mind one program in Britain that invites volunteers to participate in community projects. Viewers can telephone in to ask for details. What happens if the demand outstrips the supply? How far does the work satisfy the viewer's expectations?

Alternatively, as the BBC has shown in its literacy campaign, it is possible to use television to inform the wider public, to show that being illiterate is not something to be ashamed of (i.e., it attempted to increase the self-respect of those affected), and then to provide, within the community, the appropriate infrastructure (such as volunteer tutors carefully selected and trained). Their research, still in progress, shows both a change in the public's attitude towards illiteracy and adult illiterates as well as evidence of the success of this novel broadcasting endeavor. Also valuable would be an intensive study of the ingredients of the "package" that made for success and to chart the natural history of the influence of the series to determine at what point the potential volunteer tutor felt engaged and the potential client sufficiently "aroused" to telephone, make an appointment, keep it and/or start and continue the lessons.

The success of the Open University has meant that its broadcasting element, together with specially prepared books, are now used for teaching in many colleges of education. This situation illustrates well the interdependence of broadcasting and other institutions. In Britain, the government considers at this point in time that we have too many colleges training teachers; thus the colleges, to survive, needed to look for a new role. Equally, the Open University wishes to broaden its base through links with other institutions of continuing education. A series of novel partnerships between Open University training and the other educational institutions have been concluded, so far, with great success.

Context, the way of life of the viewers, requires more detailed and careful study. Attitude change research (Fishbein & Ajzen, 1975) and studies of broadcasting itself (Himmelweit et al., 1958) have yielded principles concerning the conditions under which broadcasting is likely to have maximal impact. It will exert least influence in spheres in which values and attitudes are firmly established through forming part of the cognitive map of the individual supported by significant reference groups that subscribe to the same values. Broadcasting will also exert little influence where the topic or the values have little relevance to the individual's life, do not relate to his general cognitive map of society, and are not reinforced by the value system of his reference groups. Influence will be maximal where the individual is consciously or unconsciously looking for information (information here denotes evaluations as well as knowledge); where the environment provides either no, or conflicting, cues, or where his or her own values are not yet firmly established. It follows that the influence of broadcasting will vary with the topic, with the viewer, and (this last is often forgotten) with the climate of opinion in society. *Consequently, findings about severity, duration, or even presence of effects must be both time-and place-bound.*

This last can be clearly seen when considering the influence of television on political attitudes and on voting. In the 50s and early 60s voters in the United States, for instance, tended to have fairly strong party identification. Studies of the influence of broadcasting showed some attitude change consequent on viewing, as in the case of the Nixon–Kennedy debate, but did not influence voting. Lang and Lang (1968) suggested that the "law of minimal consequences" might well apply. Today, in the late 70s, the situation is quite different. Party stability has decreased both in the United States and in Britain (Crewe, 1977; Crewe, Särlvik, & Alt, 1977), and the habit pattern of voting for the same party has consequently been interrupted. A longitudinal study in which we have been engaged examined the voting history of a cohort of young men from their first election to their sixth election and showed that consistency in voting was rare and that attitude change accompanied vote change (Himmelweit, Katz, Humphreys, & Jaeger, in press). We predict that, as a consequence of changes in voting habits, television as a source of influence will become more important. The conditions outlined earlier suggest that the media play a special part in a confused situation or one in which old cues and habits have lost their relevance.

It equally follows that the less educated, those living in more restricted environments (e.g., in developing countries) are on the one hand likely to be more influenced because television provides information not obtainable from the immediate surroundings. On the other hand, they might be less influenced because the values portrayed may have little relevance to the viewers' lives or to their reference groups. A reasonable and researchable question. From what has been said it is more likely that differences in degree of influence among viewers would be related to the viewers' characteristics and the social context in which they live rather than to the hours spent viewing. Yet most existing research places a great deal of reliance on time spent viewing as the relevant explanatory variable.

Social climate, social context, the relation of broadcasting to other institutions, all affect the influence of broadcasting and are, as the foregoing examples show, amenable to research.

Another area well worth study is the generalizability of the skills that children acquire to comprehend the iconic imagery of television (Salomon, 1974). He showed that such skills were to some extent transferable to other situations.Salomon emphasizes the effect of watching television generally. Hatano (1978) refers to the fact that children who have learned many Chinese characters by the time they go to kindergarten are known to retain their superiority in processing information of iconic figures for a number of years, even though their superiority over other children in dealing with Chinese characters is rapidly lost. This would suggest that the experiences of repeatedly processing iconic image information transfers to other image information-processing.

Hatano (1978) raises another point—namely, the extent to which, following Piaget, the development of conceptual skills requires active experience rather than passive absorption. If as a result of viewing television, less time is devoted to interacting with objects, are children who view a lot handicapped relative to those who view little? Such study might require cross-cultural comparison and the selection of countries that make different provisions for broadcasting to avoid contamination of the effects through self-selection of the period of viewing.

So far I have made suggestions for research dealing with the segments (F, G, H) of the model providing examples of different types of effects that might be studied, chosen either because they emphasize social context or relate to basic psychological processes.

Most research, however, should address itself to two neglected areas: the study of the broadcasting industry itself (D, E) and the study of the likely effects of new technological developments (B, D, E).

Broadcasting companies are an extremely interesting industry. It is one of the few whose influence, like that of newspapers, extends to almost every sphere of life; yet it differs from newspapers in many respects. First, newspapers aim at a particular public; television does not; it addresses itself to the public at large. Second, it is governed by regulatory provisions, whereas newspapers have no such restrictions. Third, there is far greater interest and hence concern about the broadcasting, compared with the press, in particular with regard to problems of accountability and responsiveness to the public. Also there is an extraordinary amount of information available about the viewership and its characteristics. Hence comparative studies are badly needed to determine whether or not, and to what extent, differences in regulatory provisions and mode of financing influence output and lead to differences in responsiveness to the public. In the British context, for instance, it would be inconceivable that an expensive series like "Beacon Hill," which was taken of the air in the States because it failed to please, would not in fact be shown. Where immediate success is not so financially necessary, as in the case of broadcasting systems that do not, or do not wholly, rely on advertising, the industry can build up liking for a series. It is equally possible that the demand characteristics of wishing to attract large viewerships, whether for financial reasons or self-respect, may reduce or obliterate what ought to be large differences between broadcasting systems.

Different structures may permit different learning experiences, discourage or encourage innovation, have a different relationship to its audience. Such studies need to concern themselves with the recruitment, promotion, and socialization into the broadcasting ethos of its staff as well as with the staff's perception of its public and the way it is built up and modified.

Studies of this kind may be complemented by others examining differences, if any exist, between the broadcaster's perception of the public's responses and the actual response of the public, both with regard to particular programs and to program categories. One would look for systematic bias across programs and examine whether or not the degree of difference varies with the topic, the category of broadcaster (producer, editor, research personnel), and with the length of service. If Tom Burns (1977) is right, one would expect a greater homogeneity of outlook concerning the public, with increasing length of service. That is, intergroup differences would become smaller, independent of the accuracy of perception of public's response. Organizational psychology and sociology have a particular role to play here.

New technological developments occur at such a speed and with fast decreasing cost that within the next 10 years the scene will have changed dramatically, comparable to the dramatic and qualitative change in society brought about by the introduction of television. It is here that most research effort needs to be concentrated. In the case of radio, we were too late; in the case of television almost so. There is urgency to plan now, using an interdisciplinary team that includes economists, agreeing on the collection of a minimum of standard information across studies about the locality, available media choice, and characteristics of the population.

A whole variety of research approaches should be used, including participant observation as well as the questioning of broadcast personnel and recipients at different stages of the experiment, so as to monitor participation and social influence. It may well be that information and education programs will become more attractive because more geared to the needs of particular audiences and that their revision will be speeded up through immediate and accurate feedback. Multiplex, for instance, which provides the opportunity to show the same picture with different voice-overs, can be used to determine to what extent the manner or the substance of what is said influences viewing, enjoyment, comprehension, and recall.

I see the need to study these developments as pressing and exciting. They offer, at the initial stage, a unique opportunity for social scientists, through concentrated research efforts, to influence how the technologies are used. Because applications of the new technology are in the experimental phase, researchers would be welcome, not just tolerated, and their research would lead rather than (as now) follow public interest and concerns. We know a good deal about learning, attitude development and change, and group relations; economists know about the influence of costs and finance; sociologists and political scientists know about the relations of institutions to one another; and lawyers know about the contracts, government and others, that so profoundly influence output and access. All these skills are needed in collaborative work.

I urge that we plan now on a large canvas, preferably involving more than one country. Already there are isolated studies; we need to extend them. It would be impressive if, when the next Annan Committe or its counterpart in other countries meet 10 or 15 years from now, they were able to draw on social science research that has a true bearing on policy options, as they were not able to do in the 70s.

REFERENCES

Annan, Lord. *Report of the Committee on the Future of Broadcasting,* London: Cmnd 6753, Her Majesty's Stationery Office, March 1977.

Barnouw, E. *The golden web—a history of broadcasting in the United States 1933-1953.* New York: Oxford University Press, 1968.

Birt, J. *The Times* (London), February 28, 1975 p. 14.

Blumler, J. G., & Katz, E. (Eds.), *Uses of mass communications, Sage Annual Reviews of Communication Research,* (Vol. 3). Beverly Hills: Sage, 1974.

Burns, T. *The BBC: Public institution and private world.* London: Macmillan, 1977.

Broadcasting, London: Cmnd 7294, Her Majesty's Stationery Office, July 1978.

Crewe, I. *Prospects for party realignment: An Anglo-American comparison.* Paper presented at meeting of the American Political Science Association, Washington, D.C., September 1977.

Crewe, I., Särlvik, B., & Alt, J. Partisan realignment in Britain 1964-1974. *British Journal of Political Science,* 1977, *7,* 129-190.

Fishbein, M., & Ajzen, I. *Belief, attitude, intention and behavior: An introduction to theory and research.* Reading, Mass: Addison-Wesley, 1975.

Furu, T. *The function of television for children and adolescents.* Tokyo: Sophia University Press, 1971.

Gerbner, G., & Gross, L. Living with television: The violence profile. *Journal of Communication,* 1976, *26,* 173-199.

The Glasgow Media Project. *Bad News* (Vol. 1.) London: Routledge & Kegan Paul, 1976.

Goodhardt, G. J., Ehrenberg, A. S. C. & Collins, M. A. *The television audience: Patterns of viewing.* London: Saxon, 1975.

Hall, S. *Broadcasting politics and the State—the independence-impartiality couplet.* Paper presented at the meeting of Mass Communication Research, Leicester, August 1976.

Hatano, G. *Broadcasting and learning.* Symposium on the Cultural Role of Broadcasting, Hoso Bunka Foundation, Tokyo, October 1978.

Himmelweit, H. T., Katz, M., Humphreys, P., & Jaeger, M. E. *A social psychological study of electoral choice. The voter as consumer.* In press.

Himmelweit, H. T., Oppenheim, A. N., & Vince, P. *Television and the child.* London: Oxford University Press, 1958.

Himmelweit, H. T., & Swift, B. Continuity and discontinuity in media usage and taste: A longitudinal study. *Journal of Social Issues,* 1976, *32,* 133-156.

Himmelweit, H. T., Swift, B., & Jaeger, M. E. The audience as critic: An approach to the study of entertainment. In P. Tannenbaum (Ed.), *Entertainment functions of television.* Hillsdale, N.J.: Lawrence Erlbaum Associates, 1980.

Hirsch, P. Organizational analysis and industrial sociology: An instance of cultural lag. *American Society,* 1975, *10,* 3-10.

Hornick, R. *Mass media use and the revolution of rising frustrations: A reconsideration of the theory.* Paper of the East-West Communication Institute, Honolulu, Hawaii, 1974.

Katz, E. *Social research on broadcasting: Proposals for further development.* London: BBC, 1977.

Katz, E., & Wedell, C. *Broadcasting in the Third World: Promise and performance.* Cambridge, Mass: Harvard University Press, 1978.

Kline, S. *Structure and characteristics of television news broadcasting: Their effects on opinion change.* Unpublished doctoral dissertation, University of London, 1977.

Lang, K., & Lang, G. E. *Politics and TV.* Chicago: Quandrangle, 1968.

Moscovici, S. *Social influence and social change.* London: Academic Press, 1976.

Rogers, R. S. The effects of televised sex education at the primary school level. In R. S. Rogers (Ed.), *Sex education: Rationale and reaction.* Cambridge: Cambridge University Press, 1975.

Rushton, J. P. Effects of television and film material on the pro-social behavior of children. In L. Berkowitz (Ed.), *Advances in experimental social psychology Vol 12.* New York: Academic Press, 1979.

Salomon, G. What is learned and how it is taught: The intersection between media, message, task and learner. In D. R. Olsen (Ed.), *Media and symbols: The forms of expression, communication and education.* Chicago: University of Chicago Press, 1974.

Schramm, W., Lyle, J., & Parker, E. B. *Television in the lives of our children.* Stanford: Stanford University Press, 1961.

Wheldon, H. British experience in television. The Richard Dimbleby Lecture. *The Listener, 95,* No. 2447, London: BBC Publications, March 1976.

Wilson, H. *Labour Government 1964-1970: A personal record.* London: Joseph, 1971.

8 The Influence of Television on Personal Decision-Making

Irving Janis
Yale University

POTENTIAL POWER OF TELEVISION

A recent survey of social scientists who are active in research on the behavioral effects of television has identified a number of problem areas that are regarded as warranting high priority for the next decade (Comstock, 1975). The topic of this paper cuts across two of the highest priority areas—the influence of television on socialization of young persons and behavioral effects of the "picture of the world" provided by entertainment programs.

The high priority assigned to these areas is directly or indirectly based on the assumption that in a variety of subtle ways television might prove to have considerable power in shaping the actions of large numbers of people. This assumption seems to be very much alive despite the deadening findings reported by social scientists indicating that television and other mass media have slight, if any, observable effects on the public's attitudes and behavior. As William McGuire (1969) puts it, the outcome of two decades of mass media research "has been quite embarrassing for proponents of the mass media, since there is little evidence of attitude change, much less change in gross behavior such as buying or voting [p. 227]." But there are many diehards around, not only among those responsible for spending billions of dollars each year on television advertising but also among social scientists who maintain that television is a major source of social influence in our society.

Here are a few samples of impressive assertions about the potential power of television by responsible social scientists:

George Comstock (1975), who has spent several years preparing comprehensive reviews and analyses of the extensive literature on the effects of television, speaks of the "vicarious socialization" that television provides—conveying values, norms, and taboos to maturing individuals in a way that could affect how they function in their society. He goes on to say that television has challenged the functions of parents, teachers, peers, and various social institutions that previously were the exclusive sources of socializing communications.

George Gerbner (1972), too, emphasizes the persuasive power of television when he argues that the recurrent themes of popular television dramas exert a subtle cumulative effect on viewers' beliefs and anxieties about the urban environment. He presents evidence from audience surveys indicating that the amount of exposure to television is positively correlated with grossly exaggerated expectations about the likelihood of being involved in violence—exceptions that could affect a variety of personal and political decisions, such as whether or not to move away from the center of the city and whether or not to increase appropriations for law enforcement agencies.

John Platt (1975) emphasizes the politicizing aspects of the transformations brought about by television, the instant spread of new information needed for public participation in the new policy decisions, and the general education of the public:

> The relation of television to action has not been generally appreciated by most commentators, who have tended to bewail the passive spectator character of television watching. But the instant imitation mentioned earlier shows the falseness of this view, at least for part of the audience. Any spectacular hijacking or kidnapping or television movie violence is followed by a wave of imitations, on a scale that the print media and the radio never produced. It also produces a sharing of emotions by large groups, so that by the next morning there are 100,000 telegrams pouring into Washington, of protest or support.... It is an instantaneous, spontaneous response, a chain reaction like a neutron reaction, which, when the adrenalin has been energized in this fashion by television, spreads everywhere at once. The Olympic games in Munich and the shootings of the Israeli athletes, flashed by satellite, were simultaneous events for an estimated 1.5 billion television watchers—38% of the world's human beings—and the audience response to such a medium cannot help but have international political impact.
>
> . . . .
>
> On any subject of national interest or even of interest to a small minority, television can call up within a couple of hours the best experts in the world. This is not the somewhat artificial method of consulting an indexed encyclopedia but the human method of asking direct questions of the knowers, and of face-to-face consultation or confrontation. In the best of these presentations, the viewers learn very quickly what is interesting, what is relevant, what is known or unknown, what is in dispute, who is working on the problem, and what different things can be done about it. It is not abstract knowledge but knowledge directly relevant to value choices and policy decisions [pp. 52–53, 55].

Comstock, Gerbner, and Platt and a number of other social scientists believe, contrary to all the opposing generalizations in social science monographs and textbooks, that the mass media—and television in particular—can exert a marked influence on many different types of decisions made by large numbers of viewers, not just the consumer decisions in which TV advertisers are interested.

At this point, I cannot resist the temptation of invoking a moral equivalent of Pascal's divine wager for the salvation of social science research on television effects. If the demurring monographs and textbooks are completely right, such research will be a waste of time and money. But even if the few commentators just cited are overestimating so grossly that they are only one-tenth right about the power of television to influence the viewing population throughout the world, social scientists had better try to locate samples of the tens of millions of people who might be affected, and learn whatever they can about when, how, and why.

Key Questions

The central problem on which I propose to focus pertains to the influence of television on the courses of action selected by viewers when they are confronted with major choices concerning career, marriage, life style, health, and the welfare of their families. Among which people, in what ways, and under what conditions does television influence vital personal decisions made by viewers? When and how do the words and vivid images conveyed by television become linked up in the minds of viewers as consequences of their own action, even though the producers of the television show may have had no intention of trying to influence the audience in that particular way? In discussing these questions I shall make use of conceptual schemas that have grown out of recent research, including my own research in recent years on social influence in relation to decision-making (Janis, 1972, 1975, 1977; Janis & Mann, 1976, 1977).

These questions are worth investigating, it seems to me, even if the visual medium of television does not prove to be more effective than newspapers and other mass media that rely mainly on the printed word. Television seems to differ from the press not just in the size of its daily audience throughout the world but also in the limited number of themes it presents in the relatively few programs available to regular viewers. It is important to find out exactly what those themes are and what effects they have on the behavior of viewers. Of course within one or two decades many of the findings might become obsolete, except as past history, if the monopolistic character of television programming is greatly reduced as a result of a huge increase in the variety of diverse programs made accessible by cable television, cassettes, and the proliferation of new channels catering to the special interests of different types of people. Even so, we can hope to learn something of lasting importance about the ways mass communications affect the personal choices people make and influence the procedures they use to arrive at their decisions.

AVAILABILITY OF IMAGES AND
PERSONAL SCRIPTS

Let me begin with a very simple example of the potential power of visual images on television to influence a minor decision. I have selected this obvious example because it can serve to illustrate some of the key theoretical concepts that might be brought to bear in the analysis of many more complex examples involving more important personal decisions. In New Haven, Connecticut, on the evening of January 28, 1977, a supper meeting of a parent–teachers' association, whose activities were always enthusiastically supported by the members, was poorly attended as a result of forecasts of a bad storm. Throughout the hour before the potluck supper was scheduled to start, many members phoned the organizers to say that although they were all ready to drive to the meeting and had already prepared a big dish of food as their contribution, they wanted to postpone the affair because a blizzard was predicted for that evening. Many spoke about having seen television newscasts of the blizzard that had hit Buffalo, New York, where traffic was brought to a complete standstill and some people were trapped in cars completely buried by snow drifts. They said they didn't want be caught in anything like that. Actually, at the time of these phone calls, the temperature was 44°F and the weather clear (and remained so for many hours thereafter). The organizers decided to hold the meeting despite the dire forecast. But most of the parents and teachers who had phoned to say they were worried after seeing the televised newscast about the blizzard did not show up. Many of those who came had heard about the blizzard from the radio or newspapers.

Perhaps those who rely on television to get the news are more susceptible than those who use the other media. Or perhaps the same weather forecast transmitted by radio or newspaper could not have been as effective in influencing people to decide to stay home when they felt under such a strong social obligation to attend the supper meeting. From the way the absentees spoke about it on the phone that evening and at the school the next day it seems likely that the vivid images of the blizzard they had seen on television played a crucial role in mediating their decision.

Apparently the television newscasts had made the image of a disastrous blizzard *available*, in the sense that Tversky and Kahneman (1974) use that term. These two authors have given a detailed account of various illusions, some notorious and others not yet well known, which make for errors in estimating the probability that given events will occur. Some of their empirical investigations indicate that the risks of a particular course of action are likely to be grossly overestimated if decision-makers have *vivid mental images of the possible unfavorable outcomes*, so that it is easy for them to imagine an ensuing disaster. Perhaps much of the power of television to influence the decisions of large numbers of people, whether for good or for ill, resides in its capacity to increase the availability of images of specific outcomes, desirable ones as well as undesirable. The greater availabilty of an image of any positive or negative outcome increases the probability of that particular image becoming dominant over images of other outcomes when people are trying to decide what to do. This can play a crucial role in determining which course of action they will choose.

In the brief example just given, the image of a blizzard, presented in an impersonal way on a television news program, seems to have greatly augmented the influence of the purely verbal forecast that the storm was approaching the local area. A number of viewers evidently took away a personalized message to the effect that "If I go to our parent–teacher association meeting tonight, even though the weather is fine right now, I shall encounter a very dangerous snowstorm when I try to drive home, like the one I just saw." This cognition, which represents a combination of a decision with a visual image of an ensuing outcome, is a simple example of what Shank and Abelson (1975) refer to as a "script." They use the term script to designate any specific cognitive schema or frame that represents a *"coherent sequence of events expected by the individual, involving him either as a participant or as an observer* (Abelson, 1976, [p. 33])." These psychologists propose to use scripts as a basic unit of psychological analysis in the study of language, memory, attitudes, and decision-making. They distinguish between two main types of scripts. One type, called *"situational scripts,"* pertains to stereotyped sequences of actions in well-known situations, such as what one has to do to get a room in a hotel. A second type of cognitive structure, which is more important for decision making, is what Shank and Abelson call "planning scripts." These scripts specify the choices that one can make when confronted with novel situations or personal problems. We rely on planning scripts, according to their terminology, whenever we tell ourselves what we can or ought to do in order to accomplish a goal. From the standpoint of research on television effects, I propose that we should try to learn more about the *personal* scripts—those situational and planning scripts involving individuals as participants or social actors—that they acquire from vicarious observations of events depicted on the television screen.

Do Fictitious Scenarios Become the Viewers' Personal Scripts?

The personal script just exemplified was acquired from a television newscast and involved a very limited sphere of action pertaining to an immediate source of potential danger. But the images conveyed by entertainment programs and films shown on television can also induce personal scripts in viewers, some of which may unintentionally affect their personal policies that are carried out time and again for many years, perhaps even for the rest of their lives. Let us consider next an example of an implicit script conveyed by a typical entertainment program. The opening of "Queen of the Gypsies," a Kojak episode about bank robberies that was shown on the CBS network during the winter of 1977, depicts a seemingly foolish, bumblingly incompetent young woman who, in a very pleasant way, asks a bank teller to help her out by answering her naive questions, while the bank customers in line behind her grow more and more impatient. The audience soon learns, however, that her naiveté is merely a false front that enables her to con the teller into giving her change for a fake $50 bill. And before the program is over it becomes apparent that the young woman is, in fact, an extraordinarily sophisticated con artist who can use her naive front so skillfully as to con even the great Kojak himself. An unintended message, it seems to me, could be conveyed by this fictitious episode: "Be suspicious of any seemingly helpless female who asks a lot of questions; she could be putting on an act to con you." Perhaps most viewers were too distracted by the exciting events in the drama or too absorbed in the specifics of the story to pick up any such message. But suppose that among the millions of viewers watching the program the minority who got the message included thousands of intelligent people in the business world who must deal with customers' naive questions. This little component in the television writers' scenario could conceivably reduce what little gracious behavior exists in the business world by building up personal scripts in those viewers that influence the way they respond to naive people who ask for help.

One of the problems for television research is to ascertain when and among whom such implicit messages are internalized as personal scripts and thereafter acted upon. Even simple explicit messages may be missed, especially by children, because of disracting cues, lack of opportunity for mental replaying and reflecting, and the rapid shifts of focus that characterize commercial television. As a result of these and other characteristics, television programs may often succeed in keeping the viewers' attention fixed on the screen at the expense of failing to transmit anything more than vague or garbled messages (Singer, 1976). Nevertheless a small percentage of children and a larger percentage of adults may pick up some of the unintended implicit messages that are repeated hour after hour and day after day in entertainment fare (Himmelweit, 1977). A study of school children by Dominick (1974) indicates that their viewing of crime shows on television was positively correlated not only with the belief that criminals usually get caught (which is an ideological theme intentionally introduced by the scenario writers) but also with specific knowledge about an arrested suspect's civil rights (which may be an *unintended* message that could be beneficial for some viewers, especially for anyone encountering police misconduct).

Other unintended messages picked up from those same programs may be socially detrimental because they go counter to democratic values. On the basis of an analysis of television crime shows from 1974 to 1976, Arons and Katsch (1977) conclude that Columbo, Kojak, and other television detectives regularly violate the legal rights of crime suspects, which may subtly convey the message to children and adults that there is nothing wrong with such conduct on the part of the police in their own communities.

Effects of Misleading Stereotypes of Occupational Roles

Of prime importance, from the standpoint of the potential role of television in the socialization of children and youth in our society, are the effects of explicit or implicit messages conveyed by television contents that induce personal scripts pertaining to vital decisions. Consider, for example, the complaints made by representatives of large scientific associations and technological organizations that many bright young people are not choosing careers in science or engineering because scientists and engineers are so often depicted in television stories as evil, power-hungry, and socially irresponsible. To what extent are such complaints based on factually accurate assumptions? Will a content analysis of popular entertainment programs show an imbalance in the direction of unwarranted negative portrayals of scientists and engineers? If so, do the stereotyped portrayals affect the beliefs and attitudes of the young people who see the programs? Do those beliefs and attitudes in turn affect their occupational choices?

Similar questions need to be answered by systematic content analysis, opinion surveys, and audience-response analysis of the portrayals in everyday television fare of each major type of occupation. A pioneering study by Himmelweit, Oppenheim, and Vince (1958) indicated that children who viewed television regularly knew more about prestigeful occupations than those who did not have access to television. This finding has been interpreted as suggesting that children can pick up and remember incidental information accurately from viewing television.

A number of studies cited by Leifer and Lesser (1976) indicate that children from preschool age through sixth grade absorb some information about occupations from entertainment television. There are also indications that children's attitudes can be influenced by television: children who are heavy viewers are more likely than those who are light viewers to express biased stereotypes, such as the view that professional jobs are appropriate for men but not for women. Other studies, most notably a well-known one by Lesser (1974), indicate that educational programs like "Sesame Stret" can convey to young children an accurate view of the activities involved in various forms of work, including information that they might have no other way of learning about. Although unwilling to draw any definitive conclusions from the limited set of studies now available on the effects of television, Leifer and Lesser (1976) offer the following plausible suppositions:

> Young children do develop concepts about careers very early in thier lives, but because their information and experience are fragmentary, their concepts are narrow and sterotypic.
>
> Once early stereotypes are formed, and no concerted effort is made to counter them, they persist into adolescence and young adulthood with only superficial improvements and expansions based on a child's personal experiences with employed persons.
>
> The majority of adolescents and young adults thus make career decisions while relatively uninformed of the range and variety of opportunities available to them and of the exact nature of the occupations within that range [p. 40].

The authors emphasize that there is little dependable evidence bearing on these hypotheses, except for the repeated correlational findings suggesting that children aged 4 to 10 acquire concepts and attitudes about occupational roles from television programs as well as from other sources. The main gap lies in the lack of systematic observations concerning the degree to which the concepts and attitudes acquired at an early age influence the career choices the young viewers make years later.

Content analysis studies in recent years suggest that television portrayals of occupations might be conveying some grossly distorted views of the world of work, which could lead to unrealistic levels of aspiration among many youthful viewers. For example, Seggar and Wheeler (1973) found that female occupations were grossly underrepresented in television portrayals. Whereas male professional and managerial roles were overrepresented. In presenting occupational roles and other aspects of the world that young people have little or no opportunity to learn about in other ways, television all too often promotes misconceptions and stereotypes (DeFleur, 1964; DeFleur & DeFleur, 1967; Leifer & Lesser, 1976). Physicians, lawyers, and police officers, for example, are generally presented in an overglamorized manner that grossly misinforms unsophisticated viewers about what their duties and daily routines are really like. Persons in lower status occupations are given stereotyped negative characteristics. Because blacks are so often shown in menial followship roles, white prejudices may be reinforced and black youths may acquire the impression that professional and leadership roles are out of the question for them. (See chapters 11 and 12, this volume.) Similar dysfunctional personal scripts might be acquired from the portrayals of limited occupational roles of other minority groups.

At present a national organization called Action for Children's Television has a grant of $25,000 to prepare a resource handbook on occupational role models for broadcasters, teachers, parents, and others interested in how television affects the welfare of young people. The handbook is supposed to include an account of how people in various jobs are being characterized on television and the effects of those characterizations on young people's career choices. But an extremely well-funded research project would have to be set up to obtain the essential content analysis and audience response data. Furthermore, even though the purpose of the project is educational, the findings might be misused politically.

Political Dangers of Research on Group Images. At this point, a brief digression is necessary to consider the political dangers connected with carrying out this type of research and using it to prepare handbooks of the type just described, whether they deal with occupational groups or any other kind of social or economic group. One danger is that research findings concerning unfavorable images of various groups depicted on televisoin might be used by trade unions, professional associations, and management organizations to initiate political pressures designed to push the networks into presenting the best possible image of each group. Another serious danger is the possibility that impressive evidence of distortions of reality might be used by political organizations and others who want to abridge freedom of speech by imposing new regulations on television content, in the name of protecting our youth from being misled. A related danger is that the findings might promote internal directives that reduce what little artistic freedom exists in the industry and discourage television writers and producers from preparing grossly exaggerated satires or imaginative projections of how things could be that are not an accurate reflection of the world as it is right now. If, for example, the norms required that the occupational roles portrayed on television accurately represent existing demographic characteristics of these occupations, then the television content would be used to reinforce and freeze the existing curtailments of occupational opportunities for women and various minority groups, including those curtailments that result from existing social prejudices.

In the absence of convincing reassurances, many research workers hesitate to carry out studies that might promote undesired changes in norms and regulations. It could make a difference, however, if the social science community, together with prodemocratic organizations, and political leaders were to voice their concerns about the potential misuses of this type of research and to discuss constructive ways of applying the findings. The policy implications of the research proposed need not necessarily be supportive of conservative efforts to set up new standards or to impose more restrictive controls over the television industry, even though there is a serious political danger that various groups might try to use the findings for such purposes. A pertinent suggestion by Ithiel Pool (1977), for example, is that the same research findings that some interested parties might try to use to push the industry into presenting the "right" themes could be used equally by those who favor subsidies to encourage the development of a greater variety of programs that present more diversified themes and allow more selectivity by the viewers. Obviously, no impressive guarantees can be expected, but concerned research workers might be encouraged if it becomes apparent that there are influential people around who share their misgivings and will not go gently into the dying of the light.

Assessing Misrepresentations.   Returning to some of the technical problems of the proposed type of research: It will not be easy for those research workers who decide to work on the type of project I have just described to ascertain the extent to which characterizations of each of the various jobs that are open to young people are being misrepresented or distorted. One possible approach would be to compare (a) the images that people are given of the usual positive and negative consequences to be expected if they were to pursue a given professional career or go into a given line of nonprofessional work with (b) the knowledgeable reports obtained from men and women who are actively engaged in that type of career. The questions to be answers would be whether or not the typical television characterization (if there is a fairly consistent one) influences the expectations of a sizable percentage of young viewers and differs from the central tendency of the reports obtained from experienced participants. For this purpose it might be worthwhile to use the "balance sheet' of incentives described by Janis and Mann (1977, pp. 135 *ff.*) as a schema to analyze the considerations that people consiously or unconsciously are taking into account when they decide to adopt a new course of action or to stick with the one previously chosen. In this schema the expected consequences for each alternative course of action are classified into four main categories:

1. Utilitarian gains or losses for self.
2. Utilitarian gains or losses for significant others.
3. Self-approval or disapproval.
4. Approval or disapproval from significant others.

One of the main hypotheses that has grown out of studies of the balance sheets of persons making stressful decisions (such as choosing a career, getting a divorce, giving up smoking, going on a diet, and undergoing surgery or painful medical treatments) is the following: *The more errors in the decision-maker's balance sheet at the time he or she becomes committed to a new course of action, the greater will be that person's vulnerability to negative feedback when he or she subsequently implements the decision.* Janis and Mann refer to this hypothesis as the "defective-balance-sheet" hypothesis. Errors of omission include overlooking the losses that will ensue from the chosen course, which makes the balance sheet incomplete; errors of commission include false expectations about improbable gains that are overoptimistically expected, which are incorrect entries in the balance sheet. The entries in the four categories of the balance sheet obtained from content analysis of television characterizations could be compared with the most frequent entries obtained from interviews or questionnaires given to representative samples of (a) viewers who are asked to state their expectations about the pros and cons for major occupational choices that they might face in the future and (b) persons currently occupying each major occupational role who are asked to report on the positive and negative incentives for sticking with their occupational choice. (The same comparative method could be adopted for comparing television with other media and for studying correct and incorrect conceptions about other types of decision in addition to career choices.)

Obtaining objective data on the degree of verisimilitude in television characterizations of occupational roles—and identifying the most distorted stereotypes—should be regarded as only the first step in the inquiry. What we next need to know is the extent to which the accurate or inaccurate portrayals influence the personal scripts and the level of aspiration of young people when the time comes for them to choose their own career. Similarly, we need to find out the effects of television portrayals on personal scripts that enter into the subsequent career-related decisions of adult life, such as shifts in occupational role resulting from mid-career crises.

Other Types of Personal Scripts To Be Investigated

At present we know very little about personal scripts acquired from or reinforced by television portrayals on currently popular programs like "Mork and Mindy," "Laverne and Shirley," and "Mary Hartman, Mary Hartman." Dramas, situational comedies, and other entertainment fare often portray vital aspects of adult life that might sooner or later affect some of the viewers' decisions to be married, to become pregnant, to obtain a divorce, to raise a child as a single parent, to move to or from an urban area, or to make a drastic change in life style.

Similarly, we remain quite ignorant about the ultimate impact of the repeated portrayals of health problems, not only in television soap operas, but also on talk shows and in commercials, which could have profound effects on viewers' personal scripts that enter into their health-related decisions. The increased availability of images of physical suffering and of successful and unsuccessful medical treatments undoubtedly influences the actions of many people when they become ill, sometimes when it is a matter of life or death. A study of adolescent boys by Milavsky, Pekowsky, and Stipp (1975–1976) indicates that exposure to television commercials plugging the use of over-the-counter drugs is positively correlated with the use of such drugs but is negatively correlated with the use of illicit drugs. These correlational findings need to be pursued further by other types of research to determine the causal sequences. Does viewing the television commercials exert a causal influence on the decisions of young people to take one type of drug and to avoid the other? Or are the young people who use illicit drugs too busy elsewhere to view television programs and their accompanying commercials? Of special importance are questions concerning long-range effects on drug consumption: Does viewing the commercials or other television presentations bearing on drug-taking have any observable influence on drug consumption in later adult life? If so, when, why, and among whom? Multistage studies are also needed to determine the extent to which positive or negative portrayals of smokers, overeaters, and alcoholics on television entertainment programs and in commercials affect personal scripts that are pertinent to consumption decisions made by youthful and adult viewers.

✶ Similar data are needed to learn about the personal scripts acquired in childhood. A review of 21 studies on the effects of television advertising on young children's attitudes and behavior by Adler (1977) refers to a number of studies showing that children under 8 years of age have great difficulty distinguishing commercials from programs. This may be one of the main reasons why they believe the information and misinformation in advertisements to be true. There is some evidence, for example, indicating that children who are exposed to across-the-counter drug advertising are more likely than others to believe that the advertised medicines are widely needed and effective and to be receptive to their use. Other findings, according to Adler's review, show that as children grow older they can discriminate commercials from other fare and they say that they dislike them more and more. Nevertheless, the older children show little or no decrease in wanting and requesting the advertised products. The demand created by television commercials among older as well as younger children might have some corrosive effects on family life. Survey research indicates that a substantial proportion of parents complain that their children ask them to buy the things they see advertised on the screen and become disappointed or angry when the parents do not comply.

On the positive side of the ledger, a recent development in America is the deliberate attempt by government health agencies, medical organizations, and various public interest groups to use the power of television to convey messages to adult viewers that could improve the physical health and emotional well-being of the nation. Maccoby and Breitrose (cited in Comstock & Lindsey, 1975), for example, have devised a media campaign to reduce coronary heart disease, sponsored by the National Heart and Lung Institute. The campaign includes television spot announcements, together with radio and newspaper announcements, direct mailings, and instruction classes for volunteers who want to reduce their heart disease risk by improving their diet, increasing exercise, reducing weight, and cutting down on smoking. Assessments of the effectiveness of the campaign, conducted by Maccoby, Breitrose, and others in the Stanford Heart Disease Prevention Program, may provide some useful leads as to which types of persons respond positively to which types of persuasive messages. For this type of campaign it is possible to obtain behavioral measures that are dependable indicators of viewers' decisions to volunteer for classes on personal health-related instruction and to adhere to the public health recommendations. Similar measures could be used to investigate the implicit messages about desirable health practices conveyed by soap operas and other entertainment fare, which could be much more effective for some people than campaigns that are perceived by viewers as deliberate persuasion.

CUMULATIVE EFFECTS OF EXPOSURE
TO RECURRENT THEMES

Whether or not spot television announcements alone or in combination with other media prove to be effective in inducing people to take preventative action to avoid heart disease or any other type of illness, the television medium probably exerts enormous influence on health-related decisions of large numbers of viewers. Some of the talk shows, documentary videotapes, and films presented on the television screen deliberately attempt to teach the audience personal scripts about specific ways to cope with medical emergencies, chronic illness, and a variety of other problems involving physical or mental health. For example, during the week of January 24, 1977, a daily talk show on NBC titled "Not for Women Only" concentrated entirely on the topic of breast cancer, featuring physicians who presented scientific information and knowledgeable women who described what it is like to be a breast cancer patient. On the mornings of January 26 and 17, the entire program was devoted to conversations with three patients—Rose Kushner and Betty Rollins, each of whom had written a book based on her personal experiences, and Debre Hamburger, a 25-year-old student who had recently undergone a mastectomy. These three women emphasized a number of themes that could have a direct effect on the decision-making of members of the audience who might have cancer symptoms. One theme emphasized by all three patients was that even when a trusted physician tells you that you must undergo drastic surgery to save your life, you should take an active role as a skeptical decisionmaker—you should try to get a second opinion from an independent cancer specialist as to whether or not surgery is essential, and, if so, how extensive it should be; then, if you decide to undergo the recommended operation, you should consult a plastic surgeon in advance, if possible, and arrange to have him on hand during the operation so that he can influence the life-saving surgeon to take account of the need to plan for satisfactory cosmetic repair.

Perhaps even more important from the standpoint of transmitting personal scripts to the 1 out of every 14 women in the United States who will be afflicted with breast cancer was the repeated emphasis by all three patients on the theme that you can still lead a normal, productive, and self-fulfilling life, including a satisfactory sex life, despite being mutilated by the loss of one of your breasts. A subsidiary theme, which is applicable to any male or female who undergoes any kind of body mutilation, was that there can be some advantages even though one would never choose this means for achieving one's goals if given a free choice—for example, the operation is likely to evoke an immediate increase in tender affection from one's spouse or lover; ultimately, after coming to terms with the continuing threats, the personal disaster can lead one to become a more mature, self-reliant person with a more profound sense of the values one wants to pursue during the remaining days of one's life. This theme was conveyed not just by what was being said but also by the visual images of the three impressive-looking, socially competent women, each of whom was talking frankly, in a self-possessed manner, about how her mastectomy experience had affected her and about her own current assets and liabilities.

Inadvertently, the latter theme made its apearance again in an entertainment program that was viewed by over 70 million people on the evening of each of the two days that the women had discussed their mastectomies on the morning talk show. On those evening, two installments appeared of "Roots," the serial dramatization of the ABC network of the best-selling saga about an Afro–American family, which had a larger audience than any comparable entertainment program ever had before. These two episodes of the serial dealt with the experiences of Kunta Kinte, a Mandinka warrior who in the early 18th century had been captured by slave traders outside his native village in Gambia and brought to Virginia on a slave ship. In line with one of the explicit themes of "Roots," we see Kunta asserting his manliness and his indomitable striving for freedom by attempting to escape to the north. Two sadistic slave catchers punish him by chopping off his foot with an axe. At this point in the drama, a subsidiary implicit theme makes its appearance. Kunta appears to be in a state of utter despair as he lies deathly ill from an infection in his mutilated foot, and we hear him mutter that he will never walk again. But an attractive slave woman who gives him tender care is clearly not at all repelled by his feeble, mutilated state. With her encouragement, he resolves not only to walk again but also to run and presumably to do other manly things as well.

The next episode shows Kunta functioning effectively despite his permanent mutilation, once again a strong, competent, proud man who has married his loving nurse and has impregnated her. Many of the viewers who had tuned in on the morning talk show about mastectomies would be likely to recognize that here again was essentially the same theme, encouraging anyone who undegoes body mutilation to strive to overcome the handicaps and demoralization. Some of those who had not viewed the morning talk show may well have gotten the same message. But for the ones who had watched and listened to the two morning programs, the reappearance of the theme in the entertainment program on the same two evenings may well have had an interactive or cumulative effect, establishing a more generalized personal script, with more broadly applicable categories: "If you (or anyone else in your family, male or female) are mutilated by a horrible injury or illness"—so runs the more generalized script—"you can still be a competent, well-loved person, living a full life."

Focusing on Recurrent Explicit and Implicit Themes. The foregoing example of a recurrent theme has been given in considerable detail because it calls attention to neglected aspects of television research. Most of the existing research bearing on personal scripts scquired from television is on the effects of single-shot exposures to one particular television program or even just one little snippet from one episode. Such research may have value for investigating certain limited hypotheses but can tell us nothing about sensitizing and cumulative effects on the acquisition of personal scripts that arise from recurrences of the same or similar themes on a variety of different television programs—soap operas, situation comedies, newscasts, and commercials, as well as intentionally educational programs. We need to go beyond assessing merely the overall frequency of each theme that could create or reinforce personal scripts bearing on vital personal decisions. We need to learn about the *recurrent* themes that have the most potent effects on the audience. Which are the messages that are constantly being reinforced by multiple presentations and which are the ones that are being contradicted and counteracted? In order to answer these questions, content assessments should take account of whatever is known about audience apperception in order to include the latent as well as the obvious manifest themes—Kunta's implicit modeling of emotional recovery from mutilation, as well as his explicit modeling of black manliness.

The type of research proposed here need not be deterred by the realization that we cannot hope to obtain dependable answers to sociohistorical questions regarding the extent to which the themes in daily television fare are initiating social change by introducing new ideas into our culture or by making people more aware of their latent fears and aspirations as against merely reinforcing the dominant folklore that almost everyone would learn anyhow from other socializing agencies. In any case, we need to know which are the explicit and implicit themes that are repeatedly being presented to the mass audience and to what extent each of those is being registered by the viewers and entering into their personal scripts.

Differential Effects of Explicit Versus Implicit Themes. For certain types of messages, explicit presentations might evoke so much psychological resistance that implicit presentations are more effective. Many mutilated men might reject an explicit statement of the theme about obtaining tender affection as a compensatory gain from mutilation but nevertheless incorporate the theme into a constructive personal script from seeing the implicit presentation in the mutilation episode of "Roots." It is conceivable that everyone builds up much more potent personal scripts when the themes are acquired as *tacit* knowledge. The themes may be more readily internalized from a variety of implicit presentations than from an equal number of exposures to explicit presentations that can be easily identified as deliberate efforts to influence one's personal beliefs or attitudes. These are some of the more basic research questions that need to be worked on in order to improve our understanding of the role of television as a socializing agency in our society.

Expanding the Scope of Content Analyses. Obviously, one of the first steps in pursuing some of the research questions just discussed is to make use of existing content analysis data on themes that might influence the personal scripts of at least a sizable minority, if not the majority, of viewers. Some new coding schemes may have to be developed to capture themes bearing on all the various types of vital personal decisions that should be examined. Comstock (1975) points out that the socially constructive content of television is likely to continue to be investigated by many research workers because, among other reasons, a coding scheme for content analysis of such themes has already been developed. The same consideration might make it worthwhile to invest research resources into developing specific coding schemes pertaining to depictions of the positive and negative consequences of alternative courses of action for each major type of personal decision. Ultimately, if subsequent research shows that these new coding schemes are of some value for predicting the personal scripts acquired by television viewers, they might be incorporated into a set of comprehensive cultural indicators of the type being developed by George Gerbner and Larry Gross (1976), which are intended "to present annual, cumulative, and comparative indicators of dominant cultural configurations, common conceptions, and trends relevant to issues of social health and public policy" (Comstock, 1975, p. 64).

ACQUISITION OF PERSONAL SCRIPTS

In order to investigate the acquisition of new personal scripts, some of the sophisticated methods developed for basic research on learning, memory, and high-order cognitive processes might be applied (see the section on Cognitive Psychology in Janis, 1977). These methods may be useful for research on questions about when, how, and why content themes presented on television are transformed by viewers into personal scripts, with and without distortion. Similar research questions need to be answered concerning the storage of newly acquired scripts in long-term memory and their retrieval on appropriate and inappropriate occasions when a vital personal decision has to be made.

Earlier I suggested that misleading themes on television that could have adverse effects on the making of career decisions might be detected by a two-step comparative method: First, by examining the results of a content analysis of television protrayals of the favorable and unfavorable features of various types of jobs in relation to comparisons between children who are heavy and light viewers with regard to beliefs about those features in order to see if the content themes are being assimilated by young people; and second, by comparing the assimilated beliefs with the ratings of persons who really know what the favorable and unfavorable features are because they are actually working in those jobs, in order to see if those beliefs are accurate or inaccurate. Essentially the same comparative method could be used to identify other misleading content themes presented on television that could have adverse effects on the marriages, health, or life style of many viewers—misleading in the sense that children, youth, or adults are being led to build up erroneous personal scripts involving false expectations about the consequences of the courses of action they will sooner or later be contemplating, which make for choices that are unduly costly, regrettable, and regretted.

It is especially important to find out how children respond to content themes that could build up or modify personal scripts. (See Chapter 9, this volume.) On the one hand, children usually do not give undivided, continuous attention to television but engage in active play or do their homework while half-watching the screen (Bechtel, Achelpohl, & Akers, 1972; Lyle & Hoffman, 1972). Furthermore they do not understand some of the important cues to adult intentions and miss nuances in human relationships (Singer, 1976). Consequently they may learn only the most pervasive, uncomplicated,

and attention-capturing messages, sometimes with considerable miscom-
prehension because of their failure to take account of the context. But, on the
other hand, young children are sometimes highly attentive to the television
screen and when curious they are likely to notice all sorts of little details
ignored by adults. Children may be extremely vulnerable to those fictitious
representations of the social world that repeatedly give the same messages
about the police or any other group in entertainment programs and
commercials. Correlational evidence from a study by Robert S. Frank (cited
by Comstock and Linsey, 1975, p. 62) suggests that the more they watch
television, the more likely they are to regard lying and deceitful practices as
acceptable behavior, provided that it is for a good cause. This correlational
finding, of course, does not tell us whether television plays a causal role
because other factors, such as the presence or absence of parental lying and
deceit, might account for the observed relationship. Nevertheless, it suggests
yet another type of potentially adverse effect that we probably should worry
about. Fairly high priority should be given to new research on television
effects designed to tell us something about the extent to which the ideological
components of the Watergate coverup mentality are currently being
transmitted to and acquired by children.

Exactly what kind of personal *moral* scripts are being acquired by adults as
well as children and youth who watch popular television programs?
Numerous commentators cry out against the oversimplified, self-serving,
antisocial, or corrupt moral content of prime time programs. Others say that
prime time programs are not directly antisocial or corrupt, but exert a
pernicious moral influence by promoting stereotypes, ideological beliefs, and
myths that encourage classifying all persons and social groups into simplistic
moral categories (heroes or villians, good guys or bad guys) that are
antithetical to democratic values. A few claim that television is doing a fine
job of encouraging belief in law, democratic institutions, and "good" moral
values. Columnist Benjamin Stein, for example, writing in the *Wall Street
Journal* (April 1975), takes the position that it is not at all a bad thing
that "television sells its values as well as its soaps and detergents." He
argues that shows under attack for being pointlessly violent can be defended
on the grounds that they offer counterbalancing moral content. The moral
theme that regularly runs through these programs, he says, "is that the bad,
evil people who lead criminal, antisocial lives, are punished. Thus it pays to
be a good, moral person." He adds that most adventure shows portray the
people who are on the side of the law as superior and more worthy of
imitation than the criminals. Stein concludes that insofar as television is
selling these powerfully important moral messages, along with faith in our
government's institutions, it "must be one of the primary stabilizing forces in
America today." It follows that we should prize "Kojak," "The Rookies," and
"Police Story" as national treasures.

Is there any empirical validity to Stein's ultraconservative position? Or to the opposing position? And for which programs? What we really need to find out is which of the explicit and implied moral themes in television programs are being internalized. If preliminary studies indicate that television scenarios sometimes are acquired as moral scripts that enter into the making of personal decisions by viewers, the next step should be to conduct full-scale studies to determine under what conditions this effect does and does not occur and who are the viewers most likely to be influenced. It certainly will not be an easy task to find the answers. But maybe with some ingenuity, along with intensive effort, social scientists can find ways of obtaining relevant data, especially for school children.

Counteracting the Influence of Television

The vulnerability of the very young to social modeling by the heroes and heroines in television drama is suggested by many survey results. Greenberg and Reeves (1976), for example, found that there was a marked tendency for schoolchildren to regard portrayals in television dramas as real. This attribution of reality was positively related to the amount of television viewing but was inversely related to age. Earlier I mentioned parallel findings on children's belief in the truthfulness of television commercials: Older children were found to be much less likely than younger ones to believe that commercials are truthful and display the "best" products (Ward, Wackman, Faber, & Lesser, 1974). All these findings about children's beliefs in the reality of television dramas and commercials suggest that other interacting factors offset or reduce the influence of television as a child grows older. We need to learn exactly what the counteracting factors are that decrease the influence of television on personal scripts and how they operate.

A closely related problem involves the moderating effects of socializing agents who say and do things that correct or modify the ideas children pick up from television. We can expect ongoing research to tell us something about the moderating influence of parents who watch television with their children and make comments about the prosocial and antisocial contents (Chaffee, McLeod, & Atkin, 1971; Comstock, 1975; Hicks, 1968). This topic at present is exciting considerable interest among developmental psychologists, but most of their research seems restricted to obvious pro- and antisocial themes. It could be extended to a variety of other pervasive content themes, such as those pertaining to ways of avoiding common health hazards, to determine how the moderating influence of adults can correct misconceptions or reinforce correct concepts that children might acquire from television. This brings us to a general operational research question, which could lead to advances in theoretical as well as practical knowledge: *What can parents or teachers do to prevent children from acquiring or retaining erroneous*

personal scripts that could have an adverse influence on their vital decisions,
including the ones they make currently or later in life?

New Objectives for Educational Television

Operational research is obviously needed to develop television programs that
effectively induce accurate personal scripts about means–outcome
relationships. This type of research could also be oriented toward developing
and diffusing new programs for educational television. Research workers in
this area could take advantage of findings from other fields of research and
also make use of unusual television events as opportunities to obtain some
useful leads, if not solid evidence. Here is a relatively simple example of what I
have in mind: A fourth-grade class in an innercity school (99% black) was
given a special "live" program in which black members of many different
professions came to their classroom and told them what their professional
work was like, how they prepared for it, how much money they made, and
how they felt about what they were doing. The results of questionnaires given
to the children suggested that they had changed markedly in their knowledge
about the professions represented and had shifted their aspirations upward. If
such changes are retained for many years, they could presumably affect the
personal planning scripts that will be activated when the time comes for these
young people to make career choices. Even more important might be the
increase in realistic *hope* of finding a satisfactory career, which could promote
a more vigilant pattern of decision-making rather than defensive
procrastinating or wishful thinking (see the final section of this chapter). One
obvious research problem would be to see if a similar program prepared for
television had any such effects, and, if so, if it attracted a large enough
audience, when shown on local educational television programs, to make it
worthwhile to prepare and distribute.

Here is a more complicated example: During March and April 1977,
Ingmar Bergman's "Scenes From a Marriage," on New York's WNET
(Channel 13), was accompanied by a live supplementary program during
which viewers could speak via telephone with professional counselors on
marriage and family problems. Imaginative research workers might be able to
pose a number of practical questions that could be answered by small-scale
studies concerning the impact of this combination of a provocative television
drama with telephone counseling and other innovative ways of using
television to promote effective decision-making. The findings might be useful
not just for subsequent showings of one particular program but also for a
broad range of productions that depict one or another aspect of life dilemmas
involving ubiquitous personal problems. Of course, the most valuable
projects might well be those in which research workers collaborate with
writers, producers, and others in the television industry to develop and test

new educational programs designed to build up accurate means–outcome scripts in the viewers. New programs developed to help children acquire the skills and information necessary for effective personal decision-making might prove to be especially valuable when parents watch them at the same time and discuss them with their children immediately afterward to make sure that the main points are neither missed nor misunderstood (Singer & Singer, 1977). This type of operational research might go a long way toward enabling the industry as well as the public to realize the full potential of the TV medium.

EFFECTS OF CONTENT THEMES BEARING ON DECISION-MAKING PROCEDURES

Up to this point I have been discussing research on the potential effects of television themes pertaining to means–consequence relationships—themes that could induce or reinforce personal scripts in the form of an if-then proposition: "When I have to make such and such a choice, if I select X as the best course of action, then one of the outcomes I should expect to experience is Y (with a high, moderate, or low degree of certainty)." Such scripts determine the dominant entries in the people's decisional balance sheets when they are trying to make up their minds what to do in response to a given threat or opportunity that requires a decision. But television also conveys information and presents social models of effective versus ineffective *procedures* for arriving at a satisfactory decision. Here I am referring to an entirely different type of content theme that can induce or reinforce an entirely different type of personal script pertaining to such questions as, "Where can I go for relevant information about the alternatives that are open to me and about their consequences?" "Who should I trust as an expert?" "What can I do to handle pressures from other people to make my decision before I have sufficient information about the consequences to make a sensible choice?" These and a variety of other *procedural* scripts can affect the degree to which people will explore the full range of alternatives open to them and assess the probable consequences of the alternatives that need to be taken into account in order to achieve their most important goals. If their procedures of information search and appraisal are defective, people are likely to encounter unexpected outcomes that cause unnecessary suffering and that make them regret the decision.

An example of a procedural theme explicitly conveyed by a television program has already been discussed: The three cancer patients on the "Not for Women Only" program urged the audience not be overwhelmed by the authority of any physician who recommends surgery but to regard him as a fallible expert, respecting his judgment yet delaying action until a second expert opinion is obtained. This advice and the social modeling by the three self-confident women who offered it could lead viewers to change the procedures they use in arriving at a variety of health-related decisions. Many men and women regularly deal with the appearance of incapacitating symptoms and other serious health problems by consulting their regular physician and immediately accepting whatever recommendation is given; they follow a very simple procedural rule: "Tell a qualified expert about your troubles and do whatever the expert says." For such people the message on the talk show could have a profound effect on the subsequent handling of their own health problems and those of others in their families as a result of having acquired a new procedural script.

We know very little about the extent to which television exerts this type of influence on viewers' decision-making behavior. As far as I know, we do not even have any pertinent content analysis data on the extent to which television entertainment programs and commercials present models of impulsive versus deliberative decision-making or the extent to which they portray the favorable consequences of using sound decision-making procedures when people are confronted with the necessity to make a fundamental decision concerning their marriage, career, health, or physical survival. And yet the portrayal of people making just such decisions is the stuff that television drama is made of.

I have just used a phrase—"sound decision-making procedures"—that could be faulted as a value-laden term that has no place in an objective analysis of how and why people behave as they do. It is true that we have no dependable way of assessing objectively how "good" or "successful" a decision has turned out to be. Nevertheless, we can formulate propositions specifying the conditions that have favorable or unfavorable effects on a person's decision-making activity by focusing on the *quality of the procedures* used by the decision-maker in selecting a course of action.

Criteria for Assessing the Quality of Decision Making

Suppose we want to find out if a group of viewers of a given television program show an increase or decrease, relative to an equated group of nonviewers, in the quality of their personal decisions. Or suppose we want to determine whether or not heavy versus light viewers show any differences in the quality of their personal decisions. Let us assume that we could obtain fairly valid observations of the steps each person in the equated groups has taken in arriving at recent personal decisions concerning career, health, or other vital issues. When it comes to analyzing these observations in terms of the quality of decision-making procedures, what criteria could we use?

From a review of the extensive literature on effective decision-making, Janis and Mann (1977, p. 11) have extracted seven major criteria that can be used to determine whether or not the decision-making procedures used by a person or group are of high quality. They point out that although systematic data are not yet available on this point, it seems plausible to assume that those decisions that are of high quality, in the sense of satisfying these ideal procedural criteria, have a better chance than others of attaining the decision-maker's objectives and of being adhered to in the long run. The seven procedural criteria are as follows:

The decision makers to the best of their ability and within their information-processing capabilities (1) thoroughly canvass a wide range of alternative courses of action; (2) survey the full range of objectives to be fulfilled and the values implicated by the choice; (3) carefully weight whatever they know about the costs and risks of negative consequences, as well as the positive consequences, that could flow from each alternative; (4) intensively search for new information relevant for further evaluation of the alternatives; (5) correctly assimilate and take account of any new information or expert judgment to which they are exposed, even when the information or judgment does not support the course of action they initially prefer; (6) reexamine the positive and negative consequences of all known alternatives, including those originally regarded as unacceptable, before making a final choice; (7) make detailed provisions for implementing or executing the chosen course of action, with special attention to contingency plans that might be required if various known risks were to materialize.

A plausible working assumption is that "failure to meet any of these seven criteria when a person is making a fundamental decision (one that has major consequences for attaining or failing to attain important values) constitutes a defect in the decision making process. The more defects, the more likely the decision maker will undergo unanticipated setbacks and experience postdecisional regret" (Janis & Mann, 1976, p. 11). When all seven procedural criteria are met, the decision-maker's orientation in arriving at a choice is what Janis and Mann call *vigilant information-processing*. Especially for choices involving multiple objectives, a moderate or high degree of vigilant information-processing appears to be a necessary condition for arriving at a decision that will prove satisfactory to the decision-maker in the long run.

Vigilant Versus Nonvigilant Ways of Arriving at Decisions

Coming back to priorities for research on the effects of television, we should give a high rating to the task of determining when, in what ways, and to what extent the words and images conveyed by television foster or impede a vigilant information-processing orientation. Do certain television programs and advertisements present models that promote impulsive or defensive patterns of decision-making that result in failure to meet most or all of the seven criteria? Which types of television content promote vigilant information-processing?

Aside from direct effects of modeling and explicit persuasion, messages conveyed by television, like those presented in other mass media, can indirectly influence the way a person deals with personal crises requiring vital personal decisions. The types of message that might exert an indirect influence are specified in an analysis by Janis and Mann (1976, 1977) of the crucial conditions that make for vigilant as against nonvigilant patterns of decision-making. They start with the assumption that psychological stress engendered by decisional conflict frequently is a major determinant of failure to achieve high-quality decision-making. Janis and Mann postulate that there are five basic patterns of coping with the stresses generated by any realistic challenge that confronts a person with an agonizingly difficult choice. Each pattern is associated with a specific set of antecedent conditions, which could be affected by information and images conveyed by television or other mass media. These coping patterns are derived from their analysis of the research literature on how people react to warnings that urge protective action to avert health hazards or other serious threats. The five coping patterns are:

1. *Unconflicted adherence.* The decision-makers complacently decide to continue whatever they have been doing, ignoring information about the risk of losses.

2. *Unconflicted change* to a new course of action. The decision-makers uncritically adopt whichever new course of action is most salient or most strongly recommended.

3. *Defensive avoidance.* The decision-makers escape the conflict by procrastinating, shifting responsibility to someone else, or constructing wishful rationalizations that bolster the least objectionable alternative, remaining selectively inattentive to corrective information.

4. *Hypervigilance.* The decision-makers search frantically for a way out of the dilemma and impulsively seize a hastily contrived solution that seems to promise immediate relief, overlooking the full range of consequences of the choice because of emotional excitement, perseveration, and cognitive consstriction (manifested by reduction in immediate memory span and by simplistic thinking). In its most extreme form hypervigilance is referred to as "panic."

5. *Vigilance.* The decision-makers search painstakingly for relevant information, assimilate information in an unbiased manner, and appraise alternatives carefully before making a choice, which increases the chances that they will meet the seven criteria of high-quality decision-making.

The five coping patterns are represented in Fig. 8.1, which is a schematic summary of the Janis and Mann conflict theory of decision-making.[1] This conflict model specifies the mediating psychological conditions responsible for the five coping patterns, including beliefs and expectations that might be strongly influenced by what people see on television.

As indicated in Fig. 8.1., there are three main conditions that determine whether vigilance or one of the other coping patterns will be dominant: (1) awareness of serious risks for whichever alternative is chosen (i.e., arousal of conflict), (2) hope or optimism about finding a better alternative, and (3) belief that there is adequate time in which to search and deliberate before a decision is required. Each of these three mediating conditions can be induced or changed not only by communications from the experts and advisers the person talks with but also by the words and images presented on television or in other mass media. Consider for example the messages that are likely to be extracted from television family dramas about marital discord by those viewers who are themselves contemplating breaking up their marriage. Some of the portrayals in the drama can make the viewers much more keenly aware of the risks both of trying to preserve a hopeless marriage and of breaking up one's home. The events depicted in the drama can also influence viewers to become more optimistic or pessimistic about finding a good solution for their own marital problems and can also influence beliefs pertaining to time pressures—for example, the heroine's actions might convey the message that an imminent deadline set by lawyers for working out the final details of a divorce agreement can be postponed with little cost if one really wants more time to think it over before making a final decision. Insofar as television content can build up personal scripts pertaining to these psychological conditions, it can influence the likelihood that viewers will adopt a vigilant rather than a defective coping pattern in dealing with vital decisions. If subsequent research shows that such scripts are acquired from television presentations by a sizable proportion of viewers, the next step will be to find out when and among whom.

Research on the effects of television on decisional coping patterns could also move in the direction of developing and testing out new types of television programs and new ways of using the medium to help people avoid

---

[1]Although the first two patterns (unconflicted adherence and unconflicted change) are occasionally adaptive in saving time, effort, and emotional wear and tear, especially for routine or minor decisions, they often lead to defective decision-making if the person must deal with a vital choice that has serious consequences for himself, for his family, or for the organization he represents. Similarly, defensive avoidance and hypervigilance may occasionally be adaptive but generally reduce the decision-maker's chances of averting serious losses. Consequently, all four are regarded by Janis and Mann as defective patterns of decision-making. The fifth pattern, vigilance, although occasionally maladaptive if danger is imminent and a split-second response required, generally meets the main criteria for high-quality decision-making.

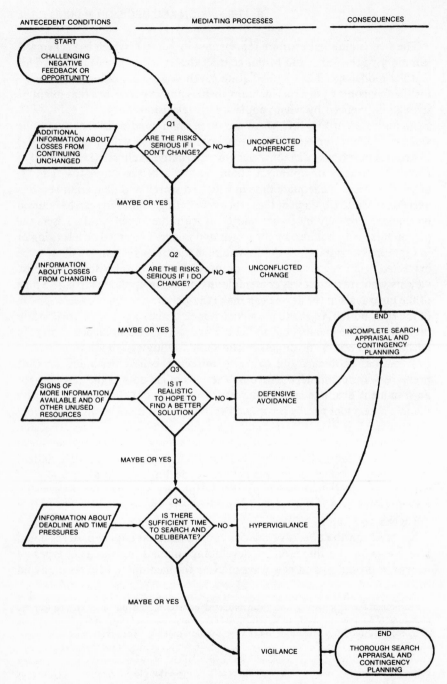

FIG. 8.1. A conflict-theory model of decision making applicable to consequential decisions. (From Janis & Mann, 1977.)

defective decision-making procedures that lead to failure and regret. Suppose that there were a marked increase in audience viewing of diversified programs on a large number of new channels, including some devoted to public service programs. Or suppose that television cassettes for home screens became popular and readily accessible in public libraries or in low-fee lending libraries in drugstores. Would it be utopian to expect that once the monopoly of current commercial programming has ended, television could take on a new role in meeting educational needs of the mass audience? Television might prove to be effective, more so than any other medium, in transmitting decision-making skills to the tens of millions of persons of all ages who never did and probably never will learn all the essential skills they need from their families, acquaintances, or classroom teachers.

When we look over what little is already known from social science research in general on how people make decisions, one thing is certain: There is plenty of room for improvement in the information-seeking and appraisal activities of most people in our society when they face the necessity of making fundamental decisions that affect their future welfare. Many of the miseries of unanticipated consequences and postdecisional regret among people who make ill-considered personal decisions might be preventable. Although hardly recognized at present as a social problem, those widespread miseries warrant a considerable increase in attention from organizations that sponsor social science investigations and from leading experts who help set the priorities and spark the research. One of the main questions that has been in the back of my mind during the writing of this chapter is this: Whether or not television is part of the problem, couldn't research on media effects make it part of the solution?

REFERENCES

Abelson, R. P. Script processing in attitude formation and decision-making. In J. S. Carroll & J. W. Payne (Eds.), *Cognition and social behavior.* Hillsdale, N.J.: Lawrence Erlbaum Associates, 1976.

Adler, R. P. *Research on the effects of television advertising on children:* Washington D.C.: U.S. Government Printing Office, 1977.

Arons, S., & Katsch, E. How TV cops flout the law. *Saturday Review,* 1977, *4,* 10–14.

Bechtel, R. B., Achelpohl, C., & Akers, R. Correlates between observed behavior and questionnaire responses on television viewing. In E. A. Rubinstein, G. A. Comstock, & J. P. Murray (Eds.), *Television and social behavior. Vol. 4. Television in day-to-day life: Patterns of use.* Washington, D.C.: U.S. Government Printing Office, 1972.

Chaffee, S. H., McLeod, J. M., & Atkin, C. K. Parental influences on adolescent media use. *American Behavioral Scientist,* 1971, *14,* 323–340.

Comstock, G. *Effects of television on children: What is the evidence?* Paper presented at the 1975 Telecommunications Policy Research Conference at Airlie House, Airlie, Virginia, April 1975.

Comstock, G., & Linsey, G. (Eds.). *Television and human behavior: The research horizon, future and present* (Vol. 2). Santa Monica: The Rand Corporation, 1975.

DeFleur, M. L. Occupational roles as portrayed on television. *Public Quarterly,* 1964, *28,* 57–74.

DeFleur, M. L., & DeFleur, L. B. The relative contribution of television as a learning source for children's occupation knowledge. *American Sociological Review,* 1967, *32,* 777–789.

Dominick, J. R. Children's viewing of crime shows and attitudes on law enforcement. *Journalism Quarterly,* 1974, *51,* 5–12.

Frank, R. S. Home–school differences in political learning: Television's impact on school children's perception of national needs. Cited in G. Comstock & G. Linsey (Eds.), *Television and human behavior: The research horizon, future and present* (Vol. 2). Santa Monica, Calif.: The Rand Corporation, 1975.

Gerbner, G. Violence in television drama: Trends and symbolic functions. In G. A. Comstock & E. A. Rubinstein (Eds.), *Television and social behavior. Vol. 1. Media content and control.* Washington, D.C.: U.S. Government Printing Office, 1972.

Gerbner, G., & Gross, L. Living with television: The violence profile. *Journal of Communication,* 1976, *26,* 173–199.

Greenberg, B. S., & R eeves, B. Children and the perceived reality of television. *Journal of Social Issues,* 1976, *32,* 86–97.

Hicks, D. J. Effects of co-observers sanctions and adult presence on imitative aggression. *Child Development,* 1968, *38,* 303–309.

Himmelweit, H. T. Yesterday's and tomorrow's television research on children. In D. Lerner & L. Nelson (Eds.), University of Hawaii Press, 1977.

Himmelweit, H. T., Oppenheim, A. N., & Vince, P. *Television and the child.* London: Oxford University Press, 1958.

Janis, I. L. *Victims of groupthink.* Boston: Houghton Mifflin, 1972.

Janis, I. L. Effectiveness of social support for stressful decisions. In M. Deutsch & H. A. Hornstein (Eds.), *Applying social psychology: Implications for research, practice, and training.* Hillsdale, N.J.: Lawrence Erlbaum Associates, 1975.

Janis, I. L. *Current trends in psychology: Readings from American Scientist.* Palo Alto: Kaufmann, 1977.

Janis, I. L., & Mann, L. Coping with decisional conflict. *American Scientist,* 1976, *64,* 657–667.

Janis, I. L., & Mann, L. *Decision making: A psychological analysis of conflict, choice, and commitment.* New York: Free Press, 1977.

Leifer, A. D., & Lesser, G. S. *The development of career awareness in young children.* NIE Papers in Education and Work: Number 1. Washington, D.C.: National Institute of Education, 1976.

Lesser, G. S. *Children and television: Lessons from Sesame Street.* New York: Random House, 1974.

Lyle, J., & Hoffman, H. R. Explorations in patterns of television viewing by preschool-age children. In E. A. Rubinstein, G. A. Comstock, & J. P. Murray (Eds.), *Television and social behavior. Vol. 4, Television in day-to-day life: Patterns of use.* Washington, D.C.: U.S. Government Printing Office, 1972.

McGuire, W. J. The nature of attitudes and attitude change. In G. Lindzey & E. Aronson (Eds.), *The handbook of social psychology* (Vol. 3). Reading, Mass.: Addison-Wesley, 1969.

Milavsky, J. R., Pekowsky, B., & Stipp, H. Television exposure and proprietary and illicit drug use. *Public Opinion Quarterly,* 1975, *39,* 457–481.

Platt, J. Information networks for human transformation. In M. Kochen (Ed.), *Information for action.* New York: Academic Press, 1975.

Pool, I. Personal communication, June, 1977.

Seggar, J. F. & Wheeler, P. World at work on TV: Ethnic and sex representation in TV drama. *Journal of Broadcasting,* 1973, *17,* 201–214.

Shank, R. C., & Abelson, R. P. *Scripts, plans, and knowledge.* Prepared for presentation at the 4th International Joint Conference on Artificial Intelligence, U.S.S.R.: Tbilisi, Mat, 1975.

Singer, J. *Television-viewing and reading in the light of current research on cognition and information-processing.* Paper presented at the annual meeting of the Magazine Publishers' Association of America, October, 1976.

Singer, D. G., & Singer, J. L. *Partners in play: A step-by-step guide to imaginative play in children.* New York: Harper, 1977.

Tversky, A., & Kahneman, D. Judgment under uncertainty. *Science,* 1974, *185,* 1124–1131.

Ward, S., Wackman, D. B., Faber, R., & Lesser, G. *Effects of television advertising on consumer socialization.* Cambridge, Mass: Marketing Science Institute, 1974.

9

When I Was a Child
I Thought as a Child

Aimée Dorr
University of Southern California

Children, busy exploring and understanding our world, seem for years to have constructed their own world out of pieces of ours. Within the changing limits of their information-processing capacity, they construct models of the world as they experience it. These models are then used to decode, encode, recall, and act. They make childhood a unique, but natural, period. The truth of this has been recognized for eons, captured in the biblical statement "When I was a child, I thought as a child," in Rousseau's admonitions (1762) to "leave childhood to ripen in your children" and to leave the mind "undisturbed till its faculties have developed," in Piaget's more than 40 years of work (e.g., Piaget & Inhelder, 1969), and in current child development research (e.g., Kohlberg, 1969a; Shantz, 1975).

Yet an alternative characterization of children as tabulae rasae upon whom we "write" also has some demonstrated accuracy. It, too, has been endorsed throughout history in philosophical writings, in advice to parents and educators (e.g., Locke, 1693; Spock, 1946), and in child development research (e.g., Bijou & Baer, 1961). If we are to progress in our understanding, this view of children as tabulae rasae must somehow be combined with the fact that they are also actively making their own sense of what we try deliberately or inadvertently to "write." Such activity may in significant ways alter the meaning of their experiences.

Much of the research on children and television to date could perhaps best be characterized as concerned with what happens when the black box transmits messages to the tabula rasa (see such recent reviews as Leifer, 1975; Leifer, Gordon, & Graves, 1974; Liebert, Neale, & Davidson, 1973; Stein &

Freidrich, 1975). We have looked at the straightforward messages of television and then examined the extent to which they appear in children's learning, attitudes, or behavior. Not many of us have undertaken the difficult task of understanding what sense children make of what they see, what constructs they use to interpret television, or what these constructs mean for the role that television plays in their development.

This is not meant to imply that such research on children and television has produced no significant findings. On the contrary, it has provided substantial evidence that children will learn attitudes, information, and behaviors from television. Television is almost certainly not the major influence on a child's development that a few researchers imply, but some of its content some of the time does go straight into the child and does come back out in almost unadulterated form. Yet, once we have demonstrated this for cognitive information and skills (e.g., Ball & Bogatz, 1970, 1973; Ball, Bogatz, Kazarow, & Rubin, 1974; Bogatz & Ball, 1971), for aggression (Goranson, 1970; Liebert et al., 1973; Stein & Friedrich, 1975), for sex roles (Frueh & McGhee, 1975; Pingree, 1975), for race roles (Graves, 1975), and for advertising (Adler, Friedlander, Lesser, Meringoff, Robertson, Rossiter, & Ward, 1977), what shall we do?[1] The answer seems to be, for research on children and television as well as for child development research in general, that we sould look at how children construct models for what they experience and what this may mean for television's role in their development.

Consider some to the "facts" about television as a medium that children must reconstruct from their experience with the world, with television, and with what people tell them about both. Television is a representational system, not little people behaving in a box. Programs are made and broadcast by other people who have a variety of motives, not the least of which is pecuniary, for doing so. There is an underlying structure of information we consider to be important, and usually deliver, in a plot. Plot lines in American television are generally stereotyped in the dilemmas presented and in the ways in which they are worked out. Finally, programs have a variety of specific production techniques for symbolizing specific content.

In addition to reconstructing these facts about the medium, children interpret the social life they see on the screen, just as they do that in "real life." They develop a sequence of changing model roles for different types of people and interpret behavior accordingly (Emmerich, 1959; Emmerich, Goldman, & Shore, 1971; Flavell, 1970). The information they use to explain and evaluate actions expands to include actions, consequences, and motives

---

[1]Although each of the cited studies and literature reviews looked primarily for straightforward stimulus–response effects, many of them explored other issues as well.

(Kohlberg, 1969b). Their ideas of what goes on in others' heads (even that something does go on) change (Selman & Byrne, 1974). Even as these and other strategies for making sense of social life develop, children continually make their own kind of sense of the television programs they see, using the constructions of social life they have derived from their own life exeperiences as well as from the media. The lack of adult constructs does not in any way diminish the child's sense-making activity!

Constructions of television as a medium and of the social life it portrays and the variations in constructions based on experience, age or stage, cognitive abilities, needs, social class, ethnicity, and unfathomable idiosyncrasies must surely have implications for the role that television plays in the lives of different children. If we unravel some of the mysteries of the constructions children make of television, perhaps we shall come closer to understanding what television can provide for them. But let me assure you, I do not intend to unravel the mysteries here. There are years of work ahead before anyone can do that well. I would, however, like to draw from child development research, from research on children and television, and from my own experiences to suggest some of the constructions children from the age of about 2 through mid-adolescence may make of the television medium and of its content[2] and to suggest implications these have for their understanding of programming and its effects on them.

In selecting such a focus for this chapter I do not mean to deny other forces that influence development. Biological and genetic factors, traditional stimulus–response and observational learning, psychodynamic processes, practice of skills at critical points in development (as in idiomotor perceptual learning), and ecological conditions all play a part in the developmental process. It is beyond the scope of this chapter and indeed beyond my abilities to present a coherent explication of the ways in which all of these factors operate as children make a place in their lives for television. I hope the focus on the constructions children make of television will provide one interesting and useful perspective on what is surely a more complex process than any one chapter can describe.

---

[2]In discussing children's constructions of the medium and its content, I may sometimes seem like a confirmed Piagetian. I am not! I have utilized Piagetian concepts when they variously seemed sensible to me, were well supported by Piagetian research, or appeared in research on children and television. I have purposely, however, avoided the use of Piagetian stage terms and ages and ascribed responsibility for variations in constructions to a number of sources (see the first sentence of the paragraph in which this footnote appears). I find Piagetian systems to be, at best, good descriptions of some of the ways in which children in Westernized, industrialized, formally educated nations make sense of their world and have tried here to use the theory and research to this end.

WHAT, REALLY, ARE THOSE THINGS IN THE BOX?

Are the People and Objects Like Those in the Real World?

Children begin watching television when they are babes in arms. They orient to its light and sound and may even be entertained by it for a while (Slaby & Hollenbeck, 1977). By the time they are 2 or 3 years old, viewing is a regular pastime for most children (Lyle & Hoffman, 1972b; Schramm, Lyle, & Parker, 1961). Much of what they see has commonalities with the "real world." There is form, movement, people, animals, objects, interaction, and linear sequencing in time. Programs with live actors in all ways except size, dimensionality, and perhaps color present experiences not at all unlike those children observe in the rest of their lives. What sense does a child make of this? Are these little people? Do they see the child?

Perhaps even the youngest child equates the television image with the real thing. Work in other areas certainly suggests that this is so. Infants as young as 5 months have been shown to recognize color photographs, black and white photographs, color drawings, and black and white line drawings of people and objects (Cohen, DeLoache, & Pearl, 1977; Dirks & Gibson, 1977; Field, 1976; Strauss, DeLoache, & Maynard, 1977). Once they become familiar with the three-dimensional person or object, they treat the representation of it as also familiar. Their visual regard, physiological responses, and reaching behavior all indicate that this is so. When they have developed one-word vocabularies (12–18 months), their behavior also indicates that they equate the two-dimensional representation with its three-dimensional equivalent (Hochberg & Brooks, 1962). This is not to say that infants do not distinguish between two- and three-dimensional presentations; they do (Fantz, Fagan, & Miranda, 1975; Ruff, Kohler, & Haupt, 1976). It does, however, indicate that they do not need to learn to recognize two-dimensional representations of their three-dimensional world. They can apparently do this in infancy.

If young children link the objects and people on television to their equivalents in the concrete world, as the research suggests they do, then perhaps they incorporate the television information into their growing conceptions of these equivalents. Television becomes only one more source of information about, for example, all the varieties of chairs in the world and all the things chairs can do. If the chair flies or talks, perhaps very young children will incorporate this capacity into their notion of what chairs are capable of. Although there is little or no evidence on this possibility, it is reinforced for me by the many stories parents and educators have related when I meet them to talk about my research. The one I remember best is an elementary school principal recounting with residual horror the time he found his 3-year-old son hanging out of the second-story window of their home. He had just watched "Superman"[3] and apparently incorporated into his conception of human capabilities that of flying.

It is not at all uncommon for children of 2 or 3 or even 4 to believe that the people they see on television can, in some mysterious way, engage in social interaction with them. Preschoolers talk to the characters of some children's programs they are viewing, and many characters talk to them. Fred Rogers ("Misterogers' Neighborhood") says "I like you just the way you are"; Bob ("Sesame Street") asks children if they got the right answer in the preceding grouping task; and the Captain ("Captain Kangaroo") says "Good" when child viewers "respond affirmatively" to his query about wanting to watch a film about the circus. More than one character in a program for young children has reported the indignation of children who, in real life encounters, find the person behaving as though they were strangers. Children will marshall all the evidence of the character talking to and looking at them each day and of them talking to the character to support their beliefs that the children and character know each other well.

My son, too, displayed such beliefs when he was 3 and saw me being interviewed on television. His father reported that he called out my name, asked me questions, and tried to show me things. He became quite angry when I continued to ignore his attempts to engage me in social interchange and finally left the room in disgust. When the medium of television is understood in this way, some of the programs the child watches may serve as simply another social interaction in which the child participates.[4] If this is true, then the young child may understand and learn from television programs in ways that are quite similar to those in other daily social encounters. We could expect information, requests, emotional displays, and orders to be responded to as they are when a live adult delivers them to a child.

---

[3]Throughout the chapter I make reference to a number of television series, all of them very popular with children and young adolescents. Because these may not be familiar to all readers, I have included the name of the relevant series at the appropriate place in the text and a brief description appears in Appendix V.

[4]Adults too engage in social interactions with television characters. I talk back to newscasters and Howard Cosell; sportsfans yell encouragement, advice, and threats to players, coaches, and officials; and Archie and Meathead come in for their share of praise and abuse. Although such activities by adults may be important indicators of involvement in or functions of viewing for them, they probably do not serve as the same kinds of social interaction via television as they do for young children.

What Does It Mean That TV Is Sometimes Like Real Life?

In addition to sometimes encouraging such character–viewer interaction, other current practices in television programming may serve to make the distinction between "real life" and "television life" less discernible. Such practices are sprinkled into discussions throughout this chapter, but I would like here to describe three programming practices that are less often considered as potential communicators of reality: (1) The lives of major characters may change in ways that parallel real life changes in the human condition; (2) characters from one series may appear—as the same characters—in episodes in other series, placed in the kinds of situations one would expect for the characters; (3) characters we first know slightly as minor participants in the lives of the main characters of one series may become better known to us when they become the main characters of their own series. Although all three changes are dictated within the industry as means to achieving larger audiences—through providing new situations to write cleverly about, maintaining or increasing the popularity of existing characters, or capitalizing on apparent interest in minor characters—for younger viewers they may simply confuse the reality issue by providing increased similarities between life in the "real world" and life on television. Such similarities might then have to be balanced off against the unrealistic aspects of television as one came to final judgments about the veracity of program content. Perhaps the following examples of the three practices will illustrate the extent to which they do mirror aspects of "real life" and therefore may increase confusion among young viewers.

In many television series, primarily situation comedies, the lives of the characters change as do the lives of ordinary people. Although there are stabilities of character and situation, as there are in real life, there are also notable transitions. Mr. Kotter ("Welcome Back, Kotter") and his wife have twins. Gloria and Mike ("All in the Family") set up their own household and present the Bunkers with a grandchild. The mother of eight children ("Eight is Enough") dies, as did the actress in real life, and the father, Tom Bradford, enters the dating world that his teenage children also inhabit. The "Good Times" children are left by their mother to survive with a neighbor's supervision. The children in "The Brady Bunch," obviously a series from another era, grow up and appear as the singing and dancing Brady Bunch on a network special. Rhoda loses weight, leaves Mary, marries Joe, struggles with the intricacies of a marriage, and finally leaves Joe (first, "Mary Tyle Moore Show" and then "Rhoda"). The momentous joys and sorrows of real life, along with the daily sameness, are there in the lives of the television characters we watch each week—even when we are watching programs other than soap operas.

Intermingling of lives also may occur both in "real life" and on television, because the characters of one television series may appear as the same characters in other series. Nancy Drew and the Hardy Boys ("Hardy Boys/Nancy Drew Mysteries") have their own lives as detectives on alternating Sunday nights during the first season of this series. Sometimes, however, the three of them fortuitously encounter each other somewhere in the world (lately in Germany and Los Angeles) and combine their talents to solve a case. In one of the more recent episodes the three of them were even joined by a number of other detectives at a detective convention. Columbo ("Columbo"), Pete ("Switch"), McCloud ("McCloud"), and Sabrina ("Charlie's Angels") are all there—as they should be if they are truly detectives—at the convention. Jaime Sommers ("Bionic Woman") and Steve Austin ("Six Million Dollar Man") also lead separate lives as bionic defenders of truth, justice, and the American Way. Yet they too sometimes work together. Steve also has been known to admire a magazine spread of his ultimate angel ("Charlie's Angels") and real life wife, Farrah Fawcett-Majors, while he was taking a rest from his latest America-saving mission. All of these interminglings of characters' lives, undoubtedly meant to reinforce the popularity of the network's series, may serve for young viewers to reinforce their beliefs that the characters must surely lead these lives. They show up in more than one series in ways that one would expect from knowledge of the multiple occurrences, but relatively consistent roles, of people in everyday life.

In addition to seeing characters' lives intermingled, we may come to know some of them more fully after a period of participating less intimately in their lives. This usually occurs when a popular series spawns spin-offs. We know more about Rhoda when she stops being Mary's friendly neighbor ("Mary Tyler Moore Show") and becomes the star of her own series ("Rhoda"). Mary's lovable, crusty boss will take on more intimate dimensions now that he has a new job and his own series ("Lou Grant"). The Jeffersons, once pesky neighbors of the Bunkers ("All in the Family") and known only because we were sharing the Bunker's life, have become better known since they moved uptown and to their own series ("The Jeffersons"). Such progressive intimacy, like changes in life circumstances and intermingling of lives, has its counterparts in most children's lives. Occasional, rather superficial social encounters may lead to more intimate sharing of lives if these first encounters are mutually rewarding.

Noble (1975), repeating an idea advanced years ago (Horton & Wohl, 1956), has argued that television viewing is appealing to children precisely because it provides them with parasocial interactions such as these. Modern children's lives are viewed as impoverished by the lack of a small, stable, multi-aged community within which to come to know one's self and the world (as the world would be defined by one whose life was lived out in the small village). Television, albeit an imperfect substitute for human encounters, makes up for some of these deficits. Accepting for the moment such a view, one could argue—as Noble does—that television should be striving to present even greater congruity between the lives of its people and the lives of ordinary citizens. Yet such congruities may also present problems for children who do not fully understand the nature of the programs they are viewing. Programs present accurate and inaccurate information about our culture to children who have not yet acquired their own stable, family- or community-based conceptions of life. The isomorphisms with real life may be sufficiently clear that young children are led to greater acceptance of the reality of television life than their families or communities view as appropriate. If this is the case, then the need increases to have children accurately understand the social significance (or insignificance) of various modes of experience, increase their knowledge and acceptance of family or community norms, decrease their acceptance of television norms, and/or refrain from consuming much if any of television's alternative construction of the social order.

Although our ability to study very early conceptions of television is severely limited, it is apparent from the preceding illustrations that television presents some congruencies with real life and that young children may have some difficulty making correct (from an adult perspective) ascriptions of reality to it. During the preschool and early elementary school years there appear to be major changes in such ascriptions. Children seem to learn to grant greater credibility to the three-dimensional world, discounting the more fantastic television capabilities of objects, animals, and people. They come to understand that most television representations are fabricated and transmitted into the box and that the realism of these fabrications varies enormously.

The Acting Is Pretend, But Can I
Ignore How Real It Seems?

By the time children are about 5 they probably understand that the television host is not Big Brother watching them. Somewhat later they may begin to understand that television programs are not glimpses into everyday social life. These understandings are, however, still a major issue for the child. In discussing what is real and pretend on television, most of the children in the early elementary grades with whom I have worked[5] focus on the fact that the actors are real people but what they do is pretend. Unlike children around 10

---

[5]Except where otherwise noted, the quotations and examples of children's, adolescents', and adults' reasoning about television are all taken from current research supported by the Office of Child Development under a grant (No. 90-C-247) originally made to Aimee Dorr Leifer, Sherryl Browne Graves, and Neal J. Gordon.

years of age, younger children do not seem to assume that what they are seeing is a performance and then go on to evaluate how realistic characters and their actions are. Their concern is still with understanding the nature of the representation on television and their major heuristic seems to be that everything, except the actors themselves, is pretend.

The heuristic—everything is pretend—is adamantly applied by children in the early elementary grades to judgments about the reality of television programming in general, of entertainment programming, and of specific actions and relationships (Dorr & Graves, 1979). Fundamentally, however, children at this age are still uncertain. The concrete, visual presentation of a television program is difficult to override with cognitive controls. This is illustrated in an interview with a second-grader who told us that the bullets used to shoot people in entertainment programs were all fake. Later, however, he added "They all wear vests just in case they try putting real bullets in there. That should help some." It is further illustrated by the kindergartener who, in discussing "The Brady Bunch," told us: "Cause people are real on television, but the stuff isn't real. What they do, that really, the husband isn't really her husband. It's really a friend of hers. But they act like it. They're not really sisters, maybe. They're just good friends." In both cases, the child's initial judgment that something was pretend on television is later somewhat modified to give more credence to what was actually seen on television—people who aren't married but behave on television as though they are must surely be close friends. It is not possible for people to behave with such intimacy and affection without at least being good friends.

The same point is illustrated by my neighbor child who, under the immediate impact of the "Six Million Dollar Man," asked why the other characters didn't behave as though they heard his bionics operating (there are always special sound effects when bionics are employed—if you are not an avid viewer). He was genuinely puzzled by this, not scornful of implausibilities in the script. Away from the set the child was much more adamant that entertainment programming, that program, and bionics were all pretend. In this case judgment could be maintained only when the child was free of the compelling force of the television presentation.

By the latter part of elementary school, children seem much more secure in their knowledge that entertainment programming is fabricated. When asked about the reality of programming in general or of entertainment programming, they are more likely to say that it is both real and pretend. They accept the fact that entertainment programming is made up, but they go on to evaluate how realistic it is and perhaps critique it technically. This is not to imply that they need learn no more about how to evaluate the reality of entertainment programming; such skills are imperfectly developed even in most adults. Nor is it meant to imply that it ever becomes consistently easy or even possible for people to distinguish when to incorporate a television representation into their conception of the "real world" (see Bogart, 1980, for a thoughtful discussion of these issues vis-à-vis television news). Still, the major task of accepting completely the fact that entertainment programs are fabricated has been accomplished for most children by the end of elementary school.

How Can I Learn About TV Reality and Why Does It Matter?

The experiences that lead children to these varying constructions of the television medium are at present undocumented. Presumably, parents, siblings, peers, teachers, and others in the child's life explain the medium at least occasionally when confusions are apparent. Most television viewing by preschool and elementary school children occurs in the presence of other people, usually parents and/or siblings, thereby providing opportunities for the younger children to learn about the medium and its content from parents' and siblings' reactions both to the content and to the children's reactions to it. Television programs are favorite topics among groups of elementary school children and among young adolescents, and such conversations may help them to evaluate the medium and its content more accurately. Perhaps repeated failures by a television character to respond to the child also contribute to a reformulation of explanations for the character's presence in the livingroom. Perhaps the repeated discontinuities in the social reality on television and elsewhere contribute.

All of these suggestions about what may lead to changing constructions of the television medium are purely speculative. Although we may know that the opportunity to help change the child's construction exists (as with co-viewing and conversations), we do not know that the instruction for or facilitation of change occurs. Although we know that discontinuities and disconfirmations abound between television and real life, we do not know that children recognize them or attempt to resolve them or if they do, when they do. My own conversations with children indicate that discontinuities do not necessarily lead to the reformulations we, as adults, expect. For example, some children have noticed that the people, animals, and objects on television are usually smaller than the same things in real life. Occasionally this is made congruent with the child's belief that all of television is real: Either people are magically made smaller so that they can behave on television or else only very small people ever appear on television. Occasionally this is taken as an indication that television content must be pretend: Even the people must be pretend, because otherwise they would have to be "squished" in order to be on television. At other times disparities in size are not even worth an explanation.

If we could determine the experiences that lead to the changing interpretations of television as a medium, we might know how to help children to acquire more accurate conceptions earlier. Failing that, we might try to learn more about the sequence of constructions children make of the medium and use this knowledge to anticipate how they will react to the variety of programming they watch. Such undertakings will demand great skill, because they require us to understand the cognitive constructs of children who are either preverbal or else verbal but largely inarticulate and unintrospective.

There are some indications in the research literature that children's judgments of the reality of program content do, in fact, mediate the influence such content has on them. Advertisements featuring women in nontraditional occupations are generally more likely to increase acceptance of such jobs for women when the character is presented as a "real" woman holding that job than when she is presented as an actress (Pingree, 1975). Aggressive scenes are more likely to lead to increased acceptance of aggression when they are believed to be newsreel footage rather than part of an entertainment program (Feshbach, 1972). Perceived reality of prosocial and antisocial aspects of entertainment programs is likely to lead children at certain ages to greater endorsement, prediction for self, and report of actual prosocial and antisocial behavior (Reeves, 1977).

All three studies I have just cited indicate that greater perceived reality of television content leads, as one would expect, to greater acceptance of it. It should, however, be noted that the results do seem to vary according to the age of the children, the specific aspect of reality measured, and some interaction of age, sex, and type of content. Moreover, two of the three studies directly manipulated the reality of the content by telling children it was either real or acted rather than measuring children's interpretations of its reality. Finally, none of the studies worked with children younger than the middle grades of elementary school, by which time most understand that entertainment programs are fabricated. Nonetheless the results of these few studies do indicate that the reality ascribed to program content may be an important mediator of the extent to which children accept it as relevant to their own lives. If further work, especially with younger children and with individual judgments of the reality of actual programming, should bear these findings out, then we would become more eager to predict how much reality specific types of viewers are likely to ascribe to specific types of content and to teach all viewers to make more accurate reality judgments.

HOW COME THOSE PROGRAMS ARE THERE?

Understanding that television programs are usually fabricated does not logically lead to questioning why they are fabricated. Nonetheless, understanding why entertainment programs are there to be seen does seem to develop as part of a child's increasingly accurate construction of television as a medium. From work that my colleagues and I have done over the past three years it appears that explanations for why programs are there to be watched develop throughout middle childhood and adolescence (Phelps, 1976). Recognition grows that programs are designed to make money by entertaining, but even adults seem to have very limited understanding of the economic structure of the American television industry. Such conceptions of why programs are produced and broadcast may be important if they mediate the effects that programs can have on the viewer.

For adults, acceptance of information depends at least partially on an evaluation of the credibility of the source of that information. Information is more believable if the source is judged to be knowledgeable in that area, to have good reasons for presenting the truth, or to achieve minimal personal gain from having the information accepted (Insko, 1967; McGuire, 1969). Our knowledge of the extent to which children and adolescents employ similar processes in evaluating information presented to them is quite limited, although there are indications that such processes are not present until middle childhood (Aronson & Golden, 1959; Roberts, 1968). At whatever point they do appear, the child's construction of why television programs are there to be seen may become an important predictor of the credibility attached to what is seen.

The earliest explanations that children seem to have for why programs are available are likely only to enhance the perceived veracity of program content—if children use these explanations in evaluating content at all. Consonant with adult beliefs and industry intentions, children in our own work often tell interviewers that entertainment programs are broadcast to amuse them:

—To entertain people when they have nothing to do. (13-year-old)

—To entertain people. And to help me think. If they didn't have TV, when you come in and you sit down and you just want to do something, you don't have anything to do. (second-grader)

—For kids to watch them. (kindergartener)

Although adults may discount content presented with a primary goal of amusement, the same may not be true for children. In fact, most of the children whom we interviewed believed that those amusing programs were also made to teach children, to tell them the truth:

—To be educational things, they try to do it to help people. (sixth-grader)

—To teach children... about how to act and all that. (second-grader)

Over the past two years my colleagues and I have collected and analyzed data from kindergarten, second- third-, and sixth-grade children about why entertainment programs are broadcast. The preceding quotations are all taken from the interviews we conducted. When we looked to see if children had any idea that entertainment programs were broadcast primarily to earn money, we found that none of the children at any grade had an accurate understanding of the economic system of audience size, advertising, and income. About one-quarter of the sixth-graders could tell us that programs were broadcast to make money, but they apparently had no concept of the factors that determined income. The rest of the sixth-graders believed that programs were broadcast to entertain, inform, or tell the truth. More than half of the second- and third-graders held similar explanations for why entertainment programs were broadcast, and the remaining half either offered no explanation for why programs were broadcast or offered one that was judged entirely wrong by our staff. A similar pattern of understanding was evidenced by the kindergarteners, although somewhat more of them offered either no explanation or incorrect explanations.

Although it is clear from these data that even at the beginning of adolescence children have only meager understanding of the economics of the television industry, it should not be assumed that complete understanding develops during adolescence. Our own interviews with adolescents and their parents indicated that even the adults had limited understanding of the relationships between audience size (and sometimes characteristics), sale of advertising time, and income, although most understood that television broadcasting was a profit-making enterprise. In fact, the economics of all the mass media are beyond the ken of most of the adult population (Hirsch, 1977).

If, as I assume, the credibility of entertainment programming may be called into question if one understands that the primary motive for broadcasting is to generate income rather than to represent a slice of life accurately, then some adults and most children even at 12 lack sufficient information to question the veracity of the programs they watch. If the 12-year-old as well as the younger child believes that entertainment programs are meant to inform (as well as entertain) he or she is less likely to question the content. This assumes that children could or would engage in an evaluation of the motives for broadcasting entertainment programs when they are evaluating their credibility. Such an assumption seems supportable for the sixth-graders, questionable for the second- and third-graders, and unsupportable for the kindergarteners. Thus in some ways we need not worry about the kindergarteners' lack of accurate information. For sixth-graders and some second- third-graders, however, we find that even though they may evaluate the source in assessing the credibility of its information, they do not have the knowledge that would lead to questioning its information. Perhaps this partially explains the finding that young children do grant relatively high credibility to television portrayals and that ascribed credibility only gradually decreases during childhood and adolescence (Lyle & Hoffman, 1972a). An increased understanding of the reasons for broadcasting entertainment programs might lead to greater questioning of the credibility of their content, which in turn might lead to either less overall acceptance or more selective acceptance of its messages.

Children's constructions of what the things on television really are and why they are put there, important as they may be, are not the only concepts necessary for "understanding" all the material that television presents. Television is an audiovisual medium that usually tells stories. If children are going to codify these aspects of television too, then it seems they will construct models for the plots they encounter on television and for the audiovisual language used to convey them. The next three sections of the paper discuss the possible constructions children may make for these aspects of television.

WHAT'S THE STORY ABOUT ANYWAY?

What Are the Parts of a Story?

Most of what children watch on television is plotted. Although the models for plot lines may differ somewhat from writer to writer, at base each contains an initiating event or problem, the internal responses of the protagonists to it, attempts at resolving the problem, the consequences of the attempts, and reactions to them (see Rumelhart, 1974, for a "grammar" of stories and Labov, 1972, for a "grammar" of well-formed, extended narratives). If children do, in fact, construct models of aspects of their world and use them to interpret it, we should expect that with time they must develop some construct, like the suggested one, for an ideal plot (which may be congruent with notions of real life cause-and-effect sequences). Each successive construction of an ideal plot model would then have implications for the sense children make of the television programs they watch.

Child development and cognitive psychologists, using experimentally constructed stories, are currently studying the ways in which children construct their notions of storybook plots and what these mean for their understanding and recall of stories (Brown & Smiley, 1977; Mandler, 1977; Mandler & Johnson, 1977; Stein, 1976, 1977; Stein & Glenn, 1975). They find that children only gradually acquire a concept of an ideal plot line, with acquisition continuing through elementary school. In the early grades children need and expect to find in a story an initiating event, an attempt at resolution, and the consequences of that resolution. When these elements are not present, they will insert them in retelling the story. They do not need, nor do they often insert, internal responses to the initiating event or reactions to the consequences of the attempted resolution. Children do best at understanding and recalling a story when the elements are presented in the ideal order. When the surface structure of the story differs from the ideal, children will often convert it to the ideal form or misunderstand it.

From an entirely different perspective sociolinguists have also recently provided evidence for developmental changes in the concept of a well-formed story. For them the analyses are likely to be of children's discourse rather than their processing of storybook plots, but some of the elements that emerge are quite similar. For example, Kernan (1977) analyzed narrative reports of personal experiences by girls in the second-third, fifth-sixth, and eighth-ninth grades. Using Labov's model for the overall structure of well-formed, extended narratives (Labov, 1972; Labov & Waletzky, 1967), he found that the youngest group of girls was quite unlikely to provide an introduction or abstract for their narratives, to devote as high a proportion of their total discourse to orientation, to give appropriate background information and/or conditions necessary to understand the action that followed, to identify the participants by characteristics and statuses (usually just name was given), and to describe mood or motivation.There was a developmental increase in the extent to which preceding and succeeding events were described so as to be interdependent, a developmental decrease in the extent to which the preceding event was described so as to be independent of or only temporally related to the succeeding event (this measure was not toally redundant to the preceding one), and a developmental decrease in the extent to which attitude was implied solely through report of action.

From studies of both children's reports of their own adventures and of their processing of storybook adventures, we see a gradual increase in the number of elements considered to be important (especially those dealing with motivation and affect) and in the expectation of connected elements of plot rather than simple actions. From both there are indications that children construct grammars for plots. From the studies of story plots there are indications that plots that conform to the ideal grammar are easiest to understand and recall and that the grammar is actively used to understand and recall. All of these findings have potential relevance for children's understanding of plotted television programs.

What About Stories on TV?

Television plots are most often like those in storybooks, so we can reasonably expect that children construct models of ideal plot lines for them, too. This has implications for what we can anticipate they might take away from different programs they watch. Prior to the development of the concept of an ideal plot, children's understanding or recall of a program is not as likely to be disturbed by randomly sequenced events, multiple plot lines, flashbacks, and inverted time sequences. One study with television provides evidence in line with this hypothesis. Collins and Westby (1975) found that randomly sequencing events from a crime drama program had little detrimental effect, compared to the correct temporal sequence, on second-graders' recall of the plot, whereas it did on fifth-graders' recall. Such a result might be explained by assuming that the second-graders had not yet developed the construct of a temporally ordered ideal plot line and that the fifth-graders had developed such a concept but were not yet adept at picking out disordered elements of a program and putting them in the right slot in the ideal format. Other studies of television programs and films also suggest that children younger than about 6 or 7 do not understand the interdependencies of events (as the study of narratives also suggests) and tend to perceive the events to be unrelated to each other (Flapan, 1968; Franck, cited in Noble, 1969; Zazzo, cited in Noble, 1969).

In addition to not being affected by temporal arrangements of elements of the plot, we might expect that children who have not yet constructed a model of an ideal plot would not be disturbed by the absence of the motivations for or the consequences of actions, nor would they necessarily include them in retelling a plot (with or without them). There are a few studies that give some credence to these expectations. In retelling plots children younger than 7 have been found to cite only the more attention-getting actions. In a crime drama this was the aggressive action, not the reason it occurred and not the consequences of it (Collins, Berndt, & Hess, 1974). In a fairy tale it was the chopping down of the massive, wish-granting tree, not the repetitive greed leading to it, the repeated granting of ever-increasing wishes, or the poverty when the tree no longer lived (Leifer, Collins, Gross, Taylor, Andrews, & Blackmer, 1971). In both these studies the youngest children did not try to supply in their retelling an initiating event or problem or the consequences of trying to resolve it, much less the internal responses of the characters to the problem or the consequences. Apparently their models of a complete story do not include all these elements.

Perhaps the lack of a concept of a plot line explains why young children have such an appetite for programming with lots of action, changes, and noise. They aren't looking for a plotted story—they don't really understand there should be one—they are looking for stimulation. Once a program meets their requirement of high activity, young children may be the most undiscriminating of viewers. Inane, disordered, partial plots may be just as appealing as sensible, linear plots with all necessary elements present—and just as understandable. If young children are, in fact, making sense of programs according to how action-packed they are rather than in accordance with some notion of plot, then perhaps this also has consequences for the possible effects of the program on them. A "message" embedded in, or demonstrated by, a plot is not what they will extract from a program, and we should not look for it.

Later, when children have first constructed the concept of an ideal plot line, what they understand of different programs and what they extract from them should reflect their changed model. Obviously, the programs truly designed for children to understand will have an initiating event, an attempt to resolve it, and the consequences of that attempt—in that order. Such programs might add the internal responses of the protagonists, but they would not add more in terms of subplots, nor would they diverge from the model order for the plot elements. So, how shall we rate the comprehensibility of the following episodes of series which are quite popular with children?

Shazam and Mentor are riding down the road in their van. They park and Shazam leaves to buy groceries. Mentor is attacked and robbed by a teenager while a younger child inadvertently watches. The younger child is then threatened with dire consequences if he speaks of what he has seen. Mentor and Shazam try to convince the child to tell. The teenager continues to intimidate the child, finally abducting him. Mentor and Shazam follow and rescue the child, who then tells all. Shazam moralizes. ("Shazam/Isis")

Steve Austin is in a submarine. A man surreptitiously sends a message out. There is a small, but damaging, explosion on the submarine. The crew escapes. An older man and younger woman in a cave monitor the explosion and escape. They send trained sharks out to intimidate the repair crew, which is sent down from the surface ship. The surface-based crew leaves and the cave-based crew begin their own repairs. Steve Austin returns to the sub and is captured and taken to the cave. Two henchmen of the leaders in the cave are alarmed at the growing dangers. They speak of all they could gain by altering some of the leaders' planned uses of the sub. Throughout Steve's stay in the cave, the leaders gradually (and only verbally) reveal that the man was once a respected naval leader whose theories about training sharks were ridiculed. His daughter proved him right with her doctoral research and now assists him in obtaining the sub and its nuclear warheads to force the public to recognize her father for the genius he is. Rudy goes down in a bathosphere to search for Steve (he did the original bionics on Steve and is devoted to him) and is stranded there when a trained shark cuts the bathosphere's support cables. Steve escapes. A shark is sent to terrrorize him until his oxygen supply runs out. Steve returns to the cave. The henchmen put Steve and the leaders under guard so they can use the sub to obtain great wealth. Steve subdues the henchmen and frees the father, who is in the process of having a heart attack, and his daughter. They rise to the surface while Steve fights a shark, this time winning, closes the cave with massive boulders, and attaches his air line to the bathosphere. The father, daughter, and Steve appear at the water's surface. The cast is in a room on the surface ship. The daughter expresses her hope that Steve will be willing to rehabilitate a female ex-con in the future, because he was willing to save her life. Steve leans over the rail of the surface ship, silently and solitarily pondering the imponderables of his life. ("Six Million Dollar Man")

Chief Robert Ironside's life has been threatened by some unknown foe at the same time that he is contemplating surgery that may restore function to the lower half of his body. Plans for the surgery, threats on his life, and attempts to identify the foe are all intertwined in a sequence of scenes. The Chief then enters the hospital. He lies in the hospital bed. His three subordinates search for the potential assailant. Ironside flashes back to the time he could walk. Another such sequence occurs with Ironside flashing back to when he was shot. Another such sequence occurs with Ironside flashing back to being hospitalized. This pattern is repeated until the full history of Ironside's original injury and his hopes for hiking after the current surgery are revealed. The potential assailant is caught. The surgery is done. Ironside still cannot walk. ("Ironside")

If children have constructed a model for plots and still can use it only in its simpler forms and manifestations, then these three programs present different problems for them. The "Shazam" episode presents one simple plot, in the model order. Although there is more than one attempt to persuade the child to tell, each attempt is similar, each attempt but the last fails, and interspersed with each attempt is another threat by the teenager. Of the three programs this is the only one in which the sense children make of it is likely to correspond to the sense the producers intended to put there. The remaining programs seem unlikely to be interpreted by younger children to be the same story older viewers would see. Each has at least two plot lines, which may lead children to intermix the plots or perhaps neglect one plot line for another. Perhaps the "Six Million Dollar Man" episode would be encoded as the submarine Steve is in blows up and/or sharks attack Steve. Steve saves the sub and attendant characters and/or subdues the sharks, the bad guys go to jail and/or the sharks die or disappear forever. Perhaps the separate plot lines would be entwined into one. The "subtleties" of wrongdoing for love of a father, of revenge on an unbelieving world, of revolt by underling henchmen, and of conversion of wrongdoers are likely to be lost amid all the plot lines, making the story less than—or at least different from—the one we as adults see. The "Ironside" episode might lead to interpretations that diverge the most from those we adults draw. Past and present are interwoven without explicit connections. Linear connections of the segments would lead to a story about a man who is alternately hunting and hunted by a sniper and repeatedly wounded, recovering, and permanently damaged. The Chief is a man who for some reason can walk and yet must also use a wheelchair, alternately. What a confusing picture of human abilities this is!

Although those of us who talk with children about the television they watch know that they understand programs differently at different ages and differently from adults, we have as yet arrived at few paradigms for predicting what their construction of a program will be. Interpreting their understanding as derived from successively more complex—and in our view more correct—models for a plot and from increasing ability to utilize plot information presented in other than ideal form may aid us in arriving at more acutate predictions for what they will understand about a television plot. It may also ultimately aid us in predicting what effect various programs will have on children of different ages if different understandings lead to the drawing of predictably different messages from a program.

WHAT DOES IT MEAN THAT THINGS ARE SO MUCH THE SAME ON TELEVISION?

In addition to almost consistently employing plots whose "deep structures" conform to an ideal model, American television programs are generally consistent in the portrayals they present of various types of people (e.g., United States Commission on Civil Rights, 1977), in the way in which people try to resolve the initial conflict situation (e.g., Gerbner & Gross, 1976), and in the final resolutions of the conflict (e.g., Larsen, Gray, & Fortis, 1968).[6] They may therefore present children with another opportunity for constructing a model of television programs. As an illustration, consider the final resolution of a program. Most dramatic programs, and most programs that children watch, are episodes in a series with a limited number of recurring, main characters. Steve, Shazam, and Ironside must somehow make it back next week. The child who watches television with any regularity must come to understand this. What happens when the child recognizes this necessity? Right now we have no inkling of the consequences of recognizing what some call consistency and others call stereotypy in entertainment programming, but let me suggest a few disparate alternatives, all drawn from my conversations with people about television.

One possibility would be that consistency of content would lead one to reject its veracity. We all know, for example, that the flesh-and-blood good guy does not always win. Knowing this and seeing that television heroes violate this reality, one might come to reject these heroes as representative of real life Ironsides ("Ironside") or Fonzes ("Happy Days"). Many of the older children, adolescents, and adults I have interviewed use just this reasoning:

—I think a lot of that is fake [portrayals of Puerto Rican women]. Like the way they are all getting beat up and this and that, you know. They are *always*

---

[6]Although these citations do not cover all possible content areas and all content analyses necessarily report data for nothing more recent than the previous season, each successive year of content analysis had indicted at best gradual declines in the consistency of whatever area is measured.

swearing and yelling and having kids and all that. I think a lot—I think most—of that is all fake. (16-year-old)

—You know it's going to have a dippy ending, but when you put the facts down, you know this happens, and they get caught. It always happens. They're always caught. (Do you think that happens in real life? [interviewer]) No. (sixth-grader)

Logical as the foregoing responses seem, I have interviewed a good many people who apparently do not reason this way. Some of them seem to use consistency of portrayals as evidence that real life people must be like those on television. As one white woman said:

I don't know much about it, but it seems that all the black shows are all the same. It seems that they're showing all black people in one way on almost all the shows, so maybe this is true... I don't know much about them, but all the shows show them and have them act the same way. But as far as white people, I know they try to make them true-to-life so people can sympathize with them.

Although at first glance I find this last an amazing flight of fancy in adults, it is altogether understandable if one assumes that viewers, like this woman, believe that television producers are generally committed to presenting accurate portrayals of our culture and if one accepts the thesis of this chapter—that people construct models for the world based on the consistencies they find in it. Viewed in this way it is not altogether unreasonable that people would interpret consistency as truth, particularly when they have little or no real life contact with those portrayed (Greenberg & Reeves, 1976; Leifer, 1976). Obviously consistency in portrayals for these viewers has entirely different implications than it does for some other viewers.

A third possibility is that consistency is recognized but dismissed in reasoning about television, because one holds some other more compelling explanation about the veracity of entertainment content. In these instances consistency may be explained in any manner that makes it consistent with the overarching explanation (if the child is old enough to believe that explanations should be logically consistent), or it may be explained in any "illogical" fashion or left unexplained. The best example I have found of explaining consistency so that it is consonant with overriding belief comes from a conversation with a 13-year-old girl. To understand her interchange with the interviewer, you must realize that she believed that Perry Mason ("Perry Mason") actually was a practicing attorney. When asked what she thought about the fact that he always won his cases, she reasoned that he used the series much like advertising for his services:

Oh, Perry Mason, he never loses a case.
(What do you think about that?)
I enjoy watching it.

(But do you think it is true?)
Yeah.
(That a man could be so good at his job that he would never lose his case?)
Well, not never. He would lose them once in awhile.
(So how come he never loses?)
It's his program. It wouldn't be too good for him if he loses.

On a different vein from that of veracity is the possibility that understanding the consistency of plots diminishes the emotional involvement of the viewer. Where is the high drama if you know that Steve Austin ("Six Million Dollar Man") must slay the shark and rise to the surface before his air runs out? This fourth possibility was suggested to me by my son, who was 7 at the time. As I grew nervous during the suspenseful part of some program in a series and asserted that I couldn't watch anymore, he scornfully asked why I was so nervous when I knew the protagonist has to be all right by the end of the program. This realization, that main characters had to be resurrected by the end of each program in a series, was a relatively new one for him. It had clearly diminished his uncertainties during a program and it seemed also to have decreased his emotional arousal. Of course, diminution of uncertainty and arousal is not a necessary consequence of understanding consistency in series plots; I, the adult facilitator of his concept, was clearly anxious. Still, I think we might want to explore the possibility of decreased arousal further; it may for a while play some role in children's reactions to programming.

We are left then with a variety of consequences and nonconsequences as possible outcomes of recognizing the consistency of television fare. Each is based on the assumptions that children construct models of television content from their experiences with it and that these models (or lack of them) may have implications for their understanding of and reactions to television programming. As yet we have no real evidence to support the assumptions or the possible outcomes for consistency in plots and portrayals, but the tantalizing tidbits gleaned from interviews and personal life suggest we may want to follow the trail further. In doing so, we will need to take account of the possibility that consistencies may not generally be recognized by viewers. Or, people may be so uncritical during viewing that the implications of consistencies over time or across programs have no impact unless discussion—or interviews like those we conducted—essentially forces examination of them.

WHAT DO THE PICTURES MEAN?

Thus far I have discussed television programming from the vantage point of my literate, print-oriented heritage, neglecting McLuhan's sage reminder that television is not a print medium (McLuhan, 1964). Although its plot lines and consistencies draw heavily on our oral and print storytelling traditions, its visual mode presents the opportunity for the development of other traditions. An increasingly sophisticated technology permits us to convey information through slow motion, quick cuts, camera angles, chroma-key, and a host of other visual techniques. Television presents us with a visual quasi-grammar that we must learn, along with the print-based grammar, if we are to understand its messages.

Because those of us heavily committed to the print tradition rarely examine the visual language of television, perhaps an example and an indication of the interpretative problems it presents for children may be helpful. Consider the use of slow motion in entertainment programming. Prior to the advent of bionics, slow motion was most often employed to indicate a dream sequence, thinking, or a flashback. The Chief ("Ironside") recollects the entry of the incapacitating bullet into his spine. In slow motion the shot is fired, he falls, pain and helplessness engulf him. The recollection adds emotional impact to the present situation of a threat on his life and of impending surgery to remove the paralysis. The slow motion, and our knowledge of the logically possible sequences of events in life, help us to identify this scene as a remembrance of things past. If we do not see slow motion as a visual device and do not understand the logically possible sequences of life events, what then are we to make of this sequence?

Although I know of no data on children's interpretation of television's visual grammar, indications of the difficulties they have abound in my informal viewing and conversations with them. For some children, slow motion sequences are accepted as the literal truth. When Steve Austin ("Six Million Dollar Man") is running slowly down the road, I think, "He's using his bionics to catch up with the bad guys," and the 5- to 7-year-old children with whom I am viewing ask how he can catch them when he's running so slowly. When slow motion is seen so literally, Ironside's shooting becomes an event that occurred at just that slow a pace—and perhaps it occurred right after he was hospitalized. The cuing property of slow motion is apparently not recognized (or utilized) by such a child viewer.

When slow motion is recognized as a symbolic form, it is probably most often given only one meaning. For children without much exposure to our current bionic heroes and heroines, this meaning is most likely to be one of thinking or dreaming. In such a case, Ironside's slow-motion shooting can be interpreted properly as a flashback or, improperly, as a dream. But a unitary interpretation of the meaning of slow motion also presents problems when the producer utilizes it differently. For example, my son learned that slow motion meant the character was dreaming. Expectedly, this construct was applied to a slow-motion insert in "Rebop." The segment focused on a young child who golfed well. As we watch him golfing on his day off from caddying, a slow motion sequence is inserted in the middle of the otherwise regularly placed segment. My son immediately asked why he was dreaming of The application of his only construct for the meaning of slow r overinterpretation of a sequence that the producer had inserted only to provide visual variety.

Similar problems of literal interpretation of visual symbolism and of unitary constructs for techniques that can have multiple meanings must surely arise for all visual forms in television. Although beginnings have been made (e.g., Huston-Stein & Wright, 1977; Salomon, 1974), we have as yet produced no comprehensive catalogue of the visual grammar of television, nor have we explicitly investigated children's understanding of any aspect of it. Because meaning may be conveyed through a variety of production techniques, as well as through the sequences of events we see and hear, we may assume that production techniques must be interpreted (and correctly) for a full and accurate understanding of many programs. Such interpretation, like that for those other aspects of television that I have already discussed, may develop through a series of constructs, the final one like ours. The application of each successive construct to any program produces an interpretation of its meaning that, though perhaps different from the one intended, may have ramifications for the information children derive from it and for its effects on them.

WHAT DO I THINK ABOUT HOW PEOPLE BEHAVE?

What Are People Thinking and Why Do They Do What They Do?

Television presents interpretative problems for a child that depend both on its characteristics as a medium, which have been discussed in the preceding sections, and on its social content, to which I now turn. The people on television and their interactions present interpretative tasks that are common to all presentations of social life, whatever the medium. They include such tasks as understanding that another person has his or her own perspective, inferring the other's perspective, and judging the morality of another's action. Within the past decade work in child development has indicated that children do indeed develop a series of constructs for interpreting these aspects of social life and that the constructs they employ at any given time have implications for their understanding of social life (e.g., Shantz, 1975).

Successful social interaction demands an awareness of the possibility that the social other may differ in significant ways from the self. Such understanding comes gradually to children—and imperfectly to adults. One formulation of the development of the ability to take the role of another has been suggested by Selman (Selman & Byrne, 1974). Before the age of about 4 the child is believed to be devoid of understanding that the social other has any perspective. Over the next eight years, approximately, the child progresses through the following stages: understanding that the other has a perspective but assuming it is the same as the child's (4–6 years), understanding that the perspective may be different and inferring what it is (6–8 or 6–10 years), understanding that the other is inferring the child's perspective just as the child infers the other's (8–10 or 8–12 years), and being able to see a social interchange in which one is participating as though one were a third-party observer—the fly on the wallpaper (after 10 years of age).

Such a progression in role-taking ability, to the extent that it may be pervasive and consistent, has implications for children's understanding of television programming. Take the simplest example, children who do not understand that another has an unique perspective. Such children will certainly miss the humor attendant in some interchanges between characters with different goals, feelings, or knowledge. They will also not understand visual attempts to portray a character's internal perspective. When, during a social interchange, the camera closes in one Joe Hardy's head ("Hardy Boys/Nancy Drew Mysteries") with a quick cut to another experience he has had and then cuts back to his head, the child will not understand that we are seeing Joe's thoughts, that those thoughts are not apparent to the other interactant, and that those thoughts are relevant to what Joe does in the social interchange. What children will make of such television sequences is not clear (although we have the right to expect to be able to make such predictions from our theories). What is clear is that children, at this stage, will not understand some parts of programs in the way producers intended. Humor is lost, and significant elements of plot development are absent.

Descriptions of role-taking often focus on children's ability to understand the motivations of another or at least on attempts to do so. Several lines of evidence indicate that motivations become an increasingly significant consideration for children during the elementary school years. Selman's formulation, for which there are supportive data (Selman & Byrne, 1974), suggests that it is not until the age of about 6 that a child will attempt to understand another's motivations (which says nothing about the ability to do so accurately). Traditional studies of moral reasoning have been interpreted as demonstrating that it is not until approximately 7 to 10 years of age that children begin to take motivations for action into account in reasoning about the moral rectitude of another's action (Kohlberg, 1969b), as has at least one study of children's judgments of kindness (Baldwin & Baldwin, 1970). More recent work in an information-integration framework suggests that children from kindergarten on (the youngest age tested) utilize both motivations and consequences in judging behavior, with increasingly more weight being given to motivations as children mature (Surber, 1977). Nonetheless, evidence for the influence of motivations is clearly present in kindergarteners' judgments.

Finally, the previous discussion of children's conceptions of plot lines in storybooks and in their own narratives also indicated that the motivations for action do not become an accepted part of the story until the early elementary grades.

Such findings, if replicated for television, present important considerations for those interested in children's understanding of television and its effects on them. From the time children are 3 or 4 they primarily watch plotted programs. For many of these programs, though surely not for all, the motivations for action are important elements for understanding the "message." Our condemnations of the rebuffed father and the shark-trainer daughter ("Six Million Dollar Man") who destroyed the sub and intimidated its crew are moderated by their motivation to revenge old wrongs and to force due recognition of genius. For anywhere from three to six years children may watch such programs and generally judge action and character by actions themselves, by their consequences, or by the standards they have learned at their home. If the effects of a television program on viewers are in any way determined by the messages they drew from it and if motivations are an important part of the message, then young children's failure to use them in interpreting the message may lead to effects we would not otherwise predict.

Such an analysis is not meant to imply that effective messages must always be imbedded in a motivation–action–consequence sequence, nor that these three elements are the only ones within a program that children may use to evaluate action and characters. Most of the plotted material that children view features one or more main characters—the "good guys"—who appear week after week.[7] Through some processes that we do not understand, even very young children soon learn who they are. Once such learning has occurred, it would certainly reduce the need for children to evaluate specific depicted motivations in order to arrive at an adult-like judgment about the morality of a character's actions. What the familiar "good guy" does must be good (his motivations usually are good), and what the usually unfamiliar opponents do must be bad (their motivations usually are bad). Even in those few series that feature the same "bad guys" each week, there are numerous cues to status other than motivations—such as relative success of each character, amount of time each is on the screen, and the reactions of other characters to him or her. Similar cues, plus the stereotyped white and black hats or their present-day equivalents, probably also operate in programs that appear only once on television rather than in a series. Thus most television programs provide multiple cues as to the moral judgments one "should" make about characters and their actions, and a complete analysis of such judgments would obviously consider as many of them as possible.

---

[7]Paul Hirsch deserves the credit for pointing this out to me, though of course I should have remembered it.

Because similar messages may for some programs be derived from analyses other than those of motivation–action–consequence, it is reasonable to assume that similar effects may also occur. If we use the message "Crime doesn't pay" and its intended effect of decreasing (or at least not encouraging) criminal acts as an example, we can easily see that this message and the attendant effect can be derived from seeing that one is stealing for personal gain, that stealing leads to a beating and incarceration, or that the unfamiliar opponent steals—or simply from believing that stealing is bad.

In some cases, in fact, deterrence may be more likely when motivations are omitted. When Pete, one of the two leads of "Switch," steals for a good cause and is temporarily incarcerated, we are less likely to condemn his theft if we include his motivations in our evaluation than if we ignore them. In such cases the younger viewer who ignores motivation (and most of the other cues discussed in the preceding paragraph) is more likely to get the message "Crime doesn't pay" than is the older, ostensibly more sophisticated viewer.

Although acknowledging that the same message may be conveyed with and without a depiction of motivations, it is important to recognize that it is the rare program that does not incorporate motivations into the important elements of the plot, and it is the rare older child, adolescent, or adult who does not consider motivation an important element in judging another's action. Within this framework, it remains important to understand the ways in which different children recognize, understand, and utilize motivation in making sense of what they see on television.

What Do I Really Think about People on TV?

A few studies suggest that young children do, in fact, tend to neglect motives in recalling and understanding television programs. Looking first at children's retelling of the plot of a television program, we find—as we would expect from studies of children's concepts of storybook plots—that kindergarten and second-grade, but not fifth- and eighth-grade, children are likely to omit motives in recounting a popular crime drama (Collins, Berndt, & Hess, 1974). In fact, only one second-grader included motives at all! The program itself had been edited down to only 11 minutes, thereby decreasing the likelihood that material intervening between motives, action, and consequences or that incidental information would seriously interfere with children's learning of the motives (see, respectively, Collins, 1973; and Collins, 1970; Hale, Miller, & Stevenson, 1968; Hawkins, 1973, for evidence for such effects). About one-half of the fifth-graders and two-thirds of the eighth-graders included motives in their retelling, but we might expect less even from them when viewing full-length crime dramas in which intervening and incidental material would be greater.

There are also indications that younger children do not understand motives well, judging by their responses to direct questions by the interviewer. In the study just described (Collins, Berndt, & Hess, 1974), almost all the subjects (96%) could, when directly questioned, describe the consequences to the characters. There was, however, a strong, linear increase with age in their ability to describe the motivations for aggression. Leifer et al. (1971) also have reported a linear increase with age (4, 7, and 10 years) in ability to describe correctly the motivations for actions. A final study compared children's recognition of motives and consequences for aggression from among four possible depicted alternatives (Leifer & Roberts, 1972). Although the curves for correct choices for motives and consequences are relatively similar to each other from kindergarten through twelfth grade, there are slightly more correct choices for consequences than motives in kindergarten and slightly more correct for motives by sixth grade. Thus consequences remain somewhat better understood than motives by children through the early elementary grades, although the differential is considerably reduced when they need only to recognize the correct motives and consequences rather than supply them themselves.

Finally, there are indications—as one would expect from the literature on the development of moral reasoning—that younger children are not as likely to use motives in evaluating television characters as are older children. In the Collins, Berndt, and Hess (1974) study, kindergarten and second-grade children were more likely than the two groups of older children to refer to aggressive acts in justifying their evaluations of characters, and the older groups were more likely than the younger to refer to motives. In the Leifer and Roberts (1972) study, kindergarten children tended to view as "bad" all characters who engaged in aggression, whereas from third grade on, children (and adults) separated aggressors into "good guys" and "bad guys," apparently based on their motives for aggression.

It should be noted in reviewing these studies that the greatest "deficits" in apparent understanding of motivations tend to occur in situations in which children are asked to produce them. Retelling plots and justifying evaluations of characters (Collins, Berndt, & Hess, 1974) both allow children to choose what they would tell the interviewer. The omission of motivations in such cases may indicate that children did not recognize or understand them, that they did not consider them important, that they did not believe the interviewer would consider them important, or that they forgot them because they occurred early in the program. Explaining motivations (Collins, Berndt, & Hess, 1974; Leifer et al., 1971) did require children to discuss motivations with the interviewer. The less adequate explanations of the younger children may, however, be due to lack of vocabulary or to forgetting of motivations, which typically are portrayed earlier than the consequences, just as much as they may be due to lack of recognition or understanding. In both the foregoing studies the performance of children younger than about third or fourth grade showed less inclusion or understanding of motivations than consequences. In the Leifer and Roberts (1972) study, in which children simply had to state how good or bad a character was, motivations seemed to influence judgments equivalently for children from third grade on, with only kindergarteners tending to ignore motivations in arriving at their judgments. Similarly when understanding of motivations and consequences was tested by presenting children with multiple choice questions supported by drawings (Leifer & Roberts, 1972), only very slight differences in understanding motivations and consequences occurred at any age from kindergarten through twelfth grade.

It is therefore possible that our estimates of young children's understanding of motivations (especially as opposed to consequences) are biased by our testing procedures. Children may understand motivations earlier than conventional tests suggest yet be unable or unwilling to verbalize them well. Certainly the research I have reviewed suggests that this is the case, although there are not yet enough data to tell us whether or not the differential in younger children's understanding of motivations as opposed to consequences is still greater than that for older children.

The available data do, however, suggest that younger children are less likely than older children to utilize motivations in evaluating people's actions (see pp. 214–215), whether or not they understand them as well. Still, even here our testing procedures influence our findings. As discussed previously (see p. 214), procedures that require children to choose the better or worse person or action generally lead to the conclusion that younger children ignore motivations in arriving at their judgments whereas procedures that require children to rate each person or action lead to the conclusion that all children utilize both motivations and consequences but that older children give more weight to motivations than do younger children. Other work has shown that judgments about stories that are relatively impoverished and are presented verbally are less likely to focus on motivations than are judgments for stories that are embellished with drawings and irrelevant detail about characters, settings, and action (Feldman, 1968) and/or are presented as mini-dramas on videotape (Chandler, Greenspan, & Barenboim, 1973). Although such work indicates that the actual age at which motivations generally become more influential than consequences may vary according to the circumstances of the testing, it does not refute the general statement that younger children are less likely than older children to employ motivations in judging another.

Does Why People on TV Do Things Matter?

In this area, in contrast to most of the others that I have discussed in this chapter, there have been a few attempts to assess the relationship between the child's understanding of television (as a medium or as content) and its effect. Let us look at what has been done with motivations for aggression, beginning with those studies in which motivations have varied in the stimulus presentation and then considering those studies in which the viewer's own understanding of the motives has been utilized.

There is some indication that the depiction of justified aggession is more likely to increase subsequent aggression by male college students than is the same aggression depicted as unjustified (Berkowitz & Rawlings, 1963). When we study children, the results are considerably more mixed—perhaps because there is more than one study to evaluate but probably because of differences in stimuli, measures, ages of children, and procedures. Of three separate studies reported by Leifer and Roberts (1972), in which the effects of motivations on children of different ages were assessed, one using stimuli nearly identical to those of Berkowitz and Rawlings found no effects for motivations for fourth-, seventh-, and tenth-graders and no interaction between motivations and age. A second, using four edited crime dramas to present good or bad motivations and good or bad consequences for aggression, found the predicted effect of justified aggression producing slightly more subsequent aggression for preschoolers, fifth-, and twelfth-graders, but there was no interaction of motivations and age. A third, using six unedited crime drama or action–adventure programs, found no clear evidence for effects of motivations on kindergarteners, third-, sixth-, ninth-, and twelfth-graders and no interaction between motivation and age.

These same three studies (Leifer & Roberts, 1972) also assessed the possibility that children's actual understanding of motivations, rather than adult judgments of depicted motivations, modified the effects of exposure to aggression. Once again the results are quite mixed. The first study found no age-related differences in understanding of motivations, but it did find that the adjudged justifiability of aggression contributed to greater subsequent aggression. The second and third studies, although finding age-related increases in understanding of motivations, found no evidence that a viewer's understanding modified his or her subsequent aggression.

What are we to make of such variegated findings? They may arise because motivations (either depicted or adjudged) do not really modify the effects of exposure to televised actions (even if there are relatively well-documented changes in understanding and utilizing motivations) or because the effects of exposure to action (in this case, aggression) are weak, variable, or nonexistent. I think the available literature makes neither of these explanations particularly likely. A third possibility is that motivations only weakly modify effects of exposure to actions. This seems arguable from the available literature, although we need more evidence if we are to accept it. My own preference is for the possibility that depicted or adjudged motivations interact with other variables (such as consequences, attractiveness of the action, age and inclinations of the viewer, the dependent measure, and the measurement situation) to produce behavior subsequent to exposure. This too must await further evidence, but it is a plausible interpretation of the present research, because none of the four studies is truly a replication of any other.

Some of the differences in the four studies may be due to the stimuli used. Two studies manipulated only the motivations, using the same presentation for both conditions (Berkowitz & Rawlings, 1963, the first study reviewed in Leifer & Roberts, 1972). The other two studies (Leifer & Roberts, 1972) manipulated both motivations and consequences, using different television programs for each of the experimental conditions. One used four programs that had been edited to present relatively unambiguous and consistent good or bad motives and consequences. The other used six unedited programs whose overall messages about motives and consequences were judged by a panel of community parents. The first three studies report the hypothesized effects for justified aggression (without finding those for age), and the last does not.

However, the first two studies also used subjects who were at least 10 years old—all old enough to use motivations to evaluate action—whereas the latter two used subjects from preschool through twelfth grade—some old enough to use motivations to evaluate action and others not. Thus the first two studies should have produced an effect for motivations, which they did for those depicted (for Berkowitz and Rawlings and for the tenth-graders in Leifer and Roberts) or for those understood (not tested in Berkowitz and Rawlings, found overall for fourth-, seventh-, and tenth-graders in Leifer and Roberts). The other two studies should have shown some interactions between motivations and age. Neither found such an interaction for depicted motivations, and neither tested for such an interaction using the children's evaluations of the motivations and consequences. Because in the first Leifer and Roberts' study it was only adjudged motivations that predicted subsequent aggression, one might hope for more support for the suggested relationship if such analyses were done in the latter two studies. Thus we have a variety of factors that may contribute to the obtained differences: stimuli that in some cases probably varied on more than the tested dimensions, subjects who varied in the extent to which they are likely to use motivations (and consequences) to evaluate actions, and analyses that considered depicted motivations only or both depicted and adjudged motivations.

Although I cannot find a single sensible explanation for the different findings of the four studies, I believe the evidence that children at different ages recognize, understand, and utilize motivations differently is strong enough and the hints that such differences may modify the effects of the (motivated) behaviors that they see on television are sufficient that the area is worth further investigation. Such research will have to select carefully the presentations, the motivations, the extent to which motivations are presented or must be inferred, measures of depicted and understood motivations, the ages of the subjects, and a host of other variables. But the results might help us better to know when and how to use television (as we now try to do) to teach children that crime does not pay, that it is wrong to use epithets intentionally to hurt, and that handicapped persons must be valued for the extent to which they choose to try to be fully competent and friendly, whether or not they succeed.

Our understanding of some of the processes children use to interpret the social world has increased considerably over the past decade. Whereas no one can pretend to understand fully the multiple processes involved in making sense of and operating in the social world (for either children or adults) or their development throughout life, the preceding discussions of role-taking and motivations indicate the potentially important role developmental differences in such processes may play in how children of different ages understand and are affected by the television programs they view. It would be possible here to review a few other such processes (e.g., inference-making and the scheme for multiple sufficient causes), but that will not be done because the theory and research support for them with children is less well developed and because the two examples serve to illustrate the point.

THE END?

At points in writing this chapter it has seemed that I could just as well have composed a general paragraph and then stop writing. The paragraph would read somewhat as follows. Children, like adults, strive to make sense of television, both as a medium and as a social world. They develop constructs and paradigms and apply them in interpreting their television-based experiences. There may be a relatively regular course of development for these constructs and paradigms. Even if their developmental course is not regular, it is clear that for many years they are not the same as those employed by adults. Different constructs and paradigms, when applied to television, lead to different understandings of the medium and its content. Such differences in understanding naturally mean that television-viewing is a different experience for viewers of different ages. They may also mean that the effects of television-viewing, where they generally exist, will differ for these viewers and that effects will occur for some viewers and not for others.

This chapter has obviously been much longer than one paragraph, even though one paragraph summarizes its theme. But the six areas in which I have tried to illustrate the general theme are not empty elaboration of it. Making sense of television requires knowledge of many different aspects of the medium and its content. We need to know where its sounds and images come from, why they are broadcast, what structures are used to tell its stories, how to interpret the consistency of its stories, what the various visual techniques signify, and how to make sense of the social world we see on it. If we do not stop to make such an analysis of what there is to understand about television, we adults may remain largely unaware of many of these areas. They are so well understood by us and taken so much for granted that we are inclined to forget that at some time one may have to learn about them. Talking with children about television and watching it with them soon lead one to realize that such understanding does not spring full-grown in the minds of yet another generation of viewers. The process of coming to understand the television medium and its content requires time and experience and learning of many things.

I have tried in this chapter to indicate what there is to learn in each of these areas, the ways in which children of various ages may make sense of them, and some approximate ages at which different constructs are likely. Table 9.1 summarizes much of the material presented. It is offered with trepidation. The formality of a table with age norms and constructs suggests far more certainty about the concepts and the ages at which they occur than is at all warranted. It also suggests greater uniformity within an age group than I would necessarily expect to find in real life. Nonetheless, it is hoped that the table will help to summarize many of the speculations I have made and provide the reader with some indications of the kinds of changes one might expect as children develop.

There are good reasons to believe that children do progress through a series of constructs for understanding television and its content and that, from an adult's viewpoint, these constructs come increasingly close to being correct— even though we are a long way from being able to specify constructs and ages. Yet it would be foolhardy to assert that "correct constructs" are fully understood and consistently applied by all adults. One should even be wary of asserting that the majority of the adult population could agree on the "correctness" of some constructs. Variations will surely occur for different social classes and ethnic groups and for those with different experiences with television, business, social life, and the social order.[8] There does, however, appear to be sufficient generality among adults in the constructs and paradigms they have for understanding television for one to use adult performance as the norm to which children's performance may be compared. Such comparison should not, however, suggest that one expects little or no variability among adults in their constructs and paradigms and in the application of them to understanding television.

---

[8]Because I believe that social class and ethnic differences are largely explained by differences in experience and opportunity, it is probably somewhat misleading to separate these locator variables from experiential variables as I have in this statement. But because social class and ethnicity often serve as convenient ways of separating people, I have let the separation stand.

Children's constructions of television are, for some, important only because they are interesting, amusing, or strange. Variations in human-kind—whether they be among adults grouped by such variables as social class, ethnicity, or experience or among children grouped by age—always intrigue us. I, too, have derived great pleasure from talking with children and glimpsing the various ways in which they can make sense of television. I believe, however, that there is a more practical reason for understanding what sense children are likely to make of it. Television, intentionally and unintentionally, does teach children. If we know how they are likely to interpret it, we can then use the medium as a more effective, positive aid in socialization. "Sesame Street" can teach cooperation better, "Mister Rogers" can better help children explore their feelings, "The Jeffersons" may not unduly influence the segregated white child's view of blacks, and young children (unlike some we have talked with) may not be so sanguine about accidents whose damages cannot really be repaired with bionics. This is not meant to imply that a "bad" television experience is likely to have substantial long-range effects on children. In my view children are remarkably resilient creatures, whose development is influenced by a multitude of people and experiences. We should try to ensure that these multiple influences—including that of television—are as good for children as possible, but we need not fear that television can in some way override all the other experiences children will have.

This chapter began with an assertion that our understanding of children and television would be enhanced if we viewed them both as tabulae rasae on whom the "real world" was imprinted and as active constructers of that world. Having devoted the chapter to a consideration of the ways in which the sense children make of the television medium and its content leads to different realities than the one we would expect to be imprinted on them, it seems wise to end by returning to the more balanced view. Different children do seem to interpret the medium and its content differently. In some cases these interpretations may lead to different effects, but in other cases one may find a more general and straightforward effect. Even the ability to interpret television content accurately does not always lead to the effects one would predict. For instance, Rossiter and Robertson (1974) have shown that children who understand very well that television commercials are designed to sell them products are generally less susceptible to their persuasive intent. Nonetheless these same children, under heavy exposure to commercials during the pre-Christmas season, still succumbed to their inducements. Here is one example, out of many I am certain we could find, in which interpretations of content matter under some conditions and not under others.

Thus in some circumstances the content to which children are exposed does have a direct effect on them (they are tabulae rasae!), whereas in others their

TABLE 9.1

Summary of Possible Constructs Employed in Understanding Television Programs and How They May Change with Age

(age in years)

Area	0–5	5–7	7–10	10–13	13 on
Reality of TV images	Images represent real thing	Actors real, but acting pretend Judgment influenced by visual representation		Evaluate realism	
Actor–viewer interaction	Actors and viewers truly interact	Actors cannot see viewers or interact with them			
Why programs are broadcast	To entertain and inform				To make money
Elements of plot	None	Initiating event, attempts to resolve, consequences		Motivations, emotions, internal responses	
		Sequences meaningfully, consequentially connected			

Order of plot elements	None	Ideal order preferred and easiest to manage	Variations in order matter
Consistency of portrayals and plots	Insufficient data even for rough age breakdown		
Visual grammar	Insufficient data even for rough age breakdown		
Role-taking (the other has):	no perspective	same perspective	own perspective — Other infers one's perspective — Watch own interaction as third party would
Use of motivation to judge character	Motivation of less importance, consequences matter more	Motivation of primary importance	

interpretation of the meaning of that content modifies the effect (they do control their fate!). Our evidence, as cited at the beginning of this chapter, is substantial that television content may have such direct effects on children. It now seems wise to expand our knowledge of television's influence on children by examining the ways in which their own understanding of the medium and its content may modify—substantially, minimally, or not at all—the nature and extent of these direct effects.

ACKNOWLEDGMENTS

I am grateful to my fellow members of the Social Science Research Council Committee on Television and Social Behavior, to Barbara Flagg, and to Marcia Chellis for their helpful comments on an earlier draft of this paper and to Donald Dorr-Bremme and Paul Hirsch, who critiqued that same draft with special care and enthusiasm, pointed me to bodies of theory and research other than those in which I usually forage, and asked questions that helped me to clarify my thoughts and the expression of them. My tenure on the Committee and the majority of the work for this chapter were completed while I was a member of the faculty of the Graduate School of Education, Harvard University. I am grateful to the faculty, staff, and students who provided substantial intellectual and physical support for this work while I was there.

REFERENCES

Adler, R. P., Friedlander, B. Z., Lesser, G. S., Meringoff, L., Robertson, T. S., Rossiter, J. R. & Ward, S. *Research on the effects of television advertising on children: Review and recommendations.* Washington, D.C.: U.S. Government Printing Office, 1977.

Aronson, E., & Golden, B. The effect of relevant and irrelevant aspects of communicator credibility on opinion change. *Journal of Abnormal and Social Psychology*, 1959, *59*, 177–181.

Baldwin, C. P., & Baldwin, A. L. Children's judgments of kindness. *Child Development*, 1970, *41*, 29–47.

Ball, S., & Bogatz, G. A. *The first year of Sesame Street: An evaluation.* Princeton: Educational Testing Service, 1970.

Ball, S., & Bogatz, G. A. *Reading with television: An evaluation of The Electric Company.* Princeton: Educational Testing Service, 1973.

Ball, S., Bogatz, G. A., Kazarow, K., & Rubin, D. B. *Reading with television: A follow-up evaluation of The Electric Company.* Princeton: Educational Testing Service, 1974.

Berkowitz, L., & Rawlings, E. Effects of film violence on inhibitions against subsequent aggression. *Journal of Abnormal and Social Psychology*, 1963, *66*, 405–412.

Bijou, S. W., & Baer, D. M. *Child development, Vol. 1. A systematic and empirical theory.* New York: Appleton, 1961.

Bogart, L. Television news as entertainment. In P. H. Tannenbaum (Ed.), *The entertainment functions of television.* Hillsdale, N. J.: Lawrence Erlbaum Associates, 1980.

Bogatz, G. A., & Ball, S. *The second year of Sesame Street: A continuing evaluation.* Princeton: Educational Testing Service, 1971.

Brown, A., & Smiley, S. Rating the importance of structural units of prose passages: A problem of metacognitive development. *Child Development,* 1977, *48,* 1–8.

Chandler, M., Greenspan, S., & Barenboim, D. Judgments of intentionality in response to videotaped and verbally presented moral dilemmas: The medium is the message. *Child Development,* 1973, *44,* 315–320.

Cohen, L. B., DeLoache, J. S., & Pearl, R. A. An examination of interference effects in infants' memory for faces. *Child Development,* 1977, *48,* 88–96.

Collins, W. A. Learning of media content: A developmental study. *Child Development,* 1970, *43,* 1133–1142.

Collins, W. A. Effects of temporal separation between motivation, aggression, and consequences: A developmental study. *Developmental Psychology,* 1973, *8,* 215–221.

Collins, W. A., Berndt, T. J., & Hess, V. L. Observational learning of motives and consequences for television aggression: A developmental study. *Child Development,* 1974, *45,* 799–802.

Collins, W. A., & Westby, S. D. *Children's processing of social information from televised dramatic programs.* Paper presented at the biennial meeting of the Society for Research in Child Development, Denver, April 1975.

Dirks, J., & Gibson, E. Infants' perception of similarity between live people and their photographs. *Child Development,* 1977, *48,* 124–130.

Dorr, A., & Graves, S. B. *Children's critical evaluation of television content.* Final report to the Office of Child Development. Harvard University, 1979.

Emmerich, W. Young children's discriminations of parent and child roles. *Child Development,* 1959, *30,* 403–419.

Emmerich, W., Goldman, K. S., & Shore, R. E. Differentiation and development of social norms. *Journal of Personality and Social Psychology,* 1971, *18,* 323–353.

Fantz, R. L., Fagan, J. F., & Miranda, S. B. Early visual selectivity. In L. B. Cohen & P. H. Salapatek (Eds.), *Infant perception* (Vol. 1). New York: Academic Press, 1975.

Feldman, J. *Children's moral judgments to traditional dilemmas presented with and without pictures and background information.* Unpublished class paper, Stanford University, 1968.

Feshbach, S. Reality and fantasy in filmed violence. In J. P. Murray, E. A. Rubinstein, & G. A. Comstock (Eds.), *Television and social behavior, Vol. 2. Television and social learning.* Washington, D.C.: U.S. Government Printing Office, 1972.

Field, J. Relation of young infants' reaching behavior to stimulus distance solidity. *Developmental Psychology,* 1976, *12,* 444–448.

Flapan, D. *Children's understanding of social interaction.* New York: Columbia University Press, 1968.

Flavell, J. H. Concept development. In P. H. Mussen (Ed.), *Carmichael's manual of child psychology* (Vol. 1). New York: Wiley, 1970.

Frueh, T., & McGhee, P. E. Traditional sex role development and amount of time spent watching television. *Developmental Psychology,* 1975, *11,* 109.

Gerbner, G., & Gross, L. P. Living with television: The violence profile. *Journal of Communication,* 1976, *26,* 172–199.

Goranson, R. E. Media violence and aggressive behavior: A review of experimental research. In L. Berkowitz (Ed.), *Advances in experimental social psychology* (Vol. 5). New York: Academic Press, 1970.

Graves, S. B. *Racial diversity in children's television: Its impact on racial attitudes and stated program preferences in young children.* Unpublished doctoral dissertation, Harvard University, 1975.

Greenberg, B. S., & Reeves, B. Children and the perceived reality of television. *Journal of Social Issues,* 1976, *32,* 86–97.

Hale, G. A., Miller, L. K., & Stevenson, H. W. Incidental learning of film content: A developmental study. *Child Development,* 1968, *39,* 69–78.

Hawkins, R. P. Learning of peripheral content in films: A developmental study. *Child Development,* 1973, *44,* 214–217.

Hirsch, P. Personal communication, December 1977.

Hochberg, J. E., & Brooks, V. Pictorial recognition as an unlearned ability: A study of one child's performance. *American Journal of Psychology,* 1962, *75,* 624–628.

Horton, D., & Wohl, R. R. Mass communication and parasocial interaction: Observations on intimacy at a distance. *Psychiatry,* 1956, *19,* 215–229.

Huston-Stein, A., & Wright, J. *Effects of formal properties of children's TV programs.* Paper presented at the biennial meeting of the Society for Research in Child Development, New Orleans, March 1977.

Insko, C. A. *Theories of attitude change.* New York: Appleton, 1967.

Kernan, K. T. Semantic and expressive elaboration in children's narratives. In S. Ervin-Tripp & C. Mitchell-Kernan (Eds.), *Child discourse.* New York: Academic Press, 1977.

Kohlberg, L. Stage and sequence: The cognitive–developmental approach to socialization. In D.A. Goslin (Ed.), *Handbook of socialization theory and research.* Chicago: Rand McNally, 1969. (a)

Kohlberg, L. *Stages in the development of moral thought and action.* New York: Holt, 1969. (b)

Labov, W. *Language in the inner city: Studies in the black English vernacular.* Philadelphia: University of Pennsylvania Press, 1972.

Labov, W., & Waletzky, J. Narrative analysis. In J. Helm (Ed.), *Essays on the verbal and visual arts.* Seattle: University of Washington Press, 1967.

Larsen, O. N., Gray, L. N., & Fortis, J. G. Achieving goals through violence on television. In O. N. Larsen (Ed.), *Violence and the mass media.* New York: Harper, 1968.

Leifer, A. D. Research on the socialization influence of television in the United States. *Fernsehen und Bildung,* 1975, *9,* 26–53.

Leifer, A. D. *Factors which predict credibility ascribed to television.* Paper presented at the annual meeting of the American Psychological Association, Washington, D.C., September, 1976.

Leifer, A. D., Collins, W. A., Gross, B. M., Taylor, P. H., Andrews, L., & Blackmer, E. R. Developmental aspects of variables relevant to observational learning. *Child Development,* 1971, *42,* 1509–1516.

Leifer, A. D., Gordon, N. J., & Graves, S. B. Children's television: More than mere entertainment. *Harvard Educational Review,* 1974, *44,* 213–245.

Leifer, A. D., & Roberts, D. F. Children's responses to television violence. In J. P. Murray, E. A. Rubinstein, & G. A. Comstock (Eds.), *Television and social behavior, Vol. 2. Television and social learning.* Washington, D.C.: U.S. Government Printing Office, 1972.

Liebert, R. M., Neale, J. M., & Davidson, E. S. *The early window.* New York: Pergamon Press, 1973.

Locke, J. *Some thoughts concerning education.* London: A. and J. Churchill, 1699. (Originally published, 1693.)

Lyle, J., & Hoffman, H. R. Children's use of television and other media. In G. A. Comstock, E. A. Rubinstein, & J. P. Murray (Eds.), *Television and social behavior, Vol. 4. Television in day-to-day life: Patterns of use.* Washington, D.C.: U.S. Government Printing Office, 1972. (a)

Lyle, J., & Hoffman, H. R. Explorations in patterns of television viewing by preschool-age children. In E. A. Rubinstein, G. A. Comstock, & J. P. Murray (Eds.), *Television and social behavior, Vol. 4. Television in day-to-day life: Patterns of use.* Washington, D.C.: U.S. Government Printing Office, 1972. (b)

Mandler, J. M. *A code in the node: Developmental differences in the use of a story schema.* Paper presented at the biennial meeting of the Society for Research in Child Development, New Orleans, March 1977.

Mandler, J., & Johnson, N. Remembrance of things parsed: Story structure and recall. *Cognitive Psychology,* 1977, *9,* 111–151.

McGuire, W. J. The nature of attitudes and attitude change. In G. Lindzey & E. Aronson (Eds.), *The handbook of social psychology* (Vol. 3). Reading, Mass.: Addison-Wesley, 1969.

McLuhan, M. *Understanding media, the extensions of man.* New York: McGraw-Hill, 1964.

Noble, G. The English report. In J. D. Halloran (Ed.), *Findings and cognition on the television perception of children and young people.* Munich: Internationales Zentralinstitut für das Jugend- und Bildungsfernsehen, 1969.

Noble, G. *Children in front of the small screen.* Beverly Hills: Sage, 1975.

Phelps, E. M. *Knowledge of the television industry and relevant first-hand experience.* Paper presented at the annual meeting of the American Psychological Association, Washington, D.C., September 1976.

Piaget, J., & Inhelder, B. *The psychology of the child.* New York: Basic Books, 1969.

Pingree, S. *A developmental study of the attitudinal effects of non-sexist television commercials under varied conditions of perceived reality.* Unpublished doctoral dissertation, Stanford University, 1975.

Reeves, B. *Children's perceived reality of television and the effects of pro- and anti-social TV content on social behavior.* Unpublished manuscript, Michigan State University, June 1977.

Roberts, D. F. *A developmental study of opinion change: source-orientation versus content orientation at three age levels.* Unpublished doctoral dissertation, Stanford University, 1968.

Rossiter, J. R., & Robertson, T. S. Children's TV commercials: Testing the defenses. *Journal of Communication,* 1974, *24,* 137–144.

Rousseau, J.-J. *Emile, or on education.* London: Dent, 1911. (Originally published in French, 1762.)

Ruff, H. A., Kohler, C. J., & Haupt, D. L. Infant recognition of two- and three-dimensional stimuli. *Developmental Psychology,* 1976, *12,* 455–459.

Rumelhart, D. E. *Notes on a schema for stories.* Paper presented at the Carbonell Memorial Conference, Pajaro Dunes, California, May 1974.

Salomon, G. *Annual report of the first year of research on cognitive effects of the media.* Submitted to the Spencer Foundation. Hebrew University, (Israel), September 30, 1974.

Schramm, W., Lyle, J., & Parker, E. B. *Television in the lives of our children.* Stanford: Stanford University Press, 1961.

Selman, R. L., & Byrne, D. F. A structural–developmental analysis of levels of role taking in middle childhood. *Child Development,* 1974, *45,* 803–806.

Shantz, C. V. The development of social cognition. In E. M. Hetherington (Ed.), *Review of child development research* (Vol. 5). Chicago: University of Chicago Press, 1975.

Slaby, R. G., & Hollenbeck, A. R. *Television influences on visual and vocal behavior of infants.* Paper presented at the biennial meeting of the Society for Research in Child Development, New Orleans, March 1977.

Spock, B. *The pocket book of baby and child care.* New York: Pocket Books, 1946.

Stein, A. H., & Friedrich, L. K. Impact of television on children and youth. In E. M. Hetherington (Ed.), *Review of child development research* (Vol. 5). Chicago: University of Chicago Press, 1975.

Stein, N. L. *The effects of increasing temporal disorganization on children's recall of stories.* Paper presented at the meeting of the Psychonomic Society, St. Louis, November 1976.

Stein, N. L. *The role of structural variation in children's recall of simple stories.* Paper presented at the biennial meeting of the Society for Research in Child Development, New Orleans, March 1977.

Stein, N. L., & Glenn, C. *A developmental study of children's recall of story material.* Paper presented at the biennial meeting of the Society for Research in Child Development, Denver, April 1975.

Strauss, M. S., DeLoache, J. S., & Maynard, J. *Infants' recognition of pictorial representations of real objects.* Paper presented at the biennial meeting of the Society for Research in Child Development, New Orleans, March 1977.

Surber, C. F. Developmental processes in social inference: Averaging of intentions and consequences in moral judgment. *Developmental Psychology,* 1977, *13,* 654–665.

United States Commission on Civil Rights. *Window dressing on the set: Women and minorities in television.* Washington, D.C.: U.S. Government Printing Office, 1977.

10 Television and Afro-Americans: Past Legacy and Present Portrayals

Gordon L. Berry
Graduate School of Education
University of California, Los Angeles

TELEVISION AS A MEDIUM AND A MEDIATOR IN SOCIETY

A major characteristic of American contemporary life is the pushing and pulling associated with the status of ethnic minorities in general and of blacks or Afro–Americans, in particular. So pervasive have been the issues of race and color in the United States that entire institutions and legal sanctions were created to constrain the cultural, educational, social, economic, and political development of Afro–Americans. These constraining strategies were almost classic in their ability to develop in a large group of black people the feeling that they were inferior and to instill in many white Americans that this feeling of inferiority was a correct perception for blacks to hold of themselves.

Historically the family, school, religious groups, and other agencies of socialization have played a major role in transmitting the values of those individuals or groups considered superior and inferior. Although these agencies are still the major transmitters of positive and negative values, contemporary society also has the phenomenon of mass communication to influence our feelings attitudes, and behavior.

One of the major media within our mass communications system that has the power to transmit meaning and values is television. The messages of television pervade our lives and establish profound mental sets that consciously or unconsciously influence our behavior. As Richard Adler (1976) pointed out:

> Television is not simply a 'medium' but a 'mediator' between fact and fantasy; between our desire to escape and our need to deal with real problems; between

our old values and new ideas; between our individual lives and the life of a nation and world. Seen from this perspective, television becomes much more interesting and much less simple than before [p. 13].

Any medium that is potentially present in the daily life of more than 95% of American households and carries words, images, and pictures that overtly and imperceptibly influence our behavior, carries a significant burden as a social force. As a force within the total environment, television's message cannot be immune from those stresses and strains associated with the race issue in our society. Many of the negative and superior attitudes held by white Americans are interwoven into the context and images of television. Similarly, the legacy of institutional racism results often in a passive acceptance by black Americans of the destructive portrayals of their customs, family life, language, and general social behavior.

However, television has the potential to become a major "mediator" in assisting American society toward a more enlightened and humane concept concerning people of different races. Television can create learning experiences that will expose our children to those values, skills, and attitudes that will open them to a better understanding of the multicultural world in which they live.

PSYCHOSOCIAL LEGACY FROM THE PAST

Historians are correct when they suggest that the events of the present are part of the legacy from the past. A good starting point in understanding today's white attitudes toward blacks is 1619, when the first 20 Africans were brought to the Virginia colony. Slavery in the United States took on a set of characteristics and was based on a rationale that was to create a special status for black people. "Slaves were denied virtually all rights, both civil and political. Perhaps the most important element in defining the slaves' status was the perpetual nature of slavery; slaves were destined to occupy this status throughout their lives and to transmit it to their children. Slavery and Afro-American became synonymous, and since slaves were defined as innately inferior, blacks were defined as inferior beings" (Pinkney, 1975, p. 3).

Whenever slavery is developed, it needs some rationale to justify the attitudes held toward the enslaved. Slavery in this country was no exception. As Herbert Aptheker (1971) pointed out, part of the rationale for slavery and the subsequent inferior status of the Afro-Americans was a series of myths— some of which were taken from the Bible and many of which had questionable "scientific" origin. The fostering of myths and the building of irrational attitudes toward Afro-Americans were so much a part of the thinking of the early colonial period that it was not unusual for the supporters of slavery to believe in the innate inferiority of black people. In addition, the myths perpetuated the belief that slaves were docile and childlike and happy in their condition—despite the many insurrections.

As a legacy of the Civil War and the Reconstruction Period, the United States created formal and informal strategies for establishing segregation as an unchallenged way of life. Much of this impetus grew out of the 1883 Supreme Court repeal of the Civil Rights Act of 1875 and out of the 1896 decision in the Plessy vs. Ferguson case that established the doctrine of "separate but equal." These legal and legislative moves, coupled with the creation of Jim Crow laws, lynchings, and other practices, helped to codify discrimination and segregation in our public life. The patterns of dehumanization in relations with black people were not accidental or haphazard but rather the products of a "culture" of discrimination and segregation.

During the period between 1900 and 1940, there was a burgeoning development of newspapers, magazines, radios, and motion pictures. Although television emerged last, it naturally took many of its portrayals from these established media. The early newspaper, radio, and other media forms traditionally failed to reflect the contributions of blacks and their culture to American society, and television perpetuated past practices.

As Margaret Just Butcher (1973) observed, "Mass media have been staffed primarily by white middle-class journalists who purposely or unconsciously communicate a bias much more respectful of whites and white culture than of nonwhite races and their cultures [p. 202]." Consequently, the early newspapers developed a record of biased reporting about the black experiences in the United States. This lack of sensitivity has continued and was especially noted by the National Commission on Civil Disorders (1968) as a contributing factor to the riots of the 1960's:

> When the white press does refer to Negroes and Negro problems it frequently does so as if Negroes were not a part of the audience. This is perhaps understandable in a system where whites edit and, to a large extent, write news. But such attitudes, in an area as sensitive and inflammatory as [racial riots and conflicts], feed Negro alienation and intensify white prejudices [p. 211].

Perhaps the most flagrant of the nationwide media in terms of negative ethnic portrayals and social sterotyping was the motion picture industry. After the *Uncle Tom's Cabin* movie debut in 1903, there appeared a variety of black presences bearing the fanciful names of the coon, the tragic mulatto, the mammy, and the brutal black buck. According to Donal Bogle (1973),

> the characters were merely filmic reproductions of black stereotypes that had existed since the days of slavery and were already popularized in American life and arts. The movies, which catered to public tastes, borrowed profusely from

all of the other popular art forms. Whenever dealing with black characters, they simply adopted the old familiar stereotypes, often further distorting them beyond recognition [pp. 3–4].

These characterizations continued in full force until around 1942, when delegates from the National Association for the Advancement of Colored People (NAACP) and the heads of several Hollywood studios met and agreed to abandon pejorative racial roles for blacks and to integrate them into several aspects of the motion picture industry (Cripps, 1977).

Another significant change of events in the middle 1950s could not help but have some impact on media as well as on society as a whole. This event was the 1954 ruling by the U.S. Supreme Court that segregated schools in the South were unconstitutional.

Following on the heels of the Brown decision came a new spirit among Afro–Americans and with it the bus boycotts in Alabama, the emergence of Dr. Martin Luther King, the continued work of the NAACP, the development of freedom marches by blacks and whites, and the push by a number of other civil rights groups. The Civil Rights Movement in this country was born and with it came a period of psychological vigor for blacks, in spite of the fact that some of the great gains were later to prove superficial and short-lived. Blacks could and did, however, move toward a stage of increased racial pride as they struggled with shaking the old images of the past in order to come to grips with a new sense of self to sustain them in the future. James Banks and Jean Grambs (1972) pointed out the internal struggle faced by the average black American when they wrote:

In modern American society we acquire identity from other human beings (significant others), and incorporate it within ourselves. A person in our society validates his identity through the evaluations of "significant others." However, the average black American has never been able to establish social or self-identity that is comparable in terms of social valuation to that of the white majority. The ideal self in America has been made synonymous with Caucasians, and particularly middle-class whites. In his quest for identity, the black man has begun to ask, "Who am I in relation to other races and ethnic groups?" [p. 7].

The 1960s were characterized by a burst of vitality in black consciousness and civil rights from which they have not retreated. It was also a sad and tragic period of assassinations., This was also a time when the cities across the country began to burn. Although the anger was unconsciously aimed at the "white power structure," the large portion of the death and destruction fell on the black community. The National Advisory Commission on Civil Disorder (1968) studied the problems and causes of the civil disorder, and concluded that "Race prejudice has shaped our history decisively; it now threatens to affect our future [p. 5]."

Unless the legacy of the past is reversed, white Americans will have succeeded in being fully socialized to the color distinctiveness of blacks. Conversely, there will remain in the experiences of black Americans this unconscious conditioning from the past, as well as the reality of the present, which tell them that there are human forces within the society that can be destructive of their accomplished gains in the present and of their future aspirations.

TELEVISION AND ITS PORTRAYALS OF BLACKS

If the early premises and myths of the superior–inferior legacy are still accepted on some level, it is necessary to work toward reversing that legacy through the major institutions in our society. The continued transmission of faulty racial attitudes and ideas can only reinforce the racist legacy and cause our institutions to maintain their retardative force on the humane growth and development of *all* of its citizens.

Consistent with the thrust of this paper, the television industry faces a special challenge in terms of its ability appropriately to communicate and portray the culture and life style of ethnic minorities in general and of blacks in particular. Television must studiously avoid the mistakes made by the film industry and other media in their early portrayal of Afro–Americans. Peter Noble (1969), drawing on the work of Dr. Lawrence D. Reddick, Curator of the Schomberg Collection of the New York Public Library, brought this point home vividly when he described the principal stereotypes of blacks that the American public has gained from the early movies and other media.

These stereotypes were as follow: the savage African, the happy slave, the devoted servant, the corrupt politician, the irresponsible citizen, the petty thief, the social delinquent, the vicious criminal, the sexual superman, the superior athlete, the unhappy nonwhite, the natural-born cook, the natural-born musician, the perfect entertainer, the superstitious churchgoer, the chicken and watermelon eater, the razor and knife "toter," the uninhibited expressionist, and the mental inferior.

There is, of course, little reason to believe that television program content will return to the stereotypes of the old cinema. Given the past and present view of Afro–Americans in our society, however, there is the potential that there will be created a host of new stereotypic characteristics that are "writ large" or "writ small" versions of old themes.

In order to establish linkages between the past legacy and present realities concerning the portrayals of blacks in television, it is necessary to consider systematically where we were and are in terms of these portrayals. Material on the portrayal of blacks in television is essentially of two types. One consists of content analyses—usually within a limited time period—of programs appearing on the major networks and of analyses of trends in content. The other consists of a large body of commentary and criticism regarding specific programs or the appearance of blacks on television. Most of the latter material is taken from the popular press and is often written by professionals in the field of television.

Content Analysis

The first body of literature—dealing with empirical content analysis over limited periods of time—provides a look at portrayals of blacks in a systematic manner. Unfortunately, content analysis was not much applied to television programs until the 1970s, and there are only a few studies before that time with which to make comparisons.

Two studies examine network television content as it existed in 1973. Hinton, Seggar, Northcott, and Fontes (1974) selected 133 programs shown between and 5 and 11 p.m. during February and March 1973. They analyzed only comedy and drama, excluding variety, news, sports, discussions, Westerns, cartoons, commercials, and quiz shows. They collected data on every black character and every ninth white character shown. Then the characters were rated as to their role attractiveness, hostility, morality, and dominance. Overall, 168 (53%) white and 149 (47%) black portrayals were analyzed. Though it was found that blacks were more often in less significant roles than whites (with black females most often ignored), blacks were portrayed somewhat favorably. Whites were more likely to be seen as violent, hostile illegal, or immoral. Hinton and colleagues (1974) conclude that charges of tokenism in television are justified and that "the favorable characteristics of blacks when portrayed in bit parts and minor roles constitute no threat to the world of the white man on television [p. 431]," because small roles showing black characters do not have much impact in changing the white–black status quo.

The second study, conducted in 1973 (O'Kelley & Bloomquist, 1976), analyzed 28 randomly selected hour-long segments aired between 7 a.m. and 12 p.m. between October 1 and 28. Each videotaped segment was coded for number of males and females, race, roles, and whether they were children or adults. Of the 2309 characters analyzed, 63% were male, 37% female, and only 4.9% nonwhite. The contents were also classified as to type of program, which revealed that nonwhites comprised 7.4% of the characters in adult shows but only 2.8% of those in commercials. One-third of the black portrayals were women. In news shows, nonwhites held 5.6% of the roles, whereas in accompanying commercials this group made up only 4.8%. The findings suggest that nonwhite groups are greatly underrepresented. The authors see this as part of television's reinforcement of societal stereotypes and conclude that "it is questionable whether... introducing more women and blacks on television would result in changes in the peoples' attitudes who watch television unless this change was accompanied by similar developments in many other aspects of life [p. 184]."

A study of "reality-type" programming was conducted by Roberts (1975) during two three-week periods in early 1972 and mid-1973. He distinguishes the news broadcasts he analyzed from the fantasy-like portrayals of blacks in entertainment programs, on which he earlier found that blacks were frequently shown with "super-human" traits (Roberts, 1971). In the news analysis, he found that blacks appeared often in stories dealing with racial matters on crime. Other frequent topics concerning blacks were the 1972 presidential primary, political activities, and Watergate. In contrast (Roberts, 1971) with the earlier study on entertainment programs, "blacks were often identified in occupations requiring technical skill or no skill but seldom in occupations relating to entertainment or sport [p. 53]." Blacks appeared in 204 of 874 news segments analyzed, with differences by network as to amount of visibility provided. Discernible occupations of blacks shown included skilled workers and technicians (21%), unskilled or semiskilled workers (18%), and occupations related to the law (12%). Roberts concludes that in the majority of newscasts (52%) blacks do not appear in speaking roles, are seen in blue-collar jobs, and often are shown in connection with civil rights issues.

Trends in television portrayals of blacks have been discerned in a larger number of studies. Northcott, Seggar, and Hinton (1975) compare the results of two studies conducted in 1971 and 1973. They place these findings in the context of past television research. A 1952 study is cited indicating that 10% of all characters at that time were nonwhite, only 3% of whom (as opposed to 16% of whites) were upper class. Also cited is a one-week monitoring of the three networks in 1970, during which blacks appeared on 46% of the programs and 10 percent of the commercials. Northcott and colleagues then reported on their own 1971 and 1973 work on drama broadcast on network prime time television. Their analyses of 96 half-hour units in April 1971 and 120 such units in February 1973 concerned frequency and occupations of black and white males and females. For males, no evidence of tokenism or stereotyping was found in 1971, but the 1973 data showed a shift away from professional and managerial portrayals to service portrayals. Females (of both races) showed a gain in appearances between 1971 and 1973, whereas frequency of black male roles declined. The authors attributed these changes to a shift in the influence from that of the Civil Rights Movement of the 1960s, reflected in the higher number of black portrayals in 1971, to that of the Women's Liberation Movement, spurring the appearance of more women in the 1973 data.

Trends in the area of television advertising were investigated by Gulley and Bennett (1976). They cite Dominick's and Greenberg's study (1970) of three years of prime time and daytime network commercials. The percentage of blacks appearing in ads rose from 4% in 1967 to 6% in 1968 to 10% in 1969. They also found, however, that those blacks pictured were in minor roles or in groups with whites. Greenberg and Mazingo (1973) found that the frequency had risen to 14% by 1973, but blacks continued to be part of groups or in minor positions. The Cully and Bennett study surveyed network commercials during January 1974. They reported 9.5% of all nonduplicate commercials contained a black character, but only 5.9% of all characters were black. They, too, found Afro-Americans mostly in background roles.

Most studies centering on trends in portrayals of black Americans do so within a somewhat limited time frame. This is due in large part to the lack of comprehensive studies analyzing the content of early television programs. Fife (1974) has drawn from both scholarly and popular sources to pull together a retrospective look at the evolution of the portrayals of blacks on television. She details the early appearances of blacks on variety shows, the 1953 debut of the stereotypic maid "Beulah," through the success of "Amos 'n Andy." She links the Civil Rights Movement to changes in programming in the mid-60s, noting that in 1962 the few blacks that were present on the screen were singers, dancers, or musicians. Bill Cosby became the first black star, but Fife suggests that the black identity of the international spy character was submerged. "Julia" debuted in 1968, but it certainly was "not the kind of thing that appeased militant minds [p. 13]."

Related Media Criticism and Commentary

A large amount of material on television content has emerged as part of scholarly criticism or comment. Cedric Clark (1969) theorizes that minority portrayals on television progress through four stages. The first, nonrecognition, is exclusion by the medium. He gives th Puerto Rican population as an example of a currently nonrecognized group. The second stage is ridicule, wherein the dominant culture's self-image is bolstered through perpetration of commonly held negative stereotypes. Regulation is the third stage. Minority group members at this stage are portrayed as defenders of the existing order—as policemen, government spies, or detectives. A final stage, respect, has not been attained by high-identity groups.

Newspaper and magazine clippings often deal with specific programs, generalizing toward the broader issues of blacks on television. These articles editorialize by criticizing or praising programs containing blacks and often use this with other examples to determine that black portrayals have either been favorable and beneficial or unrealistic and harmful. Sandra Haggerty, writing in the Los Angeles *Times* (1974), drew much reader reaction when she panned the stereotypes portrayed in "Good Times," "That's My Mama," and "Get Christie Love." A *Time* magazine article ("Commercials," 1968) derides television commercials for creating a new black stereotype. A Hollywood talent agent, Bill Cunningham, is quoted on the underlying explanation: [it] derives at least in part from the notion that white buyers won't go for actors who have very Negroid features [p. 83]."

Another person concerned with television's past use of black characters is Esther Rolle, the Florida of the series "Good Times," who was interviewed in the New York *Times* (1974) by Judy Klemesrud. She sees "Good Times" as a pioneer effort in portraying blacks realistically. She believes that "98-99%" of "Good Times" is reality, blasting the myth that black women are the backbones of their families. Rolle noted: The black man has had a hard time getting a job and his wife is told, you can't get welfare with a husband in the home... The myth is that she is the strong force in the family when the truth is she has *had* to be the breadwinner [p. 17, Sec. II, col. 1]."

Other writers have expressed views of television's role in extending the black experience. John J. O'Connor in the New York *Times* (1975) traces the recent evolution of black portrayals to the Civil Rights Movement and the discovery that "Blacks, too, purchased those soaps and detergents that white housewives had long been hawking in commercials [p. 1, Sec. II, col. 6]." He follows black performers from the early variety show appearances through dramatic themes such as the successful "Autobiography of Miss Jane Pittman." He acknowledges recent situation comedies, however, as the largest promotor of black visibility, discussing the to-the-minute topics handled in "Good Times." O'Connor plays up the positive aspects of exposure, however simplistic: "Never underestimate the power of being silly on television. Aside from being lucrative, it can be more influential than all the 'brotherhood' sermons delivered in several months of Sundays [p. 1, Sec. II, col. 6]."

Critics of television's characterizations of black Americans have contended that the absence of a male figure in some television program homes is harmful to the image of blacks. In an article defending programmers' decisions, J.K. Obatala (1974) sees enough positive characters to compensate for the lack of particular roles. He praises "That's My Mamma" for its strong images for the main characters, such as stable family conditions, a healthy attitude toward work and achievement, and a close relationship between the mother and her son, whose attitude is more respectful than emasculated. The view expressed by Obatalia of that show has not however been accepted by a number of other specialists and observers of television programs.

The advent of the television presentation "Roots" gained national attention when it aired eight consecutive nights to a growing audience that reached 36 to 38 million people, according to Nielsen figures. It triggered articles in most major newspaper and magazines throughout the country—and caused

America to think about the plight of black people in this country. In an article by Lee Margulies (1977) James W. Tisdale, head of the Writer's Guild's black writers committee, summarizes the effect "Roots" will have on ethnic stereotypes in television: "The American public has shown a willingness to accept truth—about Blacks, about American Indians, about Russian Jews, about the Irish, whatever [p. 1; 16, Part 4, col. 4]." The thrust of quotes by many industry personnel was that television can now explore previously risky subjects in a more direct manner.

Discussion of Blacks in Television

It seems there have been three major periods in the portrayal of blacks on television.

(1) The Stereotypic Age—1948-65. During this period blacks were shown on television either as guest performers on variety shows (Ed Sullivan, Arthur Godfrey, Milton Berle) or in roles consistent with the pre-Civil Rights Movement image of blacks. "Beulah" was the overweight Aunt Jemimah-type domestic servant. "Amos 'n Andy" were humorous yet ridiculous with their thick dialects and nuances of laziness and stupidity. The "Nat King Cole Show" of 1956 was popular yet unsponsored. Offending the white consumer, not support of the black consumer, loomed in the would-be sponsor's mind (Fife, 1974). The fledgling medium had not yet gained the power to take initiative; it focused instead on reflecting and placating the sensibilities of the dominant Anglo culture.

Toward the end of this period there was also a proliferation of black faces mixed into crowds in almost every commercial. Cyclops (1972) calls it "the obligatory black taken on every toothpaste and detergent commercial [p. 20]." Indeed, it appeared that programmers and ad people were compensating for their years of neglect.

(2) The New Awareness—1965-72. As the Watts riots brought black discontent to the nation's attention, television decision makers began to take the representation of blacks seriously. "I Spy" featured Bill Cosby in a costarring role, and "Julia" was the first all-black drama. In many programs blacks were mingled with white as regular supporting characters, usually with amazingly positive traits (Roberts, 1971). Blacks held such roles in "Star Trek," "Mod Squad," "Hogan's Heroes," "Peyton Place," "Mannix," and "Mission Impossible" within the period of 1965-72. Another program in the "Julia" vein, "Room 222" showed black professionals fully functioning and intellectually able. These two programs may have been an attempt to compensate for earlier stereotypes.

(3) Stabilization: The Settled Phase—1972 to the Present. As found by Northcott, and colleagues (1975) the gradual decrease in the number of blacks portrayed in shows and commercials was accompanied by the rise in women shown. The push to include a black in every group of three or more whites diminished as racial concern eased a great deal. Instead, producers concentrated on making purer all black programs. An effort was made toward being somewhat more realistic and focusing more on individual characters with personal concerns., Spawned from this attitude were Norman Lear's "The Jefferson" and "Good Times." Others include "Sanford and Son," "That's My Mama," and, substituting another ethnic group, "Chico and the Man."

There can be little doubt that the visibility of blacks in the television medium has been growing over the last 10 years. Along with that visibility have come some outstanding productions, such as "The Diary of Miss Jane Pittman" and "Roots." These programs described a phase of the black experience in America with clarity and understanding. Although it would be naive to expect that all programming involving blacks follow the genre of these presentations, they do contrast the depth involved in describing the experiences of blacks with the attitude of some other black-oriented shows that many Americans feel are descriptive of, and having something to say about, the black experience. Specifically, the several black-oriented situation comedies have a large audience of people of all races. Indeed, blacks are faithful viewers of these shows without noting that they include certain ingredients that portray blacks in ways that reflect some of the warmed-over stereotypes of old. (See chapters 11 and 12, this volume.) For example, a number of the situation comedies as well as other more dramatic presentations appear to have a predisposition for: (1) portraying blacks as coming from a single parent home, usually with the mother present; (2) portraying the black woman as an overly aggressive, dominant person able to intimidate the male through physical action as well as loud words; (3) utilizing nonstandard English with black regardless of their educational background and social experiences. Black dialect is often utilized to distinguish the "bad" characters from the "good." Thus the creators often do not show the beauty of the language but utilize it as a vehicle to highlight a special type of a character; (4) portraying the "black community" as a weak and destructive place and playing down its strengths and positive characteristics; (5) portraying the pimp, drug-pusher, and similar characters as individuals prized and even respected by the black community; (6) portraying Afro–Americans as a homogeneous group (a form of stereotyping) and failing to identify the heterogeneity of their culture and life style.

Three observations might be made to account for what appears to this writer as being some general reasons for the high level of acceptance by concerned black and white Americans of the present portrayal of blacks. (1) Viewers see that there are a large number of black professionals on television in commercials, dramatic presentations, and the so-called black-oriented shows, and they conclude that things are getting better from an exposure and employment point of view; (2) viewers no longer see the overt and heavy stereotypes of the "teeth chattering," "shoe-shine-dancing," and "feet, don't fail me" black on television, and the conclusion is that there has been great improvement in the black image; (3) Caucasians note that black viewers apparently support the present portrayals because many of the regular television black stars are held in high esteem inside and out of the black community.

One can find in the minds of both racial groups shades of the old legacy still functioning, and for blacks alone some deep-seated sociopsychological needs for identity and self-esteem. First, there is clearly an increase in the number of black actors and actresses compared with the early days of television. Looking at the general employment picture in today's terms, however, one can find some slippage in the job opportunities for minorities and women. According to Paul Bullock (Los Angeles *Times,* 10/22/76) of the UCLA's Institute for Industrial Relations, employment in the industry has disintegrated into a very dismal pattern across the board, and black employment fell from 9.3% to 6.6%. In that same article, Bernie Casey (1976) of the Screen Actors Guild's Minority Committee told the California Advisory Committee to the U.S. Commission on Civil Rights that blacks and other minority actors "don't feel we get a fair opportunity to compete for roles that don't have a specific racial origin [p. 2]," (Los Angeles *Times,*). Most black and white viewers are, however, more interested in the increased presence of certain groups on television, and this visibility is often enough to convince them that the situation of minority participation in the medium has improved a great deal.

Second, it is a fact that the early media portrayal of blacks is not part of television today. In an intergroup and racially fragile society, however, care must be taken in order that a new set of modern-day ethnic portrayals will not emerge to replace the old stereotypes. Unlike yesterday, however, new stereotypes can be communicated to 30 or 40 million people in one evening.

Third, it is easy for one to document that Afro–Americans are major supporters of black-oriented shows and of those black professionals who play roles supposedly related to the black experience. I would argue that this interest is partly related to a basic psychological need for black Americans to see, hear, and identify with people from their own racial groups who have made it as part of a prestigious medium such as television. Although many of the protrayals are not what black viewers would consider positive representation of their culture and life style, they realize that some blacks being employed and some identification with one's racial group is better than no identification at all.

However, to take a position that blacks have a need to identify with characters, places, sights, and sounds of their racial group is not to suggest that they are uncritical viewers. Indeed, black Americans listened to the radio and watched on television the "Amos 'n Andy" show until it was pushed off the air by pressure from black organizations, the young, and some social activist groups.

Black people continue today to be heavy viewers of black-oriented shows, often without fully appreciating the nature of the images they are presenting to their children and other groups. One runs a special risk criticizing some of the characteristics that emerge from black-oriented programs, because it can be demonstrated statistically that large numbers of blacks from all social classes look at and enjoy them (Yoder, 1978). Although there are some positive aspects to these programs, black and white American should not be *seduced* into believing that they are getting a true sample of the culture and life style of blacks and that therefore these programs are serving a "social good." To assume that pimps who report to the police all of the community activities; down-and-out business people who are ready to make a fast buck; loud-talking emasculating women; wealthy but bigoted uptown folk; and a group of hip-talking and finger-popping adolescents around a soda shop are portraying the experience of blacks is simply inconsistent. Indeed, many of these characters present the type of image that can increase stereotyped thinking in whites and further demonstrate to blacks that their community and life style are to be devalued and disorganized.

IMPLICATIONS FOR RESEARCH

Methodological Considerations

By its very premises this chapter has taken the liberty of utilizing broad strokes in an attempt to raise questions about television and its portrayal of Afro-Americans. These observations are being made at a time when decision-makers, artists, and other creative people in television are most likely feeling good about the improvements by the industry. Indeed, there has been progress over the years in some aspects of black portrayal. This chapter has, however, asked that we continue to look backward into history and at the present in order to ascertain why black Americans are portrayed as they are today. Feeling and hoping that the professional decision-makers in the television industry will continue to engage in self-introspection, I should like to direct some issues to the attention of social scientists.

Social scientists have traditionally turned much of their inquiry to black–white comparisons in terms of the different ways these groups use television and its impact on their aggressive behavior. A number of the methodologies employed have generally relied on classic research design. In this regard, the observation made by Bronfenbrenner (1976) about educational research is also applicable to television and social science inquiry. He stated that contemporary educational researches are characterized by experimental designs that are primarily statistical rather than scientific; that is, these designs enable us to predict the concomitants of certain combinations of conditions but not to understand the causal connections that produce the observed effects (p. 159).

Television is often spoken of as if it were a single and simple variable. However, as soon as one attempts to utilize it as a variable in research, one is confronted with the problem of accounting for the complex relationships between content, lighting, color, music, voice, and the type of character or model being used to develop the measure. Should we want to build some of these causal linkages among those factors that constitute positive and negative portrayals of minority groups, we will need to understand some of the subjects' prior social learning and behavior concerning race as well as to identify the multiple impact (color, voice, role model, etc.) the medium is having in the research effort. Children and adults are not one dimensional, but neither is television. There is a clear need, therefore, to build models capable of handling multiple factors and variability inherent in television along with the socialized learning experiences of the individuals who participate in the research efforts.

Consideration of Research Issues

In regard to television, race, and attitude formation, some issues deserving attention can be identified and a series of questions suitable as a point of departure can be posed. In a broad sense these issues relate to television's role in the formulation of ethnic identity and to the effects of ethnicity on the way the viewer perceives and evaluates the experiences provided by television (Mitchell-Kernan, 1976).

Television's role in the formulation of ethnic identity. The major question within this area is: How does television utilize cultural symbols to communicate messages about ethnicity and its meaning, as well as how do viewers interpret these messages? Some minor and related questions which follow are: What social roles and social identities are allocated to members of different ethnic groups? Can ethnic selectors be identified in terms of role allocation? What are the symbolic uses of accent and dialect features? What are the means of conveying solidarity, distance, superiority, equality, and inferiority? Are these means ethnically organized? What is the structure of hidden messages regarding the meaning of ethnicity in terms of social role, character, personality, and values? What background factors inhibit versus promote generalizations about entire ethnic groups on the basiş of information conveyed about specific individuals via television?

The effects of ethnicity as a sociocultural variable shaping the way the viewer perceives, conceptualizes, and evaluates the set of experiences provided by television. The major questions within this area are: Attention to a certain degree is stimulus compelled. Does ethnicity channel attention and dispose children and adults to attend to different stimuli and messages conveyed by television? What are the differences in reality construction within and across ethnic groups?

Another set of corollary research foci can be raised that also relate to the psychosocial development of children within a given sociocultural context. (1) What is the role of television in shaping the values, beliefs, attitudes, and actions of the minority child? (2) What information is given through television communication channels that might influence the child's feelings, attitudes, and behavior toward his or her ethnic groups and the broader community? (3) What role—subtle or overt—does this medium play in shaping the social behavior of minority children? (4) What role does television play in shaping the child's view of himself or herself? (5) How and in what ways does television convey to the child majority group attitudes toward his or her ethnic group? (6) What is the impact and importance of this medium in the lives of minority groups? (7) What are television's strengths, weaknesses, and values to ethnic minorities? (8) What improvements can or should be made in this medium for minorites?

The reader will note that a number of the broad issues and narrow questions have implications for racial attitudinal development for nonwhite and white groups. Although the focus is on Afro–Americans, such foci for research activity are consistent, because as it has been noted throughout the chapter, blacks have essentially been the receivers of what might be called a "culture of racism." and this message has had a debilitating impact. The race problem in America, however, is in a real sense a "white problem," and it is crucial that research go forward to identify how a medium such as television communicates the inferior–superior legacy of the past.

CONCLUSIONS: SHALL THE LEGACY SURVIVE?

This chapter argued that the portrayal of black Americans on television reflects and reinforces the perceptions and understandings that white Americans have of black Americans and of their culture and social status.

Similarly, the portrayals help to shape the self-images and behaviors of blacks. It is suggested that these perceptions are the results of a historical set of socializing experiences.

Similarly, the chapter has made a case for the fact that the past and present socializing experiences of black Americans predispose some of them to believe some of the television portrayals. Both the perceptions of blacks toward the way they are portrayed and the acceptance of whites of these portrayals are a result of a historical set of socializing experiences.

It is important to pause long enough to note that the issues goes beyond one race and how many blacks are shown on television. The real issue is what are we communicating to all children and what the implications of our messages are for building a more humane and just society. As Pierce (1974) points out, "Television and other forms of mass media, more often than not, see to it that blacks are portrayed in ways that continue to teach white superiority [p. 514]." He further suggests that although more blacks are presented regularly in nonmenial as well as menial roles on television, the way they are generally presented has immeasurable importance in keeping them in a reduced status (See Chapter 11, this volume.)

This observation is especially important for black children, because they gain a view about themselves from their total environment and are also influenced by their adult models. As one article ("Racism and Sexism," 1975) stated:

> Children learn their place in the world not just from home and school but from the larger society, and this knowledge is reinforced through toys and books, movies and televisions. No child is too small to pick up clues—intentional slurs and subtle suggestions—that just as surely tell what blacks, native Americans, Chicanos, or just plain men and women are and do—and thus what they can't become or can't accomplish [p. 21].

Nor are the Anglo children immune from consequences of the faulty television portrayal of blacks and other minorities. After all, they must also live in a multiracial, multicultural society, and their formulation of inappropriate racial attitudes concerning other people is merely laying the groundwork for unwholesome attitudes, proracist feelings, and future anxieties when they are adults.

Finally, another underlying message of this chapter has abeen the concept that both black and white Americans need to see successful children and adults in order to have a model for present and future social and psychological development. "Successful," in this context, does not necessarily mean highly educated and middle-class individuals. Rather, it means individuals who are strong and capable people who are taking charge of their own destiny and doing it with pride in their cultural heritage.

Both black and white children need programs that are bold enough to show the strengths of black history and a cross section of their culture and life style. They need programming aimed at showing adults and child models who have human foibles and real social problems, but the viewers also need to see fully capable, humane, and cognitively functioning black Americans. Both black and white children need to learn that this is a pluralistic country that has no "model American." They need to understand from television that to be racially, religiously, or ethnically different does not make one "disadvantaged," "deprived," or "inferior" and that it is important to take pride in one's unique customs, skin color, language, and life style.

The showing of positive characters and sound content will help young black children to identify with the positive aspects of their culture. Such television content will communicate to children that it is just as important to be able to *read* and *write* as it is to *rap*, because with such tools, they can direct the course of their lives.

Television is one of the instutitions in our society that has a chance to contribute to the demise of the superior–inferior legacy. Although it cannot be a miracle worker, it is a major reflector of the views of society as well as being able to free and shape our attitudes. Thus television's handling of Afro-Americans will only be free of destructive portrayals when the attitudes of black and white Americans are free. This writer notes that there is some light at the end of the tube. Let us see it soon.

ACKNOWLEDGMENTS

The author would like to acknowledge the assistance of Diane Elvenstar in the preparation of this paper.

REFERENCES

Adler, R. Introduction: A context for criticism. In D. Carter (Ed.), *Television as a cultural force.* New York: Praeger, 1976.

Aptheker, H. *Afro-American history: The modern era.* New York: Citadel Press, 1971.

Banks, J., & Grambs, J. *Black self-concept: Implications for education and social science.* New York: McGraw-Hill, 1972.

Bogle, D. *Toms, coons, mulattoes, and bucks: An interpretive history of blacks in American films.* New York: Viking Press, 1973.

Bronfenbrenner, U. The experimental ecology of education. *Teachers College Record,* 1976, 78(2), 157–204.

Butcher, M. J. *The Negro in American culture.* New York: Knopf, 1973.

Clark, C. Some observations on the portrayals of ethnic minorities. *Television Quarterly,* 1969, 8(2), 18–22.

Commercials: Crossing the color line. *Time,* October 25, 1968, pp. 82–83.

Cripps, T. *Slow fade to black: The Negro in American films 1900–1942.* New York: Oxford University Press, 1977.

Culley, J. D., & Bennett, R. Selling women, selling blacks. *Journal of Communication,* 1976, *26*(4), 160–174.

Cyclops. Black sit-com but too tame. *Life,* April 21, 1972.

Dominick, J., & Greenberg, B. Three seasons of blacks on television. *Journal of Advertising Research,* April, 1970, pp. 21–27.

Fife, M. D. The black image in American TV: The first two decades. *The Black Scholar,* 1974, *6*(3), 7–15.

Greenberg, B., & Mazingo, S. Racial issues in mass media institutions, communication among the urban poor. (Project C.V.P.) Report 16, Department of Communication, Michigan State University, 1973, pp. 27–33.

Haggerty, S. TV and black womanhood. Los Angeles *Times,* November 6, 1974.

Hinton, J. L., Seggar, J. F., Northcott, H. C., & Fontes, B. F. Tokenism and improving the imagery of blacks in TV drama and comedy: 1973. *Journal of Broadcasting,* 1974, *18*(4), 423–432.

Klemesrud, J. Florida finds good times in Chicago. New York *Times,* May 5, 1974.

Margulies, L. "Roots" uproots television stereotypes. Los Angeles *Times,* February 2, 1977.

Mitchell-Kernan, C. *Toward a conceptual focus for the study of television and ethnicity.* Unpublished manuscript, Center for Afro-American Studies, University of California, Los Angeles, 1976.

Noble, P. *The Negro in films.* Port Washington, N.Y.: Kennikat Press, 1969.

Northcott, H. C., Seggar, J. F., & Hinton, J. L. Trends in the portrayal of blacks and women. *Journalism Quarterly,* 1975, *52*(4), 741–744.

Obatala, J. K. Blacks on TV: A replay of Amos 'n Andy? Los Angeles *Times,* November 26, 1974.

O'Connor, J. J. Good times in the black image. New York *Times,* February 2, 1975.

O'Kelley, C. G., & Bloomquist, L. E. Women and blacks on TV. *Journal of Communication,* 1976, *26*(4), 179–184.

Pierce, C. Psychiatric problems of the black minority. In G. Caplan (Ed.), *American handbook of psychiatry.* New York: Basic Books, 1974.

Pinkney, A. *Black Americans.* Englewood Cliffs, N.J.: Prentice-Hall, 1975.

Racism and sexism and children's books. *Crisis,* January, 1975, pp. 21–23.

Report of the National Advisory Commission on Civil Disorders. Washington D.C.: U.S. Government Printing Office, 1968.

Roberts, C. The portrayals of blacks on network television. *Journal of Broadcasting,* 1971, *15*, 45–51.

Roberts, C. The presentation of blacks in television network newscasts. *Journalism Quarterly,* 1975, *52*(1), 50–55.

Yoder, J. T. *Perception of Cleveland. Blacks to network black dramatic programming.* Paper presented at the meeting of the International Communication Association, Portland, April 1978.

11 Social Trace Contaminants: Subtle Indicators of Racism in TV

Chester M. Pierce, M.D.
*Harvard Graduate School of Education
and Harvard Medical School*

It is the purpose of this chapter to draw attention to subtle but, to me, predictable presentations on television. It will be argued that these ubiquitous subtleties both mirror and mold society's behavior in the realm of racism. I pose the question whether or not television contributes to racism by teaching or reflecting what the society believes is probable, typical, or desirable in interracial behavior.

In this consideration I utilize two conceptual models from medicine. On the one hand there is a concept of "trace elements." These are elements found in very small amounts in the body, yet they are critical for survival. Minute disturbances of the ever-present trace elements can be disruptive to health and life. For example, manganese or zinc depletion in an alcoholic can abbreviate life.

Another concept is entitled "trace contaminants." In this scheme it is appreciated that there are elements which abound in microscopic quantity and ordinarily are without demonstrable influence on the organism. However, when such an element is continued in the environment *over a period of time* there is serious cumulative action. Indeed such accumulation can kill the organism. For example, a very small but continuous amount of a carbon compound will do no damage to an animal on Day 1 or even on Day 56. Yet by Day 90 the animal may abort or even die as a consequence of the ceaseless exposure to the compound.

Television provides "trace elements" that to the white viewer may aid and sustain his behavior in interracial contacts. Likewise television provides "trace contamination" to the black viewer that may be debilitating or disabling. Trace elements or contaminant analyses require sensitive chemical assays. Holding with the metaphor, similar sensitivities will be required in the analyses of television messages that have potentiating effects. If this theory of social trace contaminants has validity, it permits a new type of discussion relative to enlightened television production, informed media consumption, possible educative strategies to combat racism, and novel psychological and physiological investigations. Yet prior to elaborating the theory or its application, it is necessary to make some operational definitions of racism as it exists in the United States.

In the United States, anyone who is colored will suffer discrimination. American culture assumes not that all people are equal but that people of color or lower classes are in some way inferior, inadequate, and immature. Colored people, especially blacks, are a source of work or entertainment (and more recently social concern and distress). They are regarded normatively as physical creatures who attend primarily to physical outlets and who are by definition uncomplicated.

Despite written law to the contrary, black existence in these United States is governed by sociocultural principles that are at once comprehensive and only stubbornly yielding. As a result black–white interactions, particularly at subtle levels, can be predicted with a statistical certainty that would astonish our colleagues in the physical sciences. The gross, overriding rule is that a black must always be verified as being inferior. To accomplish this, blacks and whites must agree to operate in such a manner that both participants expect and accept as unremarkable the proposition that the black's time, space, and energy always will be controlled by the white. For example, as two people approach a glass door, from opposite directions, the white, by nonverbal signal and agreement, is statistically more likely to go through the doorway first. Thus the white's time, space, and energy are given priority. The preservation of energy, time, and space is made possible by the black at a cost to his own time, energy, and space. Millions of such contacts accumulating over years may actually contribute to health problems of the black as a result of sustained stress. Similarly, the relief from such stress afforded the white may actually promote longevity.

For the white, the door-opening example was a social trace element situation. His survival was enhanced a bit. For the black, however, the situation added another noxious unit whose cumulative weight might enhance morbidity. Both participants have expected and accepted as unremarkable this life-abbreviating phenomenon! Both have agreed to it and have not questioned its gratuitous nature or thought about the automatic preconscious assumptions required to accomplish the interchange.

Probably, on any given day there will be, by a factor of millions, more of such subtle interactions than of gross interracial violence. The remainder of this chapter describes the way that television promotes, sustains, and insists that these subtle rules of racism reign. Television, although far from unique, is obviously a major vehicle for carrying the cultural cues upon which humans learn how to behave.

HOW A BLACK WATCHES TELEVISION

The examples cited are an attempt to indicate to the reader how ubiquitous and incessant and predictable these subtle descriptions are in reflecting and shaping American racial attitudes. As will be seen, many of the descriptions could easily be subjected to objective count.

Prior to presenting examples, I remind the reader that the examples that follow may seem to discount many of the large gains made by blacks in society and in television programs. These gains have been briefly described in the previous chapter. It is their existence that makes the occurrence of the following trace elements particularly noticeable and significant. Under gross exclusion or mistreatment one is not sensitive to nuances and subtleties. The argument that television is much more sensitive to negative racial stereotypes than it was in its early commercial years is not an issue. Nor is it an issue about how many blacks or other coloreds are shown. More important, it should be realized that it is easy to confuse a "good role" (such as a black being an executive) with *how* the black is presented in the role. The role may speak loudly in terms of, say, status and dignity but still, on subtle levels, operate as an essential social trace element for a white viewer and as a social trace contaminant for a black viewer.

What does a black viewer see as he watches television? What does he think as he watches television? What does he feel as he watches television? Because I have no data my propositions may be speculative. They are based on a sample of one. The sample, though irreducibly small, does, in my view, show the general spirit and thrust of millions of colored viewers. Some typical programs will be presented. These are programs whose audience is counted in the tens of millions.

Sports Programs

The subtle, stunning, repetitive event that many whites initiate and control in their dealing with blacks can be termed a microaggression. Any single microaggression from an offender to a defender (or victimizer to victim) in itself is minor and inconsequential. However, the relentless omnipresence of these noxious stimuli is the fabric of black–white relations in America.

For instance, this year during the end-of-the-year football games, technology allowed millions to hear the referee meet the team captains at the beginning of a game. The referee selected a white tri-captain to be the spokesman for his team, to indicate "heads or tails." Then after the coin was flipped and it could be awarded as a memento—of an undoubted high point of a football player's life—the referee gave it to a white. Even when there was an extra souvenir coin awarded to the side that lost the flip, a white (not black) tri-captain received this gift. This was despite the odds, because four of the six tri-captains were black.

In some cases, therefore, millions approvingly watched a society that has moved to such an extent that blacks not only play on all teams but are captains on these teams, at both a college and professional level. But the subtlety still operated differentially, so that whites were in control and took the rewards and did the talking and thinking in front of the television audience.

Sports fans, black and white, accept as unremarkable a discussion by commentators that gives a disproportionate attention to black errors while minimizing those of whites. A reader surely will protest that this is not the case. Yet in preparing this essay I watched a basketball game. The commentators of course praised blacks (because in fact whites were present in the game only as a minority). Yet a white superstar, who was giving a mediocre performance, was the subject of most of the discussion by the announcers. Their subject matter, incredibly enough, dealt with how great the man had been in last year's playoff games. The point the audience received over and over about this man was how "intelligent" a player he was. In fact, the viewer knew that a black was the game's most consistent player on both offense and defense. Yet except for dry recitations about his performance that day, he was not lauded.

After this game there was a program of vignettes to eulogize past stars in the National Basketball Association. Two players were presented that day—both white, even though as all schoolboys know, any white who makes the team in that league has to be somewhat of an exception. An announcer was interviewing a white former superstar. Throughout the short interview the announcer asked at least 10 times if the white didn't believe a black superstar of his vintage really didn't want to play and was jealous over the money made by whites. To the credit of the white superstar, he diluted and undercut these questions, yet the announcer persisted so that viewers would assume that the black superstar was not a team player and was motivated by greed and jealousy. In a pregame collage about the National Football League a black viewer believes that blacks are shown more often getting hit, being entertaining (doing a dance at the goal line), or making comical blunders (like chasing a ball that keeps slipping out of hand).

During the last Olympics, some nights when the coverage ended and America's medal winners were summarized and pictured, black gold-medal winners would be passed over. One medal winner, unlike white medal winners, was obliged to share that supreme moment of his life with a white. In fact the white, who placed second, talked first, last, and most to the television audience explaining proudly how and what "we" did.

During another meet the heralded race was the one mile. It featured two world-class white athletes. Both were given extensive prerace television coverage about their backgrounds, practice methods, past victories. As it turned out a black beat them both, running the third-fastest time in history for that particular event. Despite this achievement the postrace show featured a white who placed second (beating the two celebrated participants). The audience never got a close look at or a word from the black even though he was available to the cameras.

As a black watches sports on television he observes over and over that blacks have difficulty appearing to do something well. If it is done well, the promise is modulated or more likely shared with a white who has not performed as well. The black thinks, therefore, that the general society makes achievement more unlikely and more strenuous for him and that the rules are applied unevenly. The feelings brought forth by these observations are chagrin, anger, and resentment.

Adventure Programs

One of our colleagues reported during our committee sessions that "Kojak" was the most popular television program in the world. What do the citizens around the world, as well as American citizens, see about race relations in the United States when they watch "Kojak"? The subtle factors are present: A black must be controlled, taught, directed, helped. Often the black is a helper to whites. He must and should be dependent, grateful for white help, and never truly in command. At other times the black is lawless and violent but can't prevail.

In the program I chanced on for this chapter, a black is a former athlete who finished his ring career because he "took a fall" dishonestly. Along with a black, a demented cowardly bully, this ex-boxer takes three white hostages. However, the white male priest and two white females are noticeably more brave and unperturbed than the men who terrorize them. The blacks are more distraught than their unarmed captives. In fact, the white captives calmly state of the blacks, "They're out of control."

A black policeman is put in charge of the case, but he too is depicted as too emotionally labile to be able to take control. A white underling is obliged to disobey him. Finally Kojak, who is of comparable rank, comes to help the black officer. Kojak gives him orders and tells him what to do.

The story demands that the black ex-boxer be killed by the black cop, without any damage being done to the white hostages. As the ex-boxer lies dying in a gutter, he begs the white man, Kojak, to undo his failure. Kojak promises to help the dying man's son, who is in prison and angry at his father. The death takes place as the black apologizes to the white women, and the white priest renders prayers.

Here countable subtleties that can be predicted in black–white behavior include that the black must get white help, that whites don't have to take commands from a black either on the job or even when a hostage, that whites can fix up the damage that the black ineptitude has wrought, and that blacks are in need of containment. In another "Kojak" episode a black athlete (who was both a dope addict and ex-convict) pimped for a white boss and even beat and killed a black woman for this boss. A white said to this black, "You belong to me."

The second adventure I report gives viewers the rules our society has about black–white, male–female relations. Simply stated, the rule says that white males should have access to both white and black females and that black males should be excluded from white females and be shown as inept with black females. This, too, is a researchable issue and could be counted in content analyses of television programs.

In this story there were three black males and three white males. Of the three blacks one was strong but stupid. A second was cuckholded (by a white) and frightened. The third was comical and dependent. All three white males talked to white females. No black talked to a white female. Only white males had success with females. In fact, the one black who had any evidence of female favors turns out to have been betrayed and duped by his beautiful black wife, in partnership with her white lover. No perceptive black man would expect a content analysis of television to indicate that a white wife ever betrayed her husband for a black lover.

The Black Everyman, in his contemplation, would realize that this program would predictably give many and constant references to how whites think in a superior way to blacks. A content analysis would show that blacks are never portrayed as out thinking whites. Indeed, if blacks were permitted to be shown thinking, it would be as in the program described. Here blacks were familiar with the devious, the criminal, the illegal, when they were shown cogitating or planning. The black characters were calculating, cunning, and artful in such areas as forgery. Or they were tipsters who could be of valuable assistance to the police.

During the program the white partner, in a detective business with a black, was the one who thought out how to get new clients, how to foil captors, and how to outwit their enemies. The black partner was useful as a person to whom the white could order, "Get the police," or as the member of the team who could be shown as less brave and more easily ruffled.

A black man in this particular story was *extremely* fearful. For a good part of the show it seemed the basic rule was being ruptured, for this frightened man seemed to have thought out a clever plan to obtain money. Yet by the middle of the show, the rule had asserted itself in orthodox fashion. The black man had not thought out the plan, but rather his beautiful black wife had made it up. Further, she had duped him in the process so that she could flee with her white lover. Other kinds of material helped reinforce society's idea of proper conduct and etiquette for a black in the presence of a white. For instance, when a black was endangered by a white, he produced a gun from his stocking but promptly gave it over to a white to do the serious protection of their lives.

Therefore in terms of objective analysis one could, presumably, make up rules that would predict black–white behavior. Then one could count the frequency and distribution of these rules. The view of blacks would be that racism (the assumption and presumption of white superiority) would be defined with arithmetic precision.

The hypothesis from this presentation is that depending on the perceptions of the viewer, the same subtlety would be a social trace element fostering confidence or a social trace contaminant with cumulative deleterious effects.

News Programs

Racism is as apparent on news programs as in any other facet of television. In the newscast I watched for the purpose of this essay, subtleties of commission and omission were exhibited.

So far, I have analyzed the programs in terms of what was presented. The news segment brings a forceful reminder that some predictable subtleties that the black viewer sees, feels, and thinks about are subtleties of omission.

On a Sunday night newscast the white anchor man was absent. In his stead was a lovely black woman. The armchair thinker catalogues quickly all the sites where blacks replace whites, so that whites can have time at more convenient locations in order to conserve their energy and be at liberty. Thus in our hospitals a disproportionate number of blacks can be seen at work at nights or weekends. The same pattern holds for even modest jobs like doormen. On weekends the doors are more likely to be manned by blacks and Puerto Ricans. The beautiful black newslady is in the same tradition.

Another omission is demonstrated by this vignette. This is that when whites choose to "advance" a black they seem more willing to push a black female than a black male. Thus a count of blacks in certain television jobs would lead the black viewer to expect that compared to whites, more black females were present than black males. The importance of this white behavior cannot be overstated. It "proves," among other things, that a black male is presumably not prepared, capable, or available. This encourages divisiveness in black family life, and it discourages black males. The rule has deep historical antecedents in the United States. It is observed with such fidelity that even on a subway at 5:30 a.m. the workers in transit are white males and black females. The black females are going to or coming from janitorial duties in skyscrapers or domiciles in the suburbs. The black male who could handle many of these jobs is enervated and reduced to joblessness, while his mother or wife earns the meager livelihood that is available.

Once the newscast begins, however, the black viewer has several other predictable omissions to negotiate. For example, the program indicates that the U.S. government has been bribing key officials in a number of countries. Several of the countries were white, and we are told that all the spokesmen from those countries had denied the charge. However, no comment is made about leaders in several black countries. Viewers must themselves ask if the blacks indeed did take the bribes.

There was one more item relating to blacks. In this segment we are shown a darling little white girl who for some reason wishes to be a boxer. Identification is built with this little girl as we watch her demolish male opponents. Finally there is a villain—a black villain. The little girl is defeated by a black boy. But the significant subtlety presented in this segment is that we are told that the boy won not by being able to take the offense, not by carrying the fight with initiative to his white opponent; we are told his victory is a result of "counterpunching." Finally, in the program that was viewed, the news terminated when a white man said it was over, despite the presence of the black lady anchorperson. It occasions surprise if a white even in a coequal situation does not give the overt signal that now the action is over.

ADVERTISING BEHAVIOR

The sorts of indicators under scrutiny are perhaps seen most clearly in advertisements. Public service advertisements as well as commercial ads can be studied. A not atypical public service ad shows several children reciting the virtues of getting their eyes examined. One child says he can read more. Another says grades are better since the examination. The black girl smiles and states that getting your eyes tested "is fun."

Several predictable subtleties are illustrated. As a social trace element the white and black viewers see that whites have positive, intellectual, "thinking" reasons to have their eyes tested. The blacks, viewers are taught, are concerned about their own health because it "is fun." Further, the black child reveals more teeth and wider-open eyes. The whole area of nonverbal cues is of huge importance in discussing television, because most interpersonal communication cues are nonverbal and kinetic.

Commercial ads provide numerous opportunites to predict subtleties. Thus, for instance, studies could be made to ascertain whether or not blacks are seen as often as animations, puppets, or animals. Do blacks in advertisements work more often for wages? Do they live as often in the suburbs? Do they serve more than they are served? Do they have positive, enjoyable contact with others as often as whites? Do they groom themselves as often and show other signs of liking themselves as much as whites? Do they mention family as much as whites do? Are they depicted consuming food or drink more frequently than whites? Do they teach whites as often as whites teach them? Do they seem to provide more entertainment for whites than whites provide for them? In the ads are blacks more likely to be "physical" in terms of such things as being sexual or athletic? Do blacks have as much knowledge and command of technology and science? Do blacks initiate action as often as whites? Do they travel as much as whites?

If these questions are answered negatively for blacks, the depictions are social trace contaminants. If such instances occurred by chance in about the same proportion as do blacks in society, they would not be microaggressions. But if they are disproportionate, then the relentness nature of the latent messages keeps blacks mobilized in constant defense against the covert content on commercial television ads.

CONCLUSIONS

It is never possible to isolate television from other influences in human existence. Just in terms of media influences it should be reinforced that television has ample help in shaping racial relations. The other media provide equal cases of interracial subtleties.

Suffice it to say that newspapers, billboards, radio, statues, paintings, murals, magazines, periodicals, all contribute the kinds of subtleties described. On manifest and latent levels, both blacks and whites have developed a long list of subtleties in the etiquette of racial relations. If these subtleties are exposed and documented they could be made known to consumer and producer alike, so that all media messages would be processed and made from a different base of understanding. All people in our society could be taught from early childhood not to expect and accept as unremarkable that these nuances in human behavior do exist. Perhaps the greatest such inquiries could have would be that they might also lead to the more refined medical study of mundane stress. In that way one of the society's social contaminants, racism, could be converted into socially useful benefits, at the same time that people are learning to live better with one another.

12 Psychological Effects of Black Portrayals on Television

Sherryl Brown Graves
New York University

INTRODUCTION

When television was first introduced, questions were asked about the extent to which time spent viewing television displaced other activities (e.g., reading, playing, and talking with family members) (Furu, 1962; Himmelweit, Oppenheim, & Vince, 1958; Schramm, Lyle, & Parker, 1961). More recent attention has focused on television as a potential socializer of children (Leifer, Gordon, & Graves, 1974; Stein & Friedrich, 1975). The primary evidence for the socialization power of television concerns the relationship between televised violence and subsequent viewer aggression (see Comstock, 1975; Liebert, Neale, & Davidson, 1973; Surgeon General's Scientific Advisory Committee on Television and Social Behavior, 1972). As a consequence of the investigation of the impact of televised violence, researchers have further documented the medium's capacity to influence positive social behaviors, such as sharing, self-control, and positive interpersonal interaction (Coates, Pusser, & Goodman, 1976; Leifer, 1974; Paulson, McDonald, & Whittemore, 1972; Shirley, 1974).

Similarly, attention is beginning to be paid to the nature and impact of television presentation of various groups within our society. Given the unique historical circumstances of the introduction of Africans into American society, and given the development of racist institutional mechanisms for relegating black Americans to a position of second-class citizenship, the media became suspect as one of society's instruments for maintaining racist perspectives.

After the riots of 1967, President Johnson's National Advisory Commission on Civil Disorders (1968) made the following indictment of the mass media:

They have not shown understanding or appreciation of—and thus have not communicated—a sense of Negro culture, thought or history.... If what white America reads in the newspapers or sees on television conditions its expectations of what is ordinary and normal in the larger society, it will neither understand or accept the black American.... Television should develop programming which integrates Negroes into all aspects of televised presentations.

Four years after this indictment, the Congressional Black Caucus (1972) issued the following statements:

1) the social and occupational progress of blacks was being hindered by the negative stereotypes appearing on television;
2) negative stereotypes teach self-hate and consequent destructive self-images to blacks;
3) negative stereotypes create and reinforce the myth of white superiority.

Finally, an independent film producer notes:

Most inportant, he doesn't gain from the one-eyed monster constructive information about himself and his people, information that will enable us to survive in a generally though often subtly hostile white environment. We are prevented from communicating among ourselves over the airwaves, a privilege lavished on White America [Greaves, 1969].

These various statements from black and white leaders clearly indicate the depth and breadth of concern over television's role in the creation and maintenance of prejudice and racism. It is this concern about television's social impact that has in large part generated research on the portrayals of black people on television.

The Nature of Current Research

The most noticeable characteristics of the available research on the effects of black portrayals on television is its paucity. In an extensive bibliography of 1341 key studies on television and human behavior (Comstock, 1975), only 51 have been labeled "Television Minorities and the Poor." This includes studies of how minorities and the poor use television, the effects of television, and analyses of programming especially relevant to them. The majority of the research is survey research about how minorities and the poor use television. There is very little on the direct effects of content featuring blacks on black and white audiences.

The available effects-research does not present a clear picture of the nature of the impact of the portrayal of blacks on television. Although studies have examined viewer variables such as race, socioeconomic status, age, and racial attitudes, other characteristics of the viewer, which have been demonstrated to relate to television's impact in other areas, have not been studied. Viewing habits, amount of television viewing, attitudes toward television, and the quantity and quality of previous experiences with blacks have been largely ignored. Characteristics of the televised message, aside from general descriptions of characterization or presence or absence of blacks, are noticeably absent. For example, type of show, status of communicator, and fantasy or reality presentation have not been studied. Further, this research has failed to examine the various dimensions along which black programming might affect the viewer. Does it simultaneously alter racial attitudes, self-concept, perceptions of reality, and behavior? Or is only a single dimension affected? Finally, the research does not adequately study how black portrayals might affect different audience groups. Is the impact different for viewers as a function of race, age, sex, socioeconomic status, and place of residence?

The effects-research that has been conducted does not suggest either the specific aspects of content that are most powerful or the nature of psychological processes that account for television's ability to influence in this area. Finally, there has been little effort to design research in this area that is relevant to policy or production considerations about how to influence the number of blacks presented or the nature of those presentations.

The paucity of research coupled with the necessarily limited scope of the research to date, strongly suggests the need for more research on the effects of televised black portrayals. In order to provide a rationale for a desired research agenda in this field, this chapter reviews in some detail the research that has been conducted and suggests some areas that could be pursued.

The chapter is organized into four sections. The first section summarizes the findings of content analysis studies about frequency and nature of the portrayals of blacks on television. The second considers the viewing patterns of both black and white viewers of programs featuring black actors and actresses. The third section examines the research relating the portrayals of blacks to changes in the viewers' attitudes, behavior, and perceptions. Effects on white, black, and other minority audiences are reviewed separately. At the end of each of these sections, suggestions for future research are presented and discussed. The final section of the paper summarizes the research that has been conducted and the suggested new directions.

TELEVISION CONTENT AND BLACK PORTRAYALS

While the motion pictures of the 1940s were increasing their inclusion of blacks as social problems, television entered the living rooms of America lily white (Fife, 1974). In the 1950s, black people were rarely presented on television. When blacks were included, it was either in minor roles or in a highly stereotypic fashion (Balsley, 1959; Smythe, 1954).

Content analysis studies conducted in the 1970s indicate that the overwhelming majority of characters on television are white, middle-class, and male (Barcus, 1971, 1972; Downing, 1974; Gerbner, 1972; Mendelson & Young, 1972; Ormiston & Williams, 1973; Sternglanz & Serbin, 1974). Further, these studies reveal that blacks are more likely to be depicted in a stereotypic manner both in terms of the occupational and social roles to which they are assigned and in terms of the personality characteristics with which they are endowed (Clark, 1972; Dominick & Greenberg, 1970; Gerbner, 1976; Gerbner & Gross, 1973; Mendelson & Young, 1972; Ormiston & Williams, 1973; Seggar & Wheeler, 1973). Finally, the explicit discussion of race is noticeably absent from television presentations (Atkin & Greenberg, 1976; Mendelson & Young, 1972; Ormiston & Williams, 1973).

In conclusion, blacks appear less frequently than whites in all aspects of television programming, and blacks are presented in a stereotypic fashion. Therefore, the television viewer is provided with a fairly consistent view of the role of blacks in television's society and, by implication, in American society.

Suggested Research Directions

There is a need to continue to monitor the way in which blacks are presented on television. This continued content analysis is necessary in order to assess the nature and extent of changes in the medium's portrayals of blacks. Over the last 10 years, there clearly has been noticeable change in the number of black characters included, in the types of role assignments given to blacks, and in the nature of the programs that feature black characters (Allsopp, 1977; Fife, 1974).

However, content analyses of black portrayals should move beyond the level of simply enumerating the number of blacks, role assignments, and personality characteristics. What is needed is a more detailed analysis of the behaviors of televised blacks, of the accoutrements that surround them, of the nature of social interactions including black characters, and of the way in which the issue of race is broached. Attention already has been focused on the gross, overt signs of stereotyping. The suggested focus is on the more subtle and perhaps more important aspects of stereotyping and racism that a visual medium is capable of providing (e.g., see Chapter 8, this volume). The content analysis framework of Atkin and Greenberg (1976) is indicative of this more comprehensive analysis of content. This framework attends to language behavior, dress, physical behaviors, discussions of race, and the context in which minorities appear in addition to the more usual categories of occupational role and personality characteristics.

More comprehensive content analyses can serve two purposes. First, the use of detailed content descriptions in experimental studies of television programs will greatly advance our understanding of the relationship between effects and content. Content analyses generally are conducted by individuals different from those who investigate the effects of television content. However, if the work of these two research approaches could be combined and coordinated, the effects-researcher would be more sensitive to the ways in which content presents stereotypic or nonstereotypic portrayals of blacks. This greater sensitivity would naturally affect the way in which the researcher assessed the impact of the black portrayals. Similarly, the content analysis researcher would be more sensitive to some of the questions about dimensions of content that emerge from the effects-researcher's experience.

Content analyses can also serve the purpose of addressing policy and production considerations. This research can provide evidence of "real" changes in black portrayals as opposed to cosmetic changes. The data of content analyses can serve as an important source of information for those who influence or make policy, whether it be legislators, regulatory agencies, network officials, or consumer advocates. More detailed content analyses that address the subtle expressions of racism can inform production groups of dimensions that deserve attention if they propose to provide qualitatively different presentations of black people. In short, such analyses can serve as educational devices for making those responsible for program production more sensitive to the various ways in which racism can manifest itself in an audiovisual medium like television.

TELEVISION VIEWING PATTERNS

To be able to understand how television's portrayals of black people may affect viewers, it is desirable to understand how often viewers are exposed to this content. A variety of independent variables have been found to relate to an individual's use of television (Lyle & Hoffman, 1972a, b; Schramm et al., 1961). In particular, race is important (Bogart, 1972; Carey, 1966; Fletcher, 1971; Greenberg, 1972; Liebert et al., 1973; Lyle & Hoffman, 1972a, b; Murray, 1972).

Among adults, blacks generally report more television viewing than whites (Bogart, 1972). Moreover, in one study, black viewers were more likely to report that they do not watch enough television (Bower, 1973). This trend toward greater viewing holds true for blacks across socioeconomic status, with upper-income blacks watching more prime-time offerings than lower-income blacks and upper-income whites (Bogart, 1972). Black adults were also more likely to view violent programs than whites (Israel & Robinson, 1972). Further, in a study of "The Black American Market in Washington, D.C." (Greeley, 1974), it was learned that the black audience had more interest in news, programs featuring black stars, action shows, and sports than did the combined multiracial audience that Nielsen generally monitored. The greater interest of black adults in programs featuring black actors and actresses has been confirmed in other studies (Greenberg & Hanneman, 1970; Lechenby & Surlin, 1975).

Among black children and adolescents, the amount of time spent in television viewing is greater than it is for white children and adolescents (Greenberg, 1972; Greenberg & Dervin, 1970; Greenberg & Dominick, 1969; Lyle & Hoffman, 1972a, b). For black children, differences in viewing time were unrelated to family income (Greenberg & Dervin, 1970).

Black and white children also show different program preferences. Although it is true that children, both black and white, have similar program preferences, the frequency and regularity of viewing of specific programs is related to race. In general, black children and adolescents exhibit a greater preference for programs featuring blacks than do their white counterparts (Fletcher, 1971; Greenberg, 1972; Greenberg & Dervin, 1970; Greenberg & Dominick, 1969; Lyle & Hoffman, 1972a, b; Surlin & Dominick, 1970).

The research on viewing patterns indicates that blacks watch more television than whites, independent of socioeconomic status. Further, black audiences are more attracted to programs featuring black characters. However, there is no evidence that white audiences shun "black" shows. In fact, Nielsen ratings indicate the generally high popularity of programs including and featuring blacks (Brown, 1971; Efron, 1973; Daily News, 1977).

Suggested Research

There is a need to continue research in the area of viewing habits so that the normal exposure of viewers to varying types of content is known. Further, it is desirable to isolate those variables that influence the extent of exposure to black portrayals. Such variables as amount of television viewing, socioeconomic status, age, use of other media, or racial attitudes may also affect the impact of such portrayals on the viewers.

EFFECTS OF CONTENT FEATURING BLACKS

The research indicates that television presents clearly stereotyped portrayals of black people that are widely viewed by both black and white audiences. Exposure to this content would be of little concern if there were no evidence that television has the capacity to influence its viewers on a variety of dimensions. Television can have some effects on the viewer after a single exposure or after many exposures. The effects occur both in the laboratory and in natural settings. Effects have been found immediately after exposure to specific content and for as long as 18 months after viewing. Children, adolescents, and adults have all been shown to be variously influenced by the content of television. Moreover, research indicates that television can affect attitudes, behavior, knowledge, perceptions, and values. Finally, as viewers are exposed to more examples of specific content, the influence of television appears to increase (Comstock, 1975; Leifer et al., 1974; Liebert et al., 1973; Stein & Friedrich, 1975).

While there is much to report on the effects of television on viewers, there are several excellent summaries available (Comstock, 1975; Leifer et al., 1974; Liebert et al., 1973; Stein & Friedrich, 1975). This chapter confines its review of effects of television to those studies that investigate programs featuring black characters and looks at the impact of televised portrayals of blacks on the attitudes, perceptions, and behaviors of white, black, and other minority audiences.

Effects of Black Portrayals on White Audiences

Research on the impact of black portrayals on white children and adults has been limited to measures of attitudes toward blacks, perceptions of blacks in general, and perceptions of television blacks. None of the research has assessed impact on the behavior of white audiences.

The earliest research on the impact of black portrayals on white children goes back to the 1930s, when studies of motion pictures gave the first indication that visual media could influence viewers' attitudes toward various groups (Charters, 1933). In particular, it was found that Hollywood-make films could alter white children's attitudes toward blacks, Germans, and Chinese (Peterson & Thurstone, 1933). For example, the pro-Chinese film resulted in more positive attitudes toward Chinese up to 18 months later. The black portrayals were the stereotyped presentations in "Birth of a Nation," which resulted in negative attitude change. That is, more negative attitudes toward black people were found among those white children who had seen the film. With the exception of one film, the anti-Chinese film, the magnitude of attitude change increased as a function of the number of films to which the subjects were exposed.

After this early research on the impact of motion pictures on racial and ethnic attitudes, a small but growing body of research emerged on television's ability to affect attitudes about various groups of people. Two studies point to television's importance as a source of information about nationality groups. In the first, Lambert and Klineberg (1967) found that white American children reported obtaining most of their information about other nationalities (e.g., Africans, Chinese, Russians, and Indians) from their parents and television. As the viewer's age increased, there was greater reliance on television as the information source about foreign groups.

In the second study, which analyzed the effect of a children's television series called "Big Blue Marble," there is further evidence that television can influence white children's attitudes about diverse national groups (Roberts, Herold, Hornby, King, Sterne, Whiteley, & Silverman, 1974). The purpose of this syndicated series was to make children aware of how children in other lands live, work, play, and grow up. After exposure to four episodes of the series, white children were more likely to perceive foreign children as healthier, happier, and better off; more likely to question whether U.S. conditions were highly superior to those in other countries; and less likely to assume the cognitive superiority of American children.

Research on White Children. Although the studies of attitudes toward national groups as a result of exposure to television portrayals suggest television's power in influencing attitudes, this research does not directly consider the impact of television portrayals of blacks on white child-viewers. Studies of the impact of televised blacks on white children are of two types: (1) studies of multi-ethnic educational programming for children; and (2) studies of commercial programs that children watch.

Research on educational programs that intentionally provide positive portrayals of blacks and other minorities attest to television's ability to influence attitudes. In summary research on the impact of "Sesame Street" (Bogatz & Ball, 1971), it was learned that white children who viewed the series for 2 years had more positive racial attitudes toward black and Hispanic people than did children who viewed the series for 1 year. However, there was no evidence that a year's worth of viewing resulted in more positive racial attitudes. Children in both the 1-year and 2-year viewing groups were subjects who were watching "Sesame Street" on a regular basis under encouragement of the research staff. It was determined that the children in both groups were regular viewers (Bogatz & Ball, 1971). This study, while pointing to the socializing power of television, suggests that it takes a sustained period of exposure (2 years) for television to positively influence attitudes toward minority groups.

A study of the Canadian version of "Sesame Street" lends further support to his show's influence on social attitudes (Gorn, Goldberg, & Kanungo, 1976). The Canadian version included two types of inserts relating to intergroup attitudes. One type of insert shows colored children, either Oriental or Indian, in one of two settings. The colored children are shown either in an ethnic nonintegrated setting or in a racially integrated setting. The other type of insert featured a Canadian boy who was identified or not identifiable as *French* Canadian by whether or not he spoke French. All of the subjects were white English-Canadian preschoolers.

Those exposed to one presentation of both types of inserts in a nursery-school setting showed a strong preference for playing with colored children as opposed to whites. This preference as measured by means of photograph selection was sharply contrasted with those selections of a control group who preferred a white playmate. Similarly, there was greater preference for the French-Canadian boy over an English-Canadian, whether or not his cultural identity was evident. In this study, ethnic or segregated settings and integrated settings were equally effective in inducing positive attitude change toward people of color. Further, only limited exposure was necessary to detect attitude change.

Although "Sesame Street" was designed to effect cognitive gains in its viewers, its self-conscious attempt to portray black people positively was also effective in producing positive attitude change. "Vegetable Soup" is another educational series targeted for preschool-age and early elementary-age children. Unlike "Sesame Street," this series was designed specifically to try to ameliorate some of the negative effects of racial isolation on children (Nelson, 1974). The series features live and animated characters, which include Euro-Americans, Afro-Americans, Asian-Americans, Native Americans, and Latino-Americans in a variety of settings, roles, and activities. A summary evaluation study of the impact of the series on children (Mays, Henderson, Seidman, & Steiner, 1975) indicates that it did alter a group's attitudes toward other racial groups. The white subjects included boys and girls between 6 and 10 years of age. After being exposed in a school setting to 16 half-hour programs over an 8-week period, experimental subjects—when compared to control group subjects—were more likely to endorse statements indicating greater acceptance of children of different racial groups, including blacks, and increased friendliness toward different groups. In addition, this series seemed to strengthen identification with one's own racial group.

One conclusion emerges clearly from this research: Educational programming that consciously includes blacks and other minorities in positive portrayals can alter positively racial attitudes, especially under conditions of prolonged or multiple exposure.

The research on commercial programming's portrayals of blacks provides additional evidence that television can influence racial attitudes in either a positive or negative direction. The increase in the visibility of blacks on television in the last 5 years resulted from their inclusion in police dramas like "Baretta" and "Police Woman," in social commentary or social satire comedy series such as "All in the Family" or "Sanford and Son," and in cartoons like "Harlem Globetrotters" or "Fat Albert and the Cosby Kids."

One study of children's reactions to television blacks seen in prime-time offerings found that white children in 4th and 5th grade from rural, suburban, and urban school settings were influenced by the presence of black actors and actresses (Greenberg, 1972). Of the white respondents, 43% named at least one black television character that they would most like to be like. Further, five shows featuring black characters ("Mod Squad," "Julia," "Flip Wilson," "Bill Cosby," and "Mannix") were mentioned among the white children's 10 favorite television programs. The most frequently mentioned source of information about blacks was television, and this source was most important for rural whites who had almost no other sources of contact or interaction. Additionally, television exposure to black people seemed to have more impact than personal contacts in the neighborhood or at school. So those white children with higher levels of exposure to blacks through television were more likely to cite black shows among their favorites and to name black characters as identification figures than those with lower levels of exposure. However, exposure to television blacks did not seem to influence the racial attitudes of the white children in this study. With the exception of a sharply negative subgroup of urban white youngsters, most whites in this study felt that blacks and whites were about the same on most personality attributes. Neither level of television exposure nor place of residence was related to attitudes toward blacks in general or toward television blacks.

In a study of commercial cartoons that featured blacks in either the numerical majority or minority, with either positive or negative portrayals, Graves (1975) found that the racial attitudes of white children between 6 and 8 years of age were altered after just one exposure. Attitude change was measured by the frequency with which children associated positive or negative evaluative adjectives with black or white stimuli. Positive portrayals of black characters produced more positive attitudes toward blacks, and conversely, whites exposed to negative portrayals were affected negatively. Further, regardless of characterization, black people presented in the numerical minority produced more positive attitude change than did black people in the numerical majority. This study suggests that integrated settings were more effective in producing positive attitude change than segregated or predominantly black settings.

The potentially stronger impact of integrated settings on racial attitudes had been found in another study of black portrayals in films developed for white high-school students (Kraus, 1962). However, a study of nonwhite characters in the Canadian version of "Sesame Street" pointed to equal effectiveness of integrated and segregated settings on positive racial attitudes (Gorn et al., 1976). These contradictory findings may reflect differences in the content of the stimulus materials as well as differences in the ages of the subject groups studied.

In a study of prime-time programming that featured black characters, there is evidence of alteration of the racial attitudes of white children. Kindergartners, second- and third-graders, and sixth-graders who were exposed to one episode of either "The Cay" (positive portrayal of a black male) or "The Jeffersons" (negative portrayals of blacks) changed their attitudes toward blacks in the same direction as the stimulus programs

(Leifer, Graves, & Phelps, 1976; Phelps, 1976). It should be noted that whites in this study were more likely to express positive attitudes toward blacks regardless of the characterization they saw.

The studies of cartoons and of prime-time programming suggest that entertainment television does affect white children's attitudes toward black people even though the stated purpose of the presentations is to entertain.

Several studies of the effect of black portrayals on white children have investigated how these programs influence children's perceptions of black people. Greenberg (1972) reported that white 4th- and 5th-graders who lived in urban settings had many opportunities for direct contact with black children, and that adults were more likely to perceive that the "black families on television are like black families in real life" than were suburban and rural white children. The author had hypothesized that rural children would attribute greater reality to television blacks because of their lack of direct contact. However, the rural children were least likely to perceive television blacks as realistic.

Another study of children and the perceived reality of television disconfirmed Greenberg's earlier findings that more contact with blacks led to greater perceived reality for television blacks (Greenberg & Reeves, 1976). In this study of white children, the greater their real-life experiences with blacks, the less perceived reality they attributed to black portrayals on television. Whites with little personal experience with blacks perceived greater reality in television's portrayals. Obviously, the combined influence of experience with blacks and exposure to television is not clear, and it requires more attention.

Although the next study is not a direct investigation of the impact of television on children, it does demonstrate if and how children receive messages about the racial group. In a study of children's reception of prosocial messages after viewing an episode of "Fat Albert and the Cosby Kids," white boys and girls were asked about the similarities and differences between themselves and the predominatly black, masculine "Fat Albert" cast (Office of Social Research, 1974). More white children referred spontaneously to the racial attributes of the cast than to sexual or age characteristics. Both middle- and lower-class whites were equally likely to refer to the blackness of the cast. Race was clearly an important message that the children received. This message, however, did not seem to affect negatively their enjoyment of the series, given its immense popularity.

Research on White Adults. Most of the research on the impact of televised black portrayals has investigated the effect on white childlren, while only four studies have studied the impact of black characters on white adult audiences. Some exploratory research with American and Canadian adolescents and adults suggests how entertainment portrayals may influence racial attitudes (Vidmar & Rokeach, 1973). Two studies indicated that viewers of "All in the

Family" did not perceive the program to be a satire on bigotry (Surlin, 1974; Vidmar & Rokeach, 1973); rather, highly prejudiced persons reported more frequent viewing of the series than did low-prejudiced persons. Moreover, more-prejudiced viewers tended to report greater identification with Archie, the bigot, and were more likely to perceive him as the victor in the program than did less-prejudiced viewers. Further, these subjects did not see anything wrong with Archie's use of racial and ethnic slurs! Despite the creator's prediction (Lear, 1971) that the series would ridicule bigotry and prejudice, this study suggested that the program is more likely to have reinforced prejudice and racism than to have combated it.

Similarly, in an exploratory study of "Sanford and Son" using Canadian adults, Vidmar and Rokeach (1973) found a similar pattern of response to the black bigot, as was found in Archie, the white bigot. The majority of the subjects saw Sanford as a more typical representation of blacks than his son, Lamont. Highly prejudiced subjects were significantly more likely than low-prejudiced subjects to name Sanford as typical than they were to name Lamont.

In another study, black and white adults from Chicago and Atlanta were queried about their perceptions of both "All in the Family" and "Sanford and Son" (Lechenby & Surlin, 1975). Respondents were asked about the perceived reality that these situation comedies presented and about the perceived reality of the racial attitudes presented. White middle-class Atlantans were least accepting of the notion that the whites presented in "All in the Family" were realistic portrayals compared to white middle- and lower-class Chicagoans and white lower-class Atlantans. The reaction to "Sanford and Son" was different. Lower-class white Atlantans expressed greater agreement with the statement that "Sanford and Son" revealed how people, particularly blacks, really behave. White middle-class Atlantans agreed with this statement least.

In general, the perceived reality of the racial attitudes presented in both series was generally neutral (Lechenby & Surlin, 1975). However, subjects rated "All in the Family" more realistic in its portrayals of whites' racial attitudes than they did "Sanford and Son" in expressing blacks' racial attitudes.

Another study of American adults (Greenberg & Hanneman, 1970) indicated how the viewers' racial attitudes may alter the perceptions of black characters on television. For example, whites who were antagonistic toward blacks were more likely to report that there were more blacks appearing on television than were more favorably disposed whites. Highly antagonistic whites were also more likely to report that television was more likely to be fair in the presentation of blacks than were whites with more favorable attitudes.

Summary. Adults' perceptions of the reality of television portrayals of blacks are affected by the viewer's race, socioeconomic status, region of residence, and initial racial attitudes. These differences in perceptions of the content probably mediate television's impact. However, there is no research on racial attitude or behavior change as it relates to differences in the perceived reality of black programming.

The effect of black portrayals on white audiences has only been measured in terms of attitude change and perceptions. No study to date has looked at the influence of such portrayals on the behavior of whites. For white children, black portrayals have altered attitudes after a single exposure or after extended periods of exposure. Type of characterization of blacks seems to be important in the direction of attitude change of white children. The impact of two independent variables on attitudes and perceptions is less clear. The research on relative efficacy of integrated or segregated settings on attitudes toward blacks and the interaction of direct experience with blacks and television exposure to blacks on the perceived reality of televised portrayals is indeterminant at the present time.

The effects on white adults have focused exclusively on perceptions of televised blacks. Racial attitudes, region of residence, and socioeconomic status are related to the perceived reality of television blacks. The effect on the racial attitudes or behavior of white adults has been ignored.

Effects of Black Portrayals on Black Audiences

Much of the psychological literature on black people's personality development indicates that a large percentage of black adults and children are more likely to reject black models and images than they are to accept or highly value them (Bayton, Austin, & Burke, 1965; Bronfenbrenner, 1967; Jones, 1972; Powell, 1973; Spurlock, 1973). This rejection of black stimuli may be so intense that black children will racially misidentify themselves at a rate between 5% and 54% (Clark & Clark, 1955; Greenwald & Oppenheim, 1968; Gregor & McPherson, 1966; Morland, 1958, 1962). Research into black self-concept finds that most blacks experience a negative or decreased sense of self because of their race (Jones, 1972; Spurlock, 1973), although some more recent studies evaluating racial attitudes and self-concept of blacks after the rise of the black power movement have noted a more positive psychological state for both black adults and children (Banks, 1976; Hraba & Grant, 1970; McAdoo, 1977; Teplin, 1976).

Given this background, televised black portrayals could have two very different types of effects depending on the nature of the portrayals. If televised blacks are presented in positive, nonstereotypic roles, blacks viewing such attractive role models might identify with them and imitate them. The modeling of such attractive blacks is presumed to influence positively self-concept and attitudes toward one's own racial group. However, if the black presentations support the negative stereotypes that already exist about blacks, exposure to such undesirable role models could increase low self-esteem and magnify negative attitudes toward blacks.

Early research on the process of modeling and imitation indicates that children are generally more likely to imitate figures who are similar to themselves than people unlike themselves (Rosekrans, 1967). However, in experimental studies of modeling, children do not necessarily imitate people who are similar in sex, age, or race (Bandura, 1969; Kunce & Thelen, 1972; Stouwie, Hetherington, & Parke, 1970; Thelen & Fryrear, 1971). Children seem more likely to imitate people who have high status, are powerful, or are warm.

One study investigated the influence of the race of a televised role model on black children (Neely, Heckel, & Leichtman, 1973). Southern black preschool children were shown televised models who were either rewarded or punished for their behavior. Some of the models in each reinforcement condition were black, while others were white. In general, the black preschoolers imitated the white models more often than they imitated equally or more attractive black models. In fact, the punished white model was imitated at approximately the same frequency as the rewarded black model. The importance of the model's race clearly outweighed what had been previously found to be the more important factor of whether the model was rewarded or punished.

Two recent studies (Heckel, 1977) suggest that young black viewers regard whites as more competent than blacks. In one situation, black children watched a television film of a group of black and white peers choosing toys to play with. After viewing, the black viewers were given the same toys to pick from. All of the black children selected toys chosen by the white children in the film. This imitation of white choices occurred even when the toys selected were smaller or inferior in quality to toys selected by blacks in the film. Heckel suggests that "On TV, the competent roles tend to go to whites, particularly young white males. Thus black children regard whites as someone to copy" (Newsweek, 1977, pp. 65–67).

While the previous research looks at behavioral manifestations through imitative behavior, the following study investigates imitation of television blacks by black children in terms of their willingness to report black shows and characters as favorites. Greenberg (1972) found the black children in 4th and 5th grade in urban schools watched shows featuring blacks more often than their white counterparts. For example, 61% of the blacks watched nine or more shows featuring blacks, compared to 21% of whites. When asked to name three television characters they would most like to be like, 75% of the black children named at least one black character compared to 43% of the white respondents.

In a study of black adolescents, it was also found that black adolescents who are heavy viewers of black programs are more likely to name black television characters as models (Clark, 1972).

Whereas the few studies on modeling of televised blacks by black children indicate that black models appear to be less effective when compared to white models who are presumed to have higher status and power, there is also evidence that black children and adolescents perceive black characters as role-models. Presumably this increased ability and willingness to identify with black role models will result in modeling the behaviors that the black characters present. Research on "Sesame Street" and "The Electric Company," which prominently feature black and Hispanic characters as sources of cognitive and social information, indicates that these shows are successful in influencing black children's levels of knowledge and information (Ball & Bogatz, 1970, 1973; Bogatz & Ball, 1971). Black children are clearly learning from these positive black models.

Several studies have investigated the impact of black portrayals on racial attitudes of black children. Just as exposure to "Sesame Street" over a 2-year period resulted in more positive racial attitudes for white viewers, it was found that black disadvantaged preschoolers who watched for 2 years expressed more positive attitudes toward minority groups than did a group who had viewed the series for 1 year (Bogatz & Ball, 1971). Similarly, the exposure of black elementary-age children to 16 episodes of "Vegetable Soup" over an 8-week period resulted in a change in their attitudes. Black children in general were more likely to report a higher level of friendliness toward the various groups presented in the series; that is, blacks selected a larger percentage of photographs of children as potential friends. Further, blacks exposed to the series were more likely to identify with their own ethnic group than black children who did not see the series. This greater identification was measured by their selection of photographs of blacks as being people who were like themselves or who could be members of their family. These various results indicate that this program with its positive portrayals of blacks and other minority groups did positively influence the black children who were exposed to it.

Positive and negative portrayals of black cartoon characters have been shown to influence the racial attitudes of black boys and girls 6 to 8 years old (Graves, 1975). Black children exposed to a single cartoon episode featuring black characters had more positive attitudes toward blacks than did a control group who saw an episode of a program with no blacks in it. Furthermore, of those exposed to black cartoon characters, black children who saw the characters in an integrated setting had more positive attitudes toward blacks than did those who saw black characters presented in a segregated setting.

In a study of the effect of black portrayals from prime-time offerings, it was found that the impact of television exposure was in the predicted direction with positive portrayals producing positive attitude change and negative portrayals producing negative attitude change (Leifer et al., 1976). However, there was an effect on attitude change simply as a function of the race of the subject (Phelps, 1976). Black children after being exposed to a positive or negative show were more likely to change their attitudes toward blacks in a negative direction. This result directly contradicts the findings of Graves (1975). It should be noted that the procedures utilized in the two studies were very different. Whereas subjects in the Graves study were randomly assigned to positive and negative portrayal groups, in the Leifer et al. study subjects who initially had negative attitudes toward blacks were assigned to the positive portrayal show and those who initially had positive attitudes toward blacks were assigned to the negative portrayal. The impact of random versus nonrandom assignments may largely account for the difference in findings. Further, the difference in the nature of the stimuli used may account for these contradictory results. The Leifer et al. research employed five stimuli with live black actors, whereas the Graves study used cartoons featuring black characters. The differences in subject selection procedures as well as the nature of the television stimuli must be considered in trying to reconcile the conflicting results of these two studies.

Finally, in a study of identification with black characters on prime-time programming, Clark (1972) found that black adolescents had more positive racial attitudes toward a black television character who conformed to societal norms than toward a black militant character. The former role was generally one in which the black character was able to get along in an integrated setting without much difficulty compared to a more conflict-ridden experience in integrated and segregated settings for the latter. Only two studies have been conducted on the impact of black portrayals on black adults. In a 1969 telephone survey, Greenberg and Hanneman (1970) tried to assess the impact of television blacks on the racial attitudes of black adults. In general, black adults perceived television as presenting a more realistic portrayal of life than did white adults surveyed. However, black adults felt that television presentations of minorities were less fair than they could be. There was no relationship between racial attitudes of blacks and their perceptions.

In a study of reactions to "All in the Family" and "Sanford and Son," Lechenby and Surlin (1975) found black adults varied in their reaction to the shows based on their place of residence and social class status. Middle-class and low-income blacks from Atlanta and Chicago were surveyed by telephone. For "Sanford and Son," blacks were very similar in their perceptions of the reality of the presentation, with black Atlantans being somewhat less accepting than black Chicagoans. The reaction to "All in the Family" was somewhat different. Whereas middle-class blacks from Chicago were most accepting of the reality of the characters in the series, lower-class black Atlantans were least accepting of the notion that "All in the Family" reveals real behavior.

The perceptions of reality of the racial attitudes presented in the two series differed. For "Sanford and Son," blacks generally disagreed that the program showed how blacks really felt about whites. The effect was true regardless of social class. Black Chicagoans were slightly more accepting of the portrayed racial attitudes than were Atlantans. For "All in the Family," black Chicagoans were most in agreement that the characters showed how whites feel about blacks. However, they were significantly different for black Atlantans' views. It is important to note that overall the respondents were neutral in their ratings of the reality of the racial attitudes expressed in both programs. "All in the Family" was rated slightly more real in expressing the racial attitudes of whites than "Sanford and Son" was in expressing black racial attitudes.

Summary. Televised black portrayals have varied effects on black audiences. Black children seem to regard black actors and actresses as models even though black televised models are imitated less when compared to white models. The presence of black characters, regardless of type of portrayal, seems to have a direct impact on black children's attitudes toward their own group. However, the direction of attitude change after exposure to such images is unclear.

Even though only two studies have evaluated the impact on black adult audiences, and even though these have been severely restricted in their focus, it seems that social class status and place of residence are related to how black adults evaluate black television portrayals.

Effects of Black Portrayals on Other Minorities

To my knowledge, there is only one study which has investigated the impact of black and other minority presentations on Hispanic, Asian, and Native American child viewers (Mays et al., 1975). The impact of 8 weeks of exposure to "Vegetable Soup" on these three groups was as follows. Although it is not possible to sort out the impact of black characters specifically, it is possible to see the impact of minority presentations on these three groups. It was found that Chicano children who viewed the series showed the greatest increase in identification with their own ethnic group, of the ethnic groups tested. In general for the other groups, there was a small but discernible increase in identification with one's own group for those exposed to the series when compared to control subjects.

When children's level of friendliness was measured by the number and type of sociometric choices they made, it was found that the Asians selected the smallest number of people as potential friends, and the television series did little to alter the magnitude of choices. That is, there was an extremely small difference between Asian experimental and control groups. For the other groups, there was a tendency for those exposed to the series to select more people as potential friends than those who had not seen it.

Despite the concern that has been raised about the impact of black portrayals, we know virtually nothing about how these characters affect the other minority group members. The impact of these portrayals on other minorities may well be magnified because of the relative dearth of exposure of Asian-Americans, native Americans, and Hispanic Americans on television.

Suggested Research: Effects of Black Portrayals

After reviewing the research on the impact of televised black portrayals, one can characterize this field in the following way. First, most of the studies have looked at the effects on white audiences, particularly children. Almost no attention has been focused on how nonblack minority audiences might be influenced.

Second, except for the few studies on the effect of the race of a model on imitative behaviors of black children, the research has concentrated on how these portrayals change attitudes and perceptions. We know little of how black characters might affect the interracial socialization of blacks and whites.

Third, the impact of this content on children has primarily focused on evaluations of educational series like "Sesame Street" and "Vegetable Soup," while the impact on adults has primarily focused on two situation comedies, "All in the Family" and "Sanford and Son." The differences in the formats, pacing, general plotlines, and purposes of these two types of programming are enormous. Therefore it is extremely difficult to try to compare children's responses to adults' responses.

Fourth, most of the research has utilized a single measure of the dependent variable. In addition, most have only looked at the impact along a single dimension, like attitude toward blacks. Multiple measures of dependent measures are missing, as well as the procedures which acknowledge that the impact could be along a variety of dimensions,

Fifth, most of the studies utilize survey research techniques to determine audience reactions. Very few studies are experimental in nature, and therefore very few have any control over the amount of viewing by the subjects of programs featuring blacks. Exposure to black programs is left to self-selection factors operating within the audience.

Finally, the relative paucity of research means that there is little opportunity for discrepant findings that have to be resolved through further research. There is little opportunity for researchers to utilize comparable measurement techniques or research paradigms because of their proven usefulness in this particular area.

SUMMARY

1. Viewer Characteristics and Perception of Black Characters. Several researchers have started to investigate how the viewer perceives black characters (Greenberg, 1972; Greenberg & Reeves, 1976; Lechenby & Surlin, 1975; Surlin, 1974; Vidmar & Rokeach, 1973). Except for the Greenberg studies, this line of investigation has focused on adult perceptions.

Although content analyses present an objective picture of the messages that television presents, they do not indicate what message the viewer perceives. The research that has been conducted has indicated that place of residence (North vs. South; urban, suburban, and rural), race (black and white), socioeconomic status, and personal experience with blacks are independent variables that help one predict the nature of a message about blacks that viewers perceive. Interestingly, none of these studies of viewers' perceptions of televised blacks has systematically compared the images that the viewer received with what was actually broadcast. This failure to relate content to perceptions is a consequence of the nature of these studies of perceived reality. All are survey in nature and therefore have no direct control over the subject's exposure to programs that feature blacks or to specific episodes of series like "All in the Family" and "Sanford and Son." In experimental studies in the laboratory or in naturalistic settings in which one can control or at least monitor television exposure to specific programs that feature blacks, it will be possible to discover the relationship between program variables and viewer perceptions. Under controlled study, one could try to relate characteristics of the message to the viewer's perceptions. Aside from dialogue and physical behavior, nonverbal behavior (facial expression or body movements), environmental settings (ghetto project vs. upper-class apartment), clothing, speech characteristics (standard American English, no dialect), and status could influence a viewer's perceptions of black characters.

It is clear from the research on persuasion that characteristics of the communicator influence the effectiveness of a message to persuade (McGuire, 1969; Zimbardo & Ebbesen, 1969). So a persuader who has high credibility, including expertise and trustworthiness, is more effective than a communicator with low credibility. In the case of black portrayals, people might attribute greater credibility to the portrayals of blacks on "Sanford and Son" because of their perception of Redd Foxx as a black man with expertise about the black community and because of their perception of him as being unshakably honest and forthright in interview settings.

Aside from communicator variables to be studied, there are additional viewer variables that are worthy of investigation: Is the age of the viewer an important variable predicting the nature of perception of black portrayals? Does amount of television viewing influence the reality attributed to black portrayals? One cannot control the self-selective factors that make one subject a heavy viewer and another a light viewer, but it would be valuable to note how varying amounts of exposure to television in general and to black programs in particular relate to perceived reality of the presentations. For example, it was found that heavy viewers of television were more likely to have incorrect notions about the prevalence and nature of violent crimes in America. Their distortions were directly traceable to the way television presents violence (Gerbner & Gross, 1973). In a similar manner, amount of television viewing may be helpful in predicting viewer perceptions of black characters. It may well be found that heavy viewing is characteristic of specific viewers or that excessive viewing necessarily results in exposure to a particular type of television diet that could influence perceptions of black characters.

Understanding what message viewers perceive relative to the characteristics of the actual message broadcast and understanding which viewer characteristics relate to perceptual differences will increase our ability to predict televised black portrayals' effect on the viewer. For example, if a viewer attributes greater reality to negative stereotypic portrayals of blacks than to positive portrayals, is the impact greater for negative portrayals than for positive ones for that viewer?

2. Integrated Versus Segregated Settings. Black characters on television can be presented in one of two settings. Either there can be a few blacks in a group of whites, illustrating the numerical minority status of blacks in this country, or it is also possible to portray blacks in a segregated or predominantly black community setting. This latter setting presents the housing situation and patterns characteristic for most blacks.

Several questions are raised by these two forms of presentations. First, do all viewers attribute similar meaning to presentations of blacks in integrated and segregated settings? To black viewers, segregated setting may indicate group solidarity and a sense of community, while to white viewers the same setting may illustrate failure to achieve in this society and a tendency toward clannishness and in-group isolation. On the other hand, blacks in integrated settings may represent successful adaptation to the cultural standards for white audiences, while it may represent a denial of one's cultural identity to black viewers. It is clear that the same setting can have very different meanings for different viewers. However, these differences in meaning need to be explored systematically and not left to speculation.

Given the possibility that integrated and segregated settings may have different meanings to various members of the viewing audience, one might wonder about the likelihood of certain portrayals being present within the format of these settings. For example, most blacks presented on television are shown in integrated settings (Atkin & Greenberg, 1976; Downing, 1974;

Mendelson & Young, 1972; Ormiston & Williams, 1973; Roberts, 1970). Moreover, these content analyses indicate that blacks presented in these settings are more likely to be shown in negative and stereotypic ways. For example, in a study of dominance and submission patterns on television, Lemon (1976) found that blacks were more likely to be dominated when they were presented in integrated settings. Blacks were dominant only in segregated presentations and were therefore dominating other blacks. From Lemon's analysis, it was very unlikely to find a portrayal of a dominant black within an integrated setting.

It is also clear from content analyses that blacks are more likely to have occupations in the field of personal service, especially public or private law enforcement (Clark, 1969, 1972; Dominick & Greenberg, 1970; Seggar & Wheeler, 1973). Further, black characters are more heavily involved in violence than are white characters (Gerbner, 1976; Gerbner & Gross, 1973). In particular, black characters are either perpetrators of violence or they are victims of violence.

One final characteristic of black presentations must be mentioned. In spite of the fact that most blacks are presented in integrated settings, the discussion of race appears to be taboo (Atkin & Greenberg, 1976; Mendelson & Young, 1972; Ormiston & Williams, 1973). In the rare instances of the discussion of race, it only occurs among blacks, never between blacks and whites (Atkin & Greenberg, 1976).

Given these limiting features of black portrayals in integrated settings, it is difficult to imagine that one will find positive, nontraditional roles for blacks in these settings. My experience in trying to find such portrayals for study is that they are almost impossible to find in commercial programming. Such positive roles for blacks in integrated settings may occasionally be found among made-for-television movies or children's afterschool specials. From children's series like "Sesame Street," "The Electric Company," and "Vegetable Soup," there is evidence that it is possible to present strong positive blacks within an integrated setting. Somehow it appears to be more difficult for the writers (predominantly white males) of prime-time programs or daytime serials to be able to provide such characterizations.

Given the possibility of different meaning attributed to the two types of settings and given the apparent unlikelihood of having positive blacks in integrated settings, what is the impact of these two types of settings on the viewer? The present research presents contradictory findings. White preschoolers' racial attitudes can be equally influenced with positive portrayals in either integrated or segregated settings (Gorn et al., 1976). Two other studies—one with white children 6 to 8 years old viewing cartoons (Graves, 1975), and one with white adolescents viewing a film (Kraus, 1962)—suggest that integrated settings are more powerful in changing attitudes toward blacks. The one study (Graves, 1975) that looked at the impact of such settings on black child viewers found that integrated settings were more likely to encourage positive attitudes among 6- to 8-year-old black children.

These findings indicate that integrated settings affect viewers of different ages and races differently. However, with only four studies we cannot clearly delineate the impact of these settings. Further, none of the studies addresses the issue of meaning for the viewers or the relationship of type of portrayal within type of setting to the impact of the viewer. What is needed is research that explicates the meaning of both positive and negative portrayals in segregated and integrated settings for various segments of the audience. Then the impact of television presentations of these two types of settings should be studied. For example, one might want to compare the portrayals on "Sanford and Son" (segregated setting) with those on "The Jeffersons" (integrated setting). Or one might want to compare "The Young Sentinels" cartoon, which features a black superheroine as one of a trio of heroes, with "Fat Albert," which features a group of black youngsters within a ghetto setting.

3. Real Versus Ideal Portrayals. Whenever one criticizes the limited range of black characters on television, the issue of the desirability of "real" versus "ideal" portrayals emerges. Essentially it is a question of presenting blacks the way they really are or presenting blacks as superheroes or ideals. The problem becomes apparent as such a discussion proceeds: The definition of what is "ideal" and what is "real" can vary at least as a function of race, socioeconomic status, and sex of the discussant. To one viewer, a black criminal may be realistic, and to another viewer so is a black neurologist because of their experience with similar individuals. Since it becomes extremely difficult to sort out what is "real" and what is "ideal" for a particular viewer, it seems that it would be desirable to evaluate a range of occupational portrayals and relate these characters to the viewers' past experience with such individuals.

Another dimension of real versus ideal portrayals is not just seen in the type of occupational roles or settings to which blacks are assigned, but in the personality characteristics that the roles are given. Ideal black portrayals can present well-mannered, highly attractive, controlled characterizations, like "Julia," or they can present a black doctor successfully overcoming bureaucratic neglect that results in widespread disease (made-for-TV movie). However, it is important to know how audiences respond to such portrayals. Is such one-dimensional goodness discounted as unbelievable if the character is black? Is such a character more likely to be imitated by black and white children? Or is a more realistic portrayal in which the character is endowed with positive attributes as well as reasonable human failings more effective with or more acceptable to the viewing audience?

A real difficulty arises when thinking about research in this area. One is generally limited to the portrayals that are readily available. To effectively conduct research on the effect of real versus ideal characters, researchers may need the production of experimental material that maintains the technical quality to which viewers are accustomed while varying personality attributes, occupational roles, and social settings along specified dimensions. At present, more diverse portrayals are available in children's programming through government-sponsored Emergency School Assistance ACT (ESAA) educational series like "Rebop," "Infinity Factory," and "Villa Allegre," and in adult programming through public television like the "Visions" series. However, there is a need for commercial broadcasters to be willing to develop experimental material and evaluate it in terms of viewers' willingness to accept new formats and roles for blacks and in terms of the impact of such material on the viewing public.

4. Relationship of Impact of Black Portrayals to Viewer and Program Characteristics. There are both viewer and program characteristics that could profitably be investigated in trying to assess the impact of black portrayals on the television audience. The viewer's race, place of residence, and socioeconomic status are the independent variables that have been studied. There are other viewer characteristics that would appear to be important.

First, there needs to be a systematic study of developmental differences in how televised black portrayals influence viewers. Only one study reviewed here (Leifer, Graves, & Gordon, 1974) includes more than one age group within the study sample. In particular, we have no knowledge of how children and adults respond to the same presentations. This is particularly relevant because the structure of television is such that widely diverse age groups are likely to be attracted to the same show. There are situation comedies featuring black actors and actresses appearing in the family viewing hour ("Good Times," "What's Happening") that are presumed to be appropriate for the entire family. These situation comedies present blacks as targets of social satire to be laughed at, yet the message received by viewers of various ages could be very different and could dramatically influence the effect of such presentations. A developmental study of the effect of such programming would more directly address the audience that actually views the program.

The viewer's initial attitudes toward blacks is an important variable to relate to the impact of televised black portrayals. From the research on persuasion and influence, it is clear that opinions of the audience influence the effectiveness of a particular communication (Zimbardo & Ebbesen, 1969). For example, Kendall and Wolf (1949) found that cartoons ridiculing either racial or religious prejudice were misunderstood more often by prejudiced than by unprejudiced individuals. This finding is very close to what was found later for perceptions of the message of "All in the Family" (Vidmar & Rokeach, 1973). The more-prejudiced viewers perceived Archie as hero and did not see the program as a satire on bigotry. Further, the literature on persuasion indicates that messages which did not fit the predisposition of an audience member were likely to be recast to fit his or her frame of reference (Klapper, 1960). As Greenberg and Hanneman (1970) found with whites' responses to televised blacks, those who were more antagonistic toward blacks in general, as compared to whites with less negative attitudes, were more likely to believe that television was very fair in its presentation of blacks. It seems that the concept of selective perception is a concept from the persuasion literature that could reasonably be applied to the relationship between initial racial attitudes and impact of televised black portrayals.

Not only has the audience's opinions been found to relate to the ability of a communication to persuade, but the source of that attitude is likewise important in determining if a particular member of the audience is persuaded (Zimbardo & Ebbesen, 1969). The strength of group affiliations affects the message's effect (Abelson, 1959; Klapper, 1960). People who have a stronger sense of group commitment are less influenced by communications that conflict with group norms (Abelson, 1959).

In trying to predict the influence or effect of black characters on television, it is helpful to have knowledge of the viewer's initial racial attitudes as well as knowledge of the presence or absence of group memberships that might be the source of that attitude. Besides group affiliation, there are other possibilities for the source of the attitude toward blacks. To what extent and under what circumstances does the viewer have contact with black people? Again, from the literature on attitude change, it would seem that those with more direct experience with blacks would rely less on television's information and thereby be less affected by its portrayals than those without opportunities for direct interaction or contact. The two studies (Greenberg, 1972; Greenberg & Reeves, 1976) that have tried to explore this idea with white children have resulted in contradictory findings. More research is needed to sort out whether the amount of contact with blacks influences the impact of televised black portrayals.

Finally, it seems crucial to relate, on the one hand, the viewer characteristics of initial attitude toward blacks and source of that racial attitude to, on the other hand, program characteristics of black portrayals. Is the black portrayal a positive one in which the character is featured as being competent, strong, concerned for others, and likeable? Or is the black role characterized by failure, incompetence, laziness, and trickery? In general, the attitude-change literature predicts that a message influences attitudes in the same direction as the presentation. The impact of type of characterization of television blacks seems to be influenced by the viewer's race. For white children, attitude change toward blacks was related to the type of characterization and in the predicted direction (Bogatz & Ball, 1971; Graves,

1975; Leifer et al., 1976). For black audiences, the impact of type of characterization is less clear. Two studies using educational television indicate that positive portrayals lead to positive attitude change in black children (Bogatz & Ball, 1971; Mays et al., 1975). However, two other studies, both using commerical television programs, found that type of characterization was not directly related to direction of attitude in black viewers (Graves, 1975; Leifer et al., 1976).

It is important to study what might account for blacks changing in positive or negative directions in their attitudes toward their own racial group regardless of the attitude presented in the television program to which they are exposed. Several factors could be operating. First, the message perceived by the black children may be qualitatively different from the message that adults receive. Second, viewer characteristics like initial racial attitude might influence the message received, but a program could influence the black viewer in a particular direction. Third, if the characterizations presented are "real," the black child viewer may be more negatively influenced than if the presentations are "ideal." The rise of slice-of-life portrayals of blacks that characterize such programs as "The Jeffersons" and "Good Times" may present so many images surrounding a prosocial plot, for example, that children are more influenced by the caricatures of black people than they are influenced by their concern for others, for example.

In order to understand how black portrayals influence television audiences, both viewer and program characteristics need to be explored.

5. *Attitude Change or Behavior Change.* Except for a small number of studies that investigated the likelihood of black children imitating black or white televised models, all of the effects-research in this area has focused on attitude change. How does exposure to black portrayals influence the behavior of blacks and whites? Is there any relationship between evidence of attitude change and subsequent behavior change after exposure to black portrayals? Is there a different impact on behavior change in the viewers if they have been exposed to black portrayals in ethnic segregated settings or in integrated settings? In pursuing the influence of black portrayals on behavior, it is important to look not only at the nature of quality of interracial socialization but also at the issues of imitation. Does a police drama featuring blacks as villains not only influence racial attitudes and behavior toward blacks but also affect knowledge of and behavior in the area of criminal activities?

In terms of research strategy, there is a need to assess not only attitude change but behavior change as well. Perhaps the use of multiple measures of a single dependent measure and, where possible, the simultaneous measure of several different types of effects would serve to fill gaps in the research in this area.

6. Attitude Change or Change in Self-Concept. Many people concerned about the portrayal of blacks as it affects black children have theorized that stereotypic presentations would not only influence the children's attitude toward their own racial group but also their attitudes toward themselves or self-concept. In spite of the fact that this effect has been hypothesized, no research has assessed the impact of black portrayals on black self-concept. Positive black characters on television are thought to enhance a black child's self-concept by presenting a more attractive image of blacks as a group. Since racial identity is a part of an individual's view of himself or herself, positive alteration of one's view of blackness should result in a positive influence on self-concept.

Self-concept change among blacks is a worthy dependent variable to be studied in relation to exposure to black actors and actresses on television. There is also the possibility that black portrayals may influence the self-concept of other minority groups, particularly if the viewer perceives some similarity between the self and his or her ethnic group and blacks.

7. Selective Viewing or Broad Viewing. The research on viewing patterns of black programs indicates that both black and white audiences view these programs. However, it has also been found that blacks are more likely to watch more programs featuring blacks and to watch those programs more frequently (Fletcher, 1971; Greeley, 1974; Greenberg, 1972; Greenberg & Dervin, 1970; Greenberg & Dominick, 1969; Lyle & Hoffman, 1972a, b; Surlin & Dominick, 1970). Given this evidence of differential viewing patterns between blacks and whites and given the range of attitudes toward blacks in the viewing audience, one wonders if a broad and diverse range of black portrayals were available, would people voluntarily expose themselves to this diversity?

Research on information campaigns indicates that the people whom one wants to influence may not be in the audience (Zimbardo & Ebbesen, 1969). People with an opinion on an issue will generally expose themselves only to those communications that advocate their own point of view (Hyman & Sheatsley, 1947/1958). *Selective exposure* is the term given to this phenomenon. The Vidmar and Rokeach (1973) research on "All in the Family" indicated that more-prejudiced viewers were more regular viewers of the series than less-prejudiced viewers. Perhaps selective exposure operates across the range of black portrayals that are available. The phenomenon of "Roots" indicated that approximately 80 million viewers watched the series, but how did exposure to the series relate to viewers' attitudes toward blacks and slavery. Certainly selective exposure is a phenomenon that should be further investigated.

8. Long-Term Versus Short-Term Exposure. Several studies have looked at long-term exposure to positive black portrayals (Bogatz & Ball,

1971; Mays et al., 1975). Most experimental research in this area has studied the impact of a single exposure to black portrayals. However, there is no research on cumulative effects of commercial television productions. Do television diets that vary in the type of black portrayals seen and amount of exposure to these portrayals affect viewers differently? What are the effects of black portrayals for viewers whose diet is heavy in cartoons compared to those whose diet is heavy in situation comedies or to those whose diet is heavy in police/action/adventure series?

Research on the effects of black characters needs to increase in a variety of areas. Following the paradigm of communications, we need to learn more about the impact of communication on the particular types of roles to which blacks are assigned. Next we need to more carefully assess characteristics of the message or role presentations and settings. Finally, we must isolate those characteristics of the receiver or viewing audience that affect their exposure to portrayals of blacks, their perception of those roles, and their likelihood of being influenced by these characterizations. This chapter has specified eight areas of research that would be both useful to pursue and likely to produce results.

REFERENCES

Abelson, H. *Persuasion.* New York: Springer, 1959.

Allsopp, R. *1963-1976: Thirteen years of black stereotyping in prime time programming:* A psychological view. Unpublished manuscript, New York University, 1977.

Atkin, C. K., & Greenberg, B. S. *Family, child and message factors mediating children's prosocial learning from television.* Paper presented at the annual convention of American Educational Research Association, San Francisco, April 1976.

Ball, S., & Bogatz, G. A. *The first year of Sesame Street: An evaluation.* Princeton, N.J.: Educational Testing Service, 1970.

Ball, S., & Bogatz, G. A. *Reading with television: An evaluation of The Electric Company.* Princeton, N.J.: Educational Testing Service, 1973.

Balsley, D. F. *A descriptive study of references made to Negroes and occupational roles represented by Negroes in selected mass media.* Unpublished doctoral dissertation, University of Denver, 1959.

Bandura, A. *Principles of behavior modification.* New York: Holt, Rinehart & Winston, 1969.

Banks, W. C. White preference in Blacks: A paradigm in search of a phenomenon. *Psychological Bulletin,* 1976, *83,* 1179-1186.

Barcus, F. E. *Saturday children's television: A report of television programming and advertising on Boston commercial television.* Boston: Action for Children's Television, July 1971.

Barcus, F. E. *Network programming and advertising in the Saturday children's hours: A June and November comparison.* Boston: Action for Children's Television, 1972.

Bayton, J. A., Austin, L. J., & Burke, K. R. Negro perception of Negro and white personality traits. *Journal of Personality and Social Psychology,* 1965, *3,* 250-253.

Bogart, L. Negro and white media exposure: New evidence. *Journalism Quarterly,* 1972, *49,* 15-21.

Bogatz, G. A., & Ball, S. *The second year of Sesame Street: A continuing evaluation.* Princeton, N.J.: Educational Testing Service, 1971.

Bower, R. T. *Television and the public.* New York: Holt, Rinehart & Winston, 1973.

Bronfenbrenner, U. The psychological costs of quality and equality in education. *Child Development,* 1967, *38,* 909–925.

Brown, L. *Television: The business behind the box.* New York: Harcourt Brace Jovanovich, 1971.

Carey, J. W. Variations in Negro/White television preference. *Journal of Broadcasting,* 1966, *3.*

Charters, W. W. *Motion pictures and youth: A summary.* New York: Macmillan, 1933.

Clark, C. Television and social controls: Some observations on the portrayal of ethnic minorities. *Television Quarterly,* 1969, *8,* 18–22.

Clark, C. Race, identification, and television violence. In *Television and social behavior* (Vol. 5). Washington, D.C.: U.S. Government Printing Office, 1972.

Clark, K. B., & Clark, M. P. Racial identification and preference in Negro children. In H. Proshansky & B. Seidenberg (Eds.), *Basic studies in social psychology.* New York: Holt, Rinehart & Winston, 1955.

Coates, B., Pusser, H. E., & Goodman, I. The influence of "Sesame Street" and "Mr. Rogers' Neighborhood" on children's social behavior in preschool. *Child Development,* 1976, *47,* 138–144.

Comstock, G. *Television and human behavior: The key studies.* Santa Monica, Calif.: The Rand Corporation, 1975.

Congressional Black Caucus. *A position on the mass communication media.* Mimeographed report, Washington, D.C., 1972.

Daily News (New York). Sunday, February 20, 1977. (Television review.)

Dominick, J. R., & Greenberg, B. S. Three seasons of blacks on television. *Journal of Advertising Research,* 1970, *10,* 21–27.

Downing, M. Heroine of the daytime serial. *Journal of Communication,* 1974, *24,* 130–137.

Efron, E. Report on minorities: Is television really integrated—or does it just look that way? *TV Guide,* Oct. 27, 1973, pp. 6ff.

Fife, M. D. Black image in American TV: The first two decades. *The Black Scholar,* 1974, *6*(3), 7–15.

Fletcher, A. D. Negro and white children's television program preferences. *Journal of Broadcasting,* 1971, *13,* 359–366.

Furu, T. *Television and children's life.* Radio and Television Cultural Research Institute, Japan Broadcasting Corp., 1966.

Gerbner, G. Violence in television drama: Trends and symbolic functions. In *Television and social behavior* (Vol. 1). Washington, D.C.: U.S. Government Printing Office, 1972.

Gerbner, G. The frightening world of the TV addict. *Psychology Today,* 1976, *9,* 41ff.

Gerbner, G., & Gross, L. *The social reality of television drama.* Unpublished manuscript, Philadelphia, Pa.: Annenberg School of Communication, 1973.

Gorn, G. I., Goldberg, M. E., & Kanungo, R. N. The role of educational television in changing intergroup attitudes of children. *Child Development,* 1976, *47,* 277–280.

Graves, S. B. *Racial diversity in children's television: Its impact on racial attitudes and stated program preferences.* Unpublished doctoral dissertation, Harvard University, 1975.

Greaves, W. Black Journal: A few notes from the executive producer. *Television Quarterly,* 1969, *8,* 3–17.

Greeley, B. Black viewers tune in soul and Cronkite. *Variety,* Sept. 18, 1974, pp. 35ff.

Greenberg, B. S. Children's reactions to TV blacks. *Journalism Quarterly,* 1972, *49,* 5–14.

Greenberg, B. S., & Dervin, B. *Use of mass media by the urban poor.* New York: Praeger, 1970.

Greenberg, B. S., & Dominick, J. R. Racial and social class differences in teen-agers' use of television. *Journal of Broadcasting,* 1969, *13,* 331–344.

Greenberg, B. S., & Hanneman, G. J. Racial attitudes and the impact of television blacks. *Educational Broadcasting Review,* 1970, *4,* 27–34.

Greenberg, B. S., & Reeves, B. Children and the perceived reality of television. *Journal of Social Issues*, 1976, *32*(4), 86–97.

Greenwald, H. J., & Oppenheim, D. B. Reported magnitude of self-misidentification among Negro children—artifact? *Journal of Personality and Social Psychology*, 1968, *8*, 49–52.

Gregor, A. J., & McPherson, D. A. Racial attitudes among white and Negro children in a deep south metropolitan area. *Journal of Social Psychology*, 1968, *68*, 95–106.

Heckel, R. V. Personal communication, 1977.

Himmelweit, H. T., Oppenheim, A. N., & Vince, P. *Television and the child.* London: Oxford University Press, 1958.

Hraba, J., & Grant, G. Black is beautiful: Reexamination of racial preference and identification. *Journal of Personality and Social Psychology*, 1970, *16*, 398–402.

Hyman, H. H., & Sheatsley, P. B. Some reasons why information campaigns fail. In E. E. Maccoby, T. M. Newcomb, & E. L. Hartley (Eds.), *Readings in social psychology.* New York: Holt, 1958. (Originally published, *Public Opinion Quarterly*, 1947, *11*.)

Israel, H., & Robinson, J. P. Demographic characteristics of viewers of television violence and news programs. In *Television and social behavior* (Vol. 4). Washington, D.C.: U.S. Government Printing Office, 1972.

Jones, J. M. *Prejudice and racism.* Reading, Mass.: Addison-Wesley, 1972.

Kendall, P. L., & Wolf, K. M. Deviant case analysis in the Mr. Biggot Study. In P. F. Lazarsfeld & F. N. Stanton (Eds.). *Communications Research 1948–1949.* Published for the American Jewish Committee. New York: Harper & Row, 1949.

Klapper, J. T. *The effects of mass communications.* Glencoe, Ill.: The Free Press of Glencoe, 1960.

Kraus, S. Modifying prejudice: Attitude change as a function of the race of the communicator. *Audiovisual Communication Review*, 1972, *10*(1).

Kunce, M., & Thelen, M. H. Effect of face of model on imitative behavior. *Developmental Psychology*, 1972, *6*.

Lambert, W. E., & Klineberg, O. *Children's views of foreign peoples: A cross-national study.* New York: Appleton-Century-Crofts, 1967.

Lear, N. "As I read how Laura saw Archie..." *New York Times*, October 10, 1971.

Lechenby, J. D., & Surlin, S. H. *Race and social class differences in perceived reality of socially relevant television programs for adults in Atlanta and Chicago.* Paper presented at The International Communication Association annual convention, Chicago, April 1975.

Leifer, A. *How to encourage socially-valued behavior.* Paper presented at the meeting of the Society for Research in Child Development, Denver, Colo., 1974.

Leifer, A. D., Gordon, N. J., & Graves, S. B. Children's television: More than mere entertainment. *Harvard Educational Review*, 1974, *44*, 213–245.

Leifer, A. D., Graves, S. B., & Gordon, N. J. *Children's critical evaluation of television content.* Unpublished manuscript, Center for Research in Children's Television, Harvard University, 1974.

Leifer, A. D., Graves, S. B., & Phelps, E. *Monthly report of Critical Evaluation of Television Project.* Unpublished manuscript, Center for Research in Children's Television, Harvard University, 1976.

Lemon, J. *Women and blacks on prime time television: An analysis of dominance patterns.* Unpublished manuscript, Center for Research in Children and Television, Harvard University, 1976.

Liebert, R. M., Neale, J. M., & Davidson, E. S. *The early window: Effects of TV on children and youth.* Elmsford, N.Y.: Pergamon Press, 1973.

Lyle, J., & Hoffman, H. R. Children's use of television and other media. In *Television and social behavior* (Vol. 4). Washington, D.C.: U.S. Government Printing Office, 1972. (a)

Lyle, J., & Hoffman, H. R. Explorations in patterns of television viewing by preschool-age children. In *Television and social behavior* (Vol. 4). Washington, D.C.: U.S. Government Printing Office, 1972. (b)

Mays, L., Henderson, E. H., Seidman, S. K., & Steiner, V. S. *An evaluation report on Vegetable Soup: The effects of a multi-ethnic children's television series on intergroup attitudes of children.* Unpublished manuscript, New York State Department of Education, 1975.

McAdoo, H. P. *A reexamination of the relationship between self-concept and race attitudes of young black children.* Unpublished manuscript, Howard University, 1977.

McGuire, W. J. The nature of attitudes and attitude change. In G. Lindzey & E. Aronson (Eds.), *Handbook of social psychology* (Vol. 3). Reading, Mass.: Addison-Wesley, 1969.

Mendelson, G., & Young, M. *A content analysis of black and minority treatment on children's television.* Boston: Action for Children's Television, 1972.

Morland, J. K. Racial recognition by nursery school children in Lynchburg, Virginia. *Social Forces,* 1958, *37,* 132–137.

Morland, J. K. Racial acceptance and preference of nursery school children in a southern city. *Merrill Palmer Quarterly,* 1962, *8,* 271–280.

Murray, J. P. Television in inner-city homes: Viewing behavior of young boys. In *Television and social behavior* (Vol. 4). Washington, D.C.: U.S. Government Printing Office, 1972.

National Advisory Commission on Civil Disorders. *Report.* New York: Bantam Books, 1968.

Neely, J. J., Heckel, R. V., & Leichtman, H. M. The effect of race of model and response consequences to the model on imitation in children. *Journal of Social Psychology,* 1973, *89,* 225–231.

Nelson, B. *The statement of goals and objectives for Vegetable Soup.* Unpublished manuscripts, University of Massachusetts at Amherst, 1974.

Newsweek. What TV does to kids. February 21, 1977, pp. 62–70.

Office of Social Research. *A study of messages received by children who viewed an episode of "Fat Albert and the Cosby Kids."* February, 1974.

Ormiston, L. H. II., & Williams, S. *Saturday children's programming in San Francisco, California: An analysis of the presentation of racial and cultural groups on three network affiliated San Francisco television stations.* San Francisco: Committee on Children's Television, 1973.

Paulson, F. L., McDonald, D. L., & Whittemore, S. L. *An evaluation of Sesame Street programming designed to teach cooperative behavior.* Monmouth, Ore.: Teaching Research, 1972.

Peterson, R. C., & Thurstone, L. L. *Motion pictures and the social attitudes of children.* New York: Macmillan, 1933.

Phelps, E. *Analysis of Phase II attitude data.* Unpublished manuscript, Harvard University, 1976.

Powell, G. J. Self-concept in white and black children. In C. V. Willie, B. M. Kramer, & B. S. Brown (Eds.), *Racism and mental health.* Pittsburgh, Pa.: University of Pittsburgh Press, 1973.

Roberts, C. The portrayal of blacks on network TV. *Journal of Broadcasting,* Winter 1970–71, *15*(1).

Roberts, D. F., Herold, C., Hornby, M., King, S., Sterne, D., Whiteley, S., & Silverman, T. *Earth's a big blue marble: A report of the impact of a children's television series on children's opinions.* Unpublished manuscript, Stanford University, 1974.

Rosekrans, M. A. Imitation in children as a function of perceived similarity to social model and vicarious reinforcement. *Journal of Personality and Social Psychology,* 1967, *7,* 307–315.

Schramm, W. M., Lyle, J., & Parker, E. B. *Television in the lives of our children.* Stanford: Stanford University Press, 1961.

Seggar, J. F., & Wheeler, P. World of work on TV: Ethnic and sex representation in TV drama. *Journal of Broadcasting,* 1973, *17,* 201–214.

Shirley, K. *Prosocial effects of publicly broadcast prosocial television.* Unpublished manuscript, Stanford University, 1974.

Smythe, D. W. Reality as presented by television. *Public Opinion Quarterly,* 1954, *18,* 143–156.

Spurlock, J. Some consequences of racism for children. In C. V. Willie, B. M. Kramer, & B. S. Brown (Eds.), *Racism and mental health.* Pittsburgh, Pa.: University of Pittsburgh Press, 1973.

Stein, A. H., & Friedrich, L. K. Impact of television on children and youth. In E. M. Hetherington (Ed.), *Review of child development research* (Vol. 5). Chicago: University of Chicago Press, 1975.

Sternglanz, S. H., & Serbin, L. A. Sex role stereotyping in children's television programs. *Developmental Psychology,* 1974, *10*(5), 710–715.

Stouwie, R. J., Hetherington, E. M., & Parke, R. D. Some determinants of children's self-reward behavior after exposure to discrepant reward criteria. *Developmental Psychology,* 1970, *3,* 313–319.

Surgeon General's Scientific Advisory Committee on Television and Social Behavior. *Television and growing up: The impact of televised violence.* Washington, D.C.: U.S. Government Printing Office, 1972.

Surlin, S. H. Bigotry on air and in life: The Archie Bunker case. *Public Telecommunications Review,* 1974, *212,* 34–41.

Surlin, S. H., & Dominick, J. R. Television's function as a third parent for black and white teenagers. *Journal of Broadcasting,* 1970, *15,* 55–66.

Teplin, L. A. A comparison of racial/ethnic preferences among Anglo, Black and Latino children. *American Journal of Orthopsychiatry,* 1976, *46,* 702–709.

Thelen, M. H., & Fryrear, J. L. Effect of observer and model race on the imitation of standards of self-reward. *Developmental Psychology,* 1971, *5,* 133–135.

Vidar, N., & Rokeach, M. *Archie Bunker's bigotry: A study in selective perception and exposure.* Paper presented at the annual meeting of the Eastern Psychological Association, Washington, D.C., May 1973.

Zimbardo, P., & Ebbesen, E. B. *Influencing attitudes and changing behavior.* Reading, Mass.: Addison-Wesley, 1969.

13 An Aerial View of Television and Social Behavior

Stephen B. Withey
University of Michigan

One achievement of the telecommunications media is their ability to produce the technology, human resources, and programs that create mass publics. These publics are so dispersed and so heterogeneous that they have little in common for the symbols and messages, the scenes and characters, that they receive from the telecommunications media. In industrialized nations, for example, television is so common that the gap between the rich and the poor, the urban and the rural, in their access to a mass medium has been all but closed. In the United States there is a television set, and sometimes more than one, in 98% of U.S. homes. Thus the potential for widespread exposure to a common set of symbols, messages, and events exists.

Whenever revolutionaries capture a radio station or whenever advertisers vie for minutes of television time, they assume that this widespread and common exposure to the media must have a significant influence upon the publics that compose society. The telecommunications media have altered the tactics of political campaigning. Business and government leaders have become acutely sensitive to the stories carried over the media. Boorstin (1961) writes about pseudoevents which are staged specifically for the media. An example is the mayor of New York handing the president of the United States a protest letter before his United Nations speech. It is the same assumption of the impact of the mass media that has been the guiding principle of the so-called "media effects" tradition of research. Researchers and others seem to be saying that the massive flow of communications certainly has some impact and that the only question is the nature and scope of that impact.

However, there is a problem. There are numerous kinds of effects that can be easily imagined. Even a single program may have a considerable effect on the attitudes or behaviors of certain individuals with particular proclivities. The reporting of a single airplane hijacking may result in imitative skyjackings. Other behaviors can be widely imitated, and fads and fashions can be established. Another possible effect is that expectations are created by the media so that people change their behaviors before an event rather than afterward. Weather forecasting, program advertising, portions of news and sports casting, and much political media usage are intended to alter the manner in which people will respond to future events. In short, there are many kinds of effects that can be hypothesized, as McLeod and Reeves showed in great detail in Chapter 3, this volume.

However, the assumption that exposure to television must have some hypothesized effect because it is so pervasive may be too simplistic. After all, the public is an aggregate of individuals, embedded in various social groups, who may well react differently to what they see. Mr. Gans explored this idea in the fourth chapter of this volume. Most events portrayed through the media do not have a contextual meaning of their own. The meaning given to them may vary from person to person, from time to time, with any given individual, depending on a host of personal, subgroup, or subcultural factors. Consequently it is just as reasonable to propose that some of the hypothetical impact of television is a myth. Television may serve merely as a time-filler and, in that regard, be similar to people-watching at a street corner. Or television programs may be regarded by the viewers as fantasies like daydreams and have no greater impact on them than a trip to an amusement park.

Although it is relatively easy to point to changes in the quantity and kind of mass media to which people have been exposed, there is considerable uncertainty and controversy as to the degree and kinds of effects that may be engendered by this exposure. In comparison to decades ago, more people have more frequent vicarious experiences of world scenes, events, and personalities; more view staged events such as concerts, sports, and theatrical productions; more are aware of public figures such as politicians and entertainment celebrities; more are exposed to audiovisual materials directly without prior filtering through symbolic forms or single sensory modalities; more are exposed to the same, simultaneous experience; and more people share more often the same fantasy and fiction than ever before. One could speculate at length on the potential effects of these changes in media exposure, but it is difficult to document their consequences either for individuals or for societies.

Certain of the difficulties are detailed in some of the preceding chapters. Other chapters focus more on the current state of knowledge of the effects of television. Still others present new challenges or describe new ideas for researchers in this area of social interactions. This volume is, therefore, a potpourri of material for the researcher on television. Lack of integration may be a disappointment but it should not be a criticism in the complex area of television and social behavior. Mass communications lack conceptual coherence and comprises principally a melange of diverse research interests. It is a borrowing area—borrowing theory and principles small and large from the social sciences generally. It is appropriate that it should do so, because communication is ubiquitous in the human experience and is therefore part of any approach to understanding the human condition whatever its disciplinary focus. But, as communication has evolved from person-to-person gestures, to the town crier, to drums and smoke signals, to print and to broadcast media, certain aspects became legitimate subject matter for any interdisciplinary perspective.

Questions of symbolism, comprehension, identification, arousal, level of abstraction, behavioral effects, and similar topics are not unique to mass media communication but are part of exposure to real events, stories, accounts, conversations, and so on. Similarly, questions of content, credibility, distortion, fantasy, and values are not unique to mass communications but take on significance for research and for social concern because of the mass media phenomena themselves.

The sheer existence of television raises questions about use of time—that is, what is done and not done as a consequence of its availability to both "producers" and "receivers." The technology of the various media—and now television—raises new potentials and restrictions. The organization and industries required to mount and maintain broadcasting operations; the systems of control that develop under governmental, commercial, or public corporation regulation; the sheer "massness" of distribution and common exposure; and the richness and distance of access to experience beyond one's directly sensed surroundings make television a *social* research problem and a *social* concern.

The limited scope of the laboratory is needed for much of the work on psychological processes in perceiving and reacting to television formats and presentations, but any general approach to television and social behavior should give heavy consideration to larger social impacts and mass research techniques, concepts, designs, and costs that would begin to improve our poor understanding of the large social problems and phenomena that go along with the "massness" of the television complex.

Limited understanding seems to be the state of the art after some 30 years of research. Perhaps, for progress to be made in the study of television and social behavior, a broader—almost aerial—view of the field is required along with new paradigms and approaches. In the remainder of this chapter, such an aerial view is scanned. First, by getting away from the detail of focused research studies that deal with individuals and specific programs, an aerial view of the territory may reveal areas of potential development.

AN AERIAL VIEW

During this same 30 years of research on television, the period was characterized by the rapid growth of the medium, many technological changes, and new organizational arrangements as well as by a stream of highly significant events and social changes. These two or three decades have fixed some of these studies in a cultural setting that is long gone or in a phase of technological and industrial development that is over. The experimental pioneers of the industry are now old hands, and the audiences have become people who have known television all their lives. The characteristics of the environment that provided context for some of these projects have turned out to be transitory (De Sola Pool, 1974).

Several recent reviews and a recent compilation of the research on television and human behavior (Comstock & Fisher, 1975) show content analyses of programs grown to the status of "cultural indicators"; extended investigations of audience attitudes, motivation, information, and viewing behavior; some work on the industry itself; and diligent efforts to ferret out the educational, political, and social correlates of television exposure in persuasion, modeling, behavioral change, and informational learning. Researchers have shown special interests in certain program content such as violence, and attention has focused heavily on certain audience members, particularly children.

Except for some production research and pretesting and occasional special studies, most research by the industry has entailed audience demographics and motivational and response studies for commercial interests and "research" that schedules programs and then evaluates them by the criteria of absolute size and share of the audience.

In spite of the limited interest of many studies, a body of knowledge on television and social behavior is emerging that has gaps and problems but is beginning to have some stature. A scanning of the work that has been done reveals areas that have not been worked or developed—sometimes for lack of conceptualization, sometimes because of the difficulties of measurement and research design, and sometimes because of the large budgets and time commitments that would be required for extended projects involving repeated assessments over a lengthy period.

If the available reviews and overviews are regarded as aerial photographs, the areas well wooded with research are not as conspicuous as the open spaces either untouched or sparsely marked with the occasional growth of isolated studies. Most of the well-grown areas have dealt with laboratory studies of particular programs or program content with specially selected audiences, most often children and youth. Reactions of interest (dependent variables) have usually been very restricted. In some field studies the viewing has been in a natural setting, but very few effects over extended periods of time have been explored. There are good reasons for a lack of full coverage in topics and designs. It is easier to do research with small and contrived or captive audiences, and short-term effects can be more convincingly related under controlled conditions.

Discussion of the effects of television (or movies, comic books, etc.) have often foundered over the problem of multiple determinants of behavior. For instance, "Evidence that exposure to televised violence increases the probability that viewers may engage in aggressive behavior under certain circumstances does not mean that repeated portrayal of violent methods is either a sufficient, or a necessary, condition of aggression. The same can be said for every other determinant of aggression (Bandura, 1973)": frustration, fear, broken homes, rejecting parents, lax discipline, poverty, oppression, provocative police, opportunity, and so forth. Aggression was not invented with television, but some mix of determinants may be more influential, more often, than others (Scherer, Abeles, & Fischer, 1975).

Discussion also founders over the consideration of shifts in the importance of the same factors for different individuals and different circumstances. Averaged or aggregated influences may almost disappear because in some situations they are powerful but in others negligible. To assign an average value to television effects obscures their powerful effects in particular combinations of circumstances. We may well learn from television how to be a better consumer, how to carry out certain crimes, how to be helpful, how to vote, and what to discuss, but such behaviors are likely to be evoked only under a particular set of coexisting conditions. In addition to multiple causes of behavior we must recognize that this behavior derives from particular patterns of precipitating conditions.

Sometimes we develop and accept hypotheses that contribute to our understanding of particular areas and are blind to what they might contribute to our understanding of similar phenomena. For instance, sellers of commercial time on television argue for the power of television to influence the behavior of viewers at least enough to merit the price of commercial time. It is somewhat illogical not to assume that the programs between the commercials share some of the same efffectiveness, even though they were produced with different objectives. Advertising is often regarded as successful if it influences the information, attitudes, or behavior of a very small percentage of viewers. The number of purchasers of an advertised new car in any one year is a miniscule proportion of television viewers. It would also seem reasonable to regard television as similarly influential if a noncommercial message were "bought" by relatively small minorities, say, less than 1%. The behavior of advertisers strongly suggests that they hold the conviction that this influence is cumulative. It seems illogical then not to assume that other television messages repeated in story plots, news formats, and so on, are also cumulative.

Television as an institution also has an influence on the behavior of other institutions that must deal with television or take it into account. What students and tacticians of political campaigns have learned about the use of television must surely have some bearing on and relevance to other uses and effects of the medium. If elections are won by firming up one's supporters and swinging a small minority of voters who are "changers" or independents, this process may be a model for a number of other steps in social change where small shifts lead to significant social events. Where the audience is not the whole body politic but rather a more limited one of children, young people, the elderly, or some other minority, small shifts in the proportions of those smaller aggregates may also have significant social consequences for them and for the larger society some time later. In short, the public education model, in which the entire public is the target, may not be very applicable to much of television's impact on its audiences.

It is reasonable to propose that our understanding might be enhanced if the scope of what we tried to encompass in our insights were larger. Certainly some policy issues plead for a larger understanding. Though the bulk of what we know is in the area of selected program content and certain audience responses to that content, our knowledge should include some understanding of how program content and program mixes and fads come to be. This pushes us into studies of the industry, its organization, its rewards, its market, its handling of creative enterprise, and its circumscribed legislation. Also, we need to go beyond certain (immediate) audience responses into subsequent and sustained effects and into the dynamics of mass publics. How does television "addiction," popular taste, and mass behavior—which is often the behavior of minorities—feed back into the industry through elites, special interests, investors, and the like?

Through the 30 years research on television the inconsistent performance of our communications effects models has become apparent (McLeod & O'Keefe, 1972). The motif of some very early models was the transmission of the enlightening, enriching, or pernicious content of the media to differentially receptive individuals. This theme matched the broader notion that external forces molded and shaped a relatively hapless individual. Concern over the input from the media was equally great among those who saw an impelled, driven, active individual who needed guidance, restriction, and control. Neither extreme carried the day, and more sophisticated models of the complex interplay among individual and societal forces found their counterpart in ideas about the relations between audiences and media content. Recognition rapidly developed that the media have limited effects and that the social and interpersonal context of individuals makes a difference in the impact of communications from the media. Models were forced to include more than demographic factors and general social conditions as buffers against media effects. Step or cascade flows of information were discovered, and audiences were fractioned into elites, opinion leaders, the sensitized, the indifferent, the resisters, and so forth. The idea of the obstinate audience was devised (Bauer, 1971). It would seem that covert, subtle, and slowly accreting influences of the media must be accomodated into the communications model along with a recognition of the filtering, adaptive, and self-directing capabilities of an individual and the complex social and idiosyncratic variables that intervene between television and its audiences.

An example of a fairly well-worked area is that of television fare in relation to a child's development and the maturing process of perceptual, cognitive, learning, and character development. A much less developed area is that of role stereotypes, modeling, and identification involving often subtle aspects of the portrayal of minorities, occupations, and so forth. An area that has been scarcely touched is the immunization, desensitization, or habituation effects of continued exposure to television fare. It has been proposed that continued reporting of violence, for instance, results not so much in behavioral resort to violence as in lessening concern over its occurrence.

Among the many variables that have heavily engaged research interest, two have received especially widespread attention. One is the link between acquired attitudes and behavior (if television fare changes attitudes, what is the implication for behavior?), and the other is the motivations and gratifications associated with media usage. On the first, the traditional and seemingly obvious perspective has been to regard information as influencing attitudes and these in turn guiding behavior. However, in the literature there are various propositions and associated data that suggest that attitudes lead to behavior, that behaviors lead to attitudes, or that there is some causal interaction; attitudes have also been regarded as independent from behavior (some psychoanalytic studies). These issues raise many conceptual as well as methodological questions.

There is no intrinsic reason why attitudes and behavior could not be in some causal association and also be related as coeffects of still another factor. It is quite possible that attitudes and behavior would tend to have a negative relationship via one set of dynamics and a positive relationship via another. It may be fair to suggest that a variety of dynamics underlie the confusing relationships. To the extent that low relationships are often found, it may reflect not so much the fact that there are no dynamics relating them as the fact that there are several and complex associative patterns involving structural hierarchies. Thus we should not expect simple, direct relationships between media exposure and attitudes and behavior. Nor should we claim that becuse a *simple* study finds no relationships, none exist.

Human motivations have characterized another approach to communications in which various gratifications from use of the mass media are proposed (Katz, Blumler, & Gurevitch, 1974). There is some similarity between this approach and D. Katz's (1960) functional approach to the study of attitudes. McGuire (1974) has developed a structure for these approaches that leads to 16 paradigms in a 4 by 4 table. Gratifications can be active or passive behaviors and internal or external in orientation. They can be in cognitive or affective modes and directed toward having and preserving a gratifying state or a process of growth and development. Thus an active, internally oriented preservation of a cognitive picture is gratifying in the sense of maintaining feelings of "consistency." An active, externally orientated, cognitive growth gratification becomes, in simpler terms, "stimulation." These paradigms may elaborate some of the many complex relations that are possible between various people and the media offerings to which they are exposed.

At the same time models have been devloped that are less individually oriented and more concerned with broad social processes in populations and audience aggregates. The Ball-Rokeach and DeFleur model (1976), which encourages a tripartite treatment of the media and its audiences as integral parts of a larger social system, is an example of several approaches that stress interdependencies among social units and particularly their interactions in times of social change and social conflict.

As students of the media searched for the causes of particular responses by various members of audiences, the simple models have been revised into models of multiple inputs or multiple causes with complex interactions and devious patterns of causes and effects. Similarly, as researchers groped for the effects of particular material, their search broadened to include not one or two effects but multiple and diverse consequences. One could not keep such developments isolated from each other for long, and matrix models of multiple inputs to a system with multiple effects became the appropriate model for the data and phenomena under investigation. Similar methods and models were being developed in studies of the impacts of technology, changes in the economy, and in studies of ecological interactions in the environment.

Most studies have to focus on single or on a few cells of such a matrix. A limited number of aspects of television programs can be studied in relation to a limited set of data about a limited portion of the audience for a limited period. It is not very helpful to point to other interactions in other cells that might have been studied or might have yielded more enlightening results. However, the tactics and strategies of a *program* of research require such comparisons, cross-pollinations, replications, and integrations.

The variety of content in television fare and the diverse tastes of various audiences generate a large number of potential program–audience interactions and relationships. Among the various appeals of television is the variety of its offerings and the mixture of functions it can serve to viewers with various characteristics and living in dissimilar situations. There is no reason to assume that any simple hypothesis about program–audience relationships will hold up substantially well, but there is possibly good reason to imagine that it might work well somewhere in the matrix of television–societal interactions.

On the other hand, there may be some general characteristics of television programming so widespread that the viewer comes to believe that they characterize most of his social environment. The fact that we know crises right away from television news may well generate an impatient feeling that they should be dealt with right away. "Right away," for responsible officials, may then turn out to mean no more than dealing with their treatment in the media. The neat, compact packaging of problems and events in television news may lead one to expect equally neat, compact policies, and easily handled ways of coping with the changing scene. The ease of simply turning a knob to get time-filling entertainment may inhibit more active involvement. Television's emphasis on the spectacle and importance of the winner may have sunk into public consciousness as much as our improved understanding of the dynamics of weather. Certainly a great deal of television constructs for viewers the picture of what is significant and meaningful in our culture and society. Viewers can reinforce this picture by behaving as if it were a valid portrayal. Gerbner and Gross (1976) report that avid television viewers exaggerate the incidence of violence in their areas. Trend data may reveal these effects, but they would be so common in short-term studies that one could not discriminate or isolate these effects from the general background of social conditions and current culture. Some effects may cumulate with an extended but moderate exposure to television fare so that the message sinks in. If so, it may be impossible to differentiate frequent from moderate viewers of television time in terms of such accrued impacts.

Individual differences may be important in how a message is used or how it finds expression in the somewhat idiosyncratic circumstances of an individual's life. But individual differences may be rather inconsequential compared to some of the near-univeral effects of advertising (an image of desirable life style) or the effects of repeated messages supporting the advantages of mood-altering experiences by buying better products, experiencing new interpersonal situations, hearing a new musical group, or joining a new movement. An inclination for these experiences and their ready obtainability may well be inculcated without contributing directly to the use of mood-altering substances (Milavsky, Pekowsky, & Stripp, 1976).

One impetus for larger and more comprehensive models of television and social behavior has come from efforts to put into effect ideas about new forms of television offerings, functions, and services. The kinds of thinking and activity that led to the development of public broadcasting, the staging of "Roots," the organization of Operation Prime Time, the education television network, pay television, audience participation in television tests, and other new developments required more than an idea for a new program and more than production research on its content and format. Different forms of financing, organization, and management were required to create the intended changes in television's product.

Normative research on alternative ways of organizing and structuring society's communications systems is part of the approach of policy research to deal with cable, satellites, political and business uses, educational and economic development, links to computer systems, new forms of marketing, international relations, and shifts in the ecology of human settlements. Aspects of telecommunications policy are developing in most division of government and in most of the governments of major countries. In the United States a total revision of the Communications Act of 1934 is being discussed. If not social behavior in every sense, these changes and preoccupations are certainly changes in the behavior of society.

New and developing technologies for production, transmission, storage, retrieval, recording, duplicating and reproducing, printing, displaying, and so on are linking and interweaving the various media into an increasingly commonly shared web. The forces and conditions of production, organization, financing, marketing, and product development are turning out to be equally interrelated. The market for popular taste industries creates certain forms of organization and management. The market for news operates differently, and the market for educational products presents yet new variations and patterns. In the television industry they seem to mix and merge.

Any aerial view of the research scene for studying television and social behavior provides only distant views of the actual terrain where research is done. The various chapters of this volume have given a much closer view of specific areas of interest to the authors. They elaborate in more detail difficulties that are encountered and provocative ideas for making these problems tractable. They were selective. They will not, if pieced together, fill in the whole area like the snapshots from a satellite. But they have provided close views of major features of this extensive and changing landscape.

REFERENCES

Ball-Rokeach, S. J. & DeFleur, M. L. A dependency model of mass-media effects. *Communication Research, 3,* 1976, 3–21.

Bandura, A. *Aggression: A social learning analysis.* Englewood Cliffs, N.J.: Prentice-Hall, 1973.

Bauer, R. A. The obstinate audience: The influence process from the point of view of social communication. In W. Schramm & D. F. Roberts (Eds.), *The process and effects of mass communications* (Red. ed.). Urbana Ill.: University of Illinois Press, 1971.

Boorstin, D. J. *The image, or what happened to the American dream.* New York: Atheneum, 1961.

Comstock, G., & Fisher, M. *Television and human behavior: A guide to the pertinent scientific literature.* Santa Monica: Rand, 1975.

De Sola Pool, I. The rise of communications policy research. *Journal of Communication,* 1974, *24,* 31–42.

Gerbner, G., & Gross, L. Living with television: The violence profile. *Journal of Communication,* 1976, *26,* 172–199.

Katz, D. The functional approach to the study of attitudes. *Public Opinion Quarterly,* 1960, *24,* 163–204.

Katz, E., Blumler, J. P., & Gurevitch, M. Utilization of mass communication by the individual. In J. G. Blumler & E. Katz, (Eds.), *The uses of mass communications. Sage Annual Reviews of Communication Research,* (Vol. 3). Beverly Hills: Sage, 1974.

McGuire, W. J. The uses of mass communications: Current perspectives on gratifications research. In J. G. Blumler & E. Katz (Eds.), *The uses of mass communications. Sage Annual Reviews of Communications Research,* (Vol. 3), Beverly Hills: Sage, 1974.

McLeod, J. M., & O'Keefe, G. J. The socialization perspective and communication behavior. In F. G. Kline & P. J. Tichenon (Eds.), *Current perspectives in mass communications research. Sage Annual Reviews of Communication Research* (Vol. 1). Beverly Hills: Sage, 1972.

Milavsky, J. R., Pekowsky, B., & Stripp, H. TV drug advertising and proprietary and illicit drug use among teenage boys. *Public Opinion Quarterly,* 1975–1976, *39,* 457–481.

Scherer, K. R., Abeles, R. P., & Fischer, C. S. *Human aggression and conflict: Interdisciplinary perspectives.* Englewood Cliffs, N.J.: Prentice-Hall, 1975.

APPENDIXES

Appendix I:
Members and Staff of the Committee on Television and Social Behavior

Stephen B. Withey
Chairman
University of Michigan

Leo Bogart
Newspaper Advertising
    Bureau, Inc.

Aimée Dorr
University of Southern California

Herbert J. Gans
Columbia University and Center
    for Policy Research

Hilde T. Himmelweit
London School of Economics
    and Political Science

Irving L. Janis
Yale University

Eleanor E. Maccoby
Stanford University (1972–1973)

Jack M. McLeod
University of Wisconsin

Chester M. Pierce
Harvard University

Percy H. Tannenbaum
University of California, Berkeley

Staff:

Ronald P. Abeles (1974–1978)
David A. Statt (1972–1974)

Appendix II:
A Profile of Televised Violence[1]

INTRODUCTION

Background

The effect of television on social behavior, especially on the behavior of children, has long been an area of shared concern by the general public, its elected representatives, and the scientific community. This concern has centered around several fears: First, the portrayal of violence through direct imitation and by conveying the message that violent solutions to problems are frequent, acceptable, and even preferable. Thus viewers might be influenced by television to perform violent acts.

Second, televised violence might influence people's social values and beliefs about society. Constant exposure to violence might make viewers indifferent to the suffering of others and accept violence in real life. On the basis of television programming, people may develop an inaccurate view of the prevalence of violence in society and of social relationships. This might lead to unrealistic fears and inappropriate or maladaptive behaviors.

Finally, the ill effects of viewing television violence may occur over a long period and may grow slowly in magnitude, even if short-term or immediate effects are negligible. The negative consequences might not appear immediately but many years later. This might particularly be true in the case of impressionable children and youths.

These concerns resulted recently in the Surgeon General's Scientific Advisory Committee on Television and Social Behavior's Report (*Television and Growing Up: The Impact of Televised Violence*, December 1971) that

---

[1]Report submitted by the Committee on Television and Social Behavior of the Social Science Research Council, July 1975.

307

appraised the evidence for a relationship between violence on television and children's aggressive behavior. In summarizing its findings the committee concluded: "While the data are by no means wholly consistent or conclusive, there is evidence that a modest relationship does exist between the viewing of violence and aggressive behavior [p. 9]."

As a result of hearings on the report before the Senate Subcommittee on Communications (March 1972), Senators Warren B. Magnuson and John O. Pastore wrote to the Secretary of Health, Education, and Welfare, Elliot L. Richardson, requesting that "[his] Department proceed in consultation with the Federal Communications Commission to develop a measurement of violence on television so that a report can be submitted annually to this Committee on the level of violence entering American homes [p. 374]." In response to this request, staff from the National Institute of Mental Health and the Federal Communications Commission met with social and behavioral scientists on June 2, 1972 to discuss the development of measures of televised violence. The consultants agreed that the development of such measures was feasible and recommended that a nongovernmental group participate in the planning and instigation of further research on measures of televised violence and on television and social behavior.

In January 1973 the Social Science Research Council submitted a proposal to NIMH for support of a Committee on Television and Social Behavior. The Committee was funded and established in September 1973 with the overall objective of planning and stimulating new research on television and social behavior. The committee set as one of its aims to conceptualize and give scientific context to the research required for the development of a multidimensional profile of violence in television programming. The committee has met six times between October 1973 and April 1975 to discuss the problems associated with developing a profile of television and to plan new research on television and social behavior. The remainder of this report consists of the committee's concerns about and recommendations for the development of a television violence profile.

Considerations

Despite the controversy surrounding the Surgeon General's Report, the committee agrees with the assessment that there are sufficient grounds for concern about the relationship between televised violence and violent behavior. However, several considerations should be kept in mind while developing a profile of televised violence.

First, the relative importance of television content as a cause of societal violence in comparison to other potential causes is unresolved. Television is only one medium, and violence in motion pictures and printed matter is often more vividly portrayed than in television. Beyond the media, there are many causes of violence, and a constellation of factors over a lengthy period would have to change to create a significant reduction in societal violence. Thus, reducing television violence may have little effect. Moreover, the consequences of television violence are not so well understood as to suggest any reduction in research on these effects. This does *not* imply that no efforts should be made to reduce the effects of television violence, only that the relative importance of different causes of violence and the relative ease of countering them should be considered before counter measures are implemented.

Second, by itself the development or publication of a television violence profile will not be effective in reducing violence on television. This depends on the use to which the profile is put and, in the last resort, on the policy decisions of the television industry. If a violence profile is developed, its uses by and impact on the television industry and the public should be evaluated carefully.

Finally, any set of indices may be used and abused. The uses may be coercive and restrictive or monitorial and educative. Efforts should be made to guard against the dangers of censorship that may accompany the implementation of a violence profile. This is particularly true in regard to any monitoring of nonfictional programming such as the "news." The committee recommends strongly that news and documentaries be excluded from the measurement domain of a violence profile.

Purpose of a Television Violence Profile

The committee assumes that the function of a violence profile is to inform viewers, producers, advertisers, policy-makers, legislators, scientists, and others about the content of television. However, the committee realizes that the major options for significantly altering television programming rest with the television industry.

The committee's problem is to propose measures that adequately inform about the amount and kind of violent content in television programming. The measures should enable consumers and other groups, including television industry personnel, to differentiate various forms of violence on television. The committee sees an educative function in making the public and the television industry aware of what is being shown. Changing television fare is a large and complex problem but not one that the committee is addressing. Whether parents then restrict television-viewing or minimize some of the effects by ridicule or disapproval or some other tactic is a question of parental policy. Whether the industry is monitored by consumers or by other groups and how the industry might itself use such indicators is also a question of policy.

No resolution of questions about the character and format of a profile of television violence is possible without specifying concerns of the profile's users and the resources of the profile's developers. If the information contained in a profile is to be given wide distribution to the general public, it should be relatively simple for reporting purposes and adaptable to the needs of individual families. Profiles for use within the television industry could be more detailed to monitor progress in implementing various changes in programming. If profiles were developed for research purposes, they would need to be even more detailed and to include information on the relevance of content to various characteristics of viewers. In sum, the construction of a violence profile is highly dependent on the uses for which it is intended.

PROBLEMS IN DEFINING VIOLENCE

Commonalities in Definitions

The concept of violence is complex, and any attempt to define it seems to lead to dispute and controversy. However, some aspects of violence are usually discussed in most definitions of violence.

First, violence is frequently defined in terms of the intensity of the harm-doer's behavior. The term "violence" is often employed when the behavior is viewed as excessive, with the implication of violent activity as being unrestrained or wanton. Frequently, a value judgment about the harm-doer and his justifications for his behavior is associated with labeling his behavior as "violent." People tend to regard behavior as "violent" when they believe it is "bad," employed for "evil" or "antisocial" ends, or "unjustified."

Within this perpetrator-oriented approach, distinctions can be made between instrumental and expressive violence and between intentional and unintentional violence. *Instrumental violence* is done to achieve some goal, while *expressive violence* occurs without planning in a state of hate or rage. Thus the latter is often a goal in itself, while the former is a means to another goal.

Intentional violence refers to incidents where the perpetrator acts with the knowledge and desire that his behavior will have dire consequences for his victim. He acts purposely to harm another person. *Unintentional violence* involves cases where the harm-doer does *not* know or desire that his behavior harms someone else. Indeed, the harm-doer may have even attempted to benefit his unintended victim or he may have acted without knowing his behavior would have any consequences, good or bad, for another person. Both intentional and unintentional violence can be combined with instrumental and expessive violence. Finally, the concept of unintentional violence distinguishes "violence" from "aggression," which is usually defined in terms of the actor's intentions to do harm. Thus unintentional violence allows one to categorize as violent those harmful acts caused by nonhuman agents, such as animals or natural events. Floods, earthquakes, hurricanes, and the like can be considered to be violent incidents, since they result in harm to those in their path. The concept of unintentional violence draws attention to the need to consider both the perpetrator's behavior and the consequences of that behavior for their victims.

A second perspective in defining violence considers killing, maiming, and inflicting pain as "violence" because of the magnitude of the effects from the victim's point of view. This would be true regardless of the human or nonhuman origin of the effect or regardless of the perpetrator's intentions, self-control, or degree of exerted effort. The Surgeon General's Scientific Advisory Committee on Television and Social Behavior (1971) had this victim-oriented definition in mind when it wrote:

> In order to define violence as realistically, as ethically, and with as much psychological accuracy as possible, the definition should be broadened to include the experience of its victims. The physical and psychological violence experienced under circumstances ... such as pollution, unsafe drugs, accidents resulting from defective equipment, restrictive or war-oriented legislation, etc. ... have been caused by the acts of other human beings ... When a society legislates and institutionalizes the definition of violence in terms of victims, then all violent experience becomes a matter of concern. When the definition reflects only accountable destructive behavior, much, if not most, violent experience may not even be acknowledged [p. 9].

Three Kinds of Definitions

Besides these general considerations that might be applied in defining violence, three kinds of definitions or criteria might be used in identifying and profiling televised violence.

First, one might include in a profile of television violence only those televised portrayals that are known to have specific ill effects upon the audience. Such a specific definition of televised violence based on careful empirical research would very probably be the most acceptable basis for a profile. But there are problems in assessing "ill effects," their probability, their timing, the proportion of the population affected, the types of individuals influenced, and the seriousness of their behaviors. Although such an empirically based definition might be ideal, the current body of empirical studies on the effects of television is not sufficiently clear or inclusive to serve as a basis for the labeling of television content as violent.

Second, one might base a profile on the judgments of television viewers as to what aspect of television programming are undesired portrayals of violence. People can be asked to record their perceptions of violence, but people would not always agree. In addition, Nielson ratings already show that many violent programs are not labeled "unpopular" or "undesired." It is probably true that most observers would have a fair degree of agreement among themselves on the relative levels of violence to be found in a set of television programs. However, differences in judgment would be clearly apparent and differences would be most common in the coding of individual segments and episodes (Blumenthal, Kahn, Andrews, & Head, 1972). In addition, our focus of concern is the relatively unsophisticated, young viewer whose judgments on these matters would be most suspect. Most adults would be reluctant to leave the judging of programs to children, and adults would have to do the judging for the children. This raises the problem of who should be permitted the privilege or the responsibility of making these value-laden decisions.

Third, violence could be defined in specific content terms with clear enough cues so that the indicators could be readily used by anyone with a little training who observes and can code a screened presentation. One could identify weapons or observe the use of physical force, and it would not be difficult to identify willing or unwilling victims and so forth.

It is impossible to formulate a single definition of violence that results in agreement and meets all purposes of a violence profile. The specific content approach seems most feasible. The committee therefore proposes a rather broad approach to identifying violent content and proposes that other specific characteristics of such episodes depicting violence be included in the indicator profiles.

Too Narrow A Definition

One could list a number of events and obtain widespread agreement that *all* of them could be categorized as displaying or portraying "violence":

A bully drowns a young child
A woman is sadistically tortured by an intruder
A pedestrian is knifed by a mugger
A street fight among adolescents using chains, bats, and blades
Ward attendants kick an old man into insensibility
A policeman is brutal

They are all violent crimes, and there is no commonly accepted justification for any of the behaviors. A person perpetrates the crime, and there is clear injury and pain to another, apparently innocent, person. The events involve death, serious injury, or extreme pain, and the violence is physical (not psychological, economic, etc.). They all involve some awareness, and perhaps planning, of what is going on (none of them are accidents).

If we had included "armed hijacking of an airplane with passengers and crew," many people would have accepted the categorization of "violence" even though death or injury is only threatened. Also, if we had included "a woman handing out poisoned Halloween candy," many observers would see "violence," even though no threat was expressed and no death or pain was observable—the consequence is highly probable.

A core definition, to maximize agreement, could be limited to completed violence. But this would omit the consideration of serious and compelling threats resulting in almost certain pain and injury off screen.

A Moderately Broad Definition

Dr. George Gerbner in his monitoring of television content suggests a broader definition of violence. He defines violence without regard to its justification or condemnation, involving overt action and serious actual or potential consequences and excluding voluntary participations or willingness on the parts of the victim (which excludes violence in sports events). His definition (Gerbner & Gross, 1974) is: "The overt expression of physical force against other or self, compelling action against one's will on pain of being hurt or killed or actually hurting or killing [p. 5]." According to this definition, if a comedian says, "I'm going to kill you," this is not violent. But if a cartoon character slams the door on another's nose, that is violent. If a woman is raped and the television audience hears the sounds of the struggle but does not see it happen, that is violent. However, by Gerbner's definition just hearing a report about the rape is not violent.

Under such criteria as Dr. Gerbner's, there is very little recourse to inferential judgment in coding television material. It might be added that his definition does not necessarily require a human agent. It encompasses unintentional violence, as defined above, and therefore includes such acts of nature as a burning skyscraper, an earthquake, or a traffic accident.

It may be instructive to know that by this definition one study showed that television has aired from 5 to 9 violent incidents an hour, some stations going much higher, and that densities are particularly high in cartoons and on Saturday mornings and after school hours. Cartoons, for instance, may go as high as 25 to 30 incidents per hour (Gerbner & Gross, 1974). This definition could be as readily used with duration measures as with episode counts. Gerbner has used this kind of approach and has increased his categories over the years, so that he now reports various tabulations for killing versus lesser violence, with various categories of program themes and settings and characteristics of committers and sufferers by sex, age, marital status, socioeconomic class, nationality, race, and role. Much of it is tabulated over a seven-year span. His is unquestionably the most comprehensive research that carries out an accounting of these aspects of television content.

PROBLEMS IN THE DEVELOPMENT OF
A VIOLENCE PROFILE

Multiple Indicators Required

Assuming agreement can be reached on a definition of what events to include in the category of "violent," the question still remains as to what should be indexed about those events. The argument that it is practically or politically useful to have only a single index figure of televised violence is unconvincing because it completely overlooks the nature of the material to be indicated. The committee proposes that the purposes for a television measure of violence can only be met if the information carried by the indicators is considerably richer and more detailed than a single index would allow.

The need for multiple indicators of television violence is exemplified by the inadequacy of a single crime rate index of actual violence. A crime rate index that added cases of larceny, murder, fraud, and malpractice would be nothing more than the total tally of known law-breaking and would be insensitive to various changes in crimes or to changes in their definitions. If burglaries went down and murders went up the same amount, the index would show no change even though the crime scene would have changed radically. Weather-reporting also provides a suggestive example. No one wants a single figure to index the weather. Even the causal consumer of weather information would like to know temperature ranges, and some people use pollution level, pollen count, and precipitation reports. Men in fishing fleets might check wind speeds and direction, and outside workers might be concerned with humidity and wind–chill factors. A traveler wants more than local weather data, and a plane pilot would check ground level weather and visibility and also the whole complex of weather conditions at various altitudes.

These examples from crime and weather reporting suggest that a single index would be an inadequate indicator of television violence. Consequently the committee proposes that a profile consisting of measures of different aspects of televised violence be developed as opposed to a single index.

What To Count

After an acceptable definition of violence is achieved, one indicator of violence could be a simple counting of the number of violent acts per program or per unit of time. This would provide information about the *density* or *frequency* of violence on television. While informative, a measure based solely on the frequency of violence would not be adequate by itself. Such an indicator would make programs appear less violent if episodes were reduced in number but drawn out in time.

A complementary indicator to density would be to clock the minutes devoted to violence as a proportion of total program time (excluding commercials). But such a *duration* measure would not be without its difficulties. For example, there is some ambiguity as to when violence begins. A shot or stabbing takes only a second, but the violent "episode" surely includes part of the build up and preparation, the period of threat and stalking during which the violence is expected. In addition, a duration measure would not be infallible by itself. Programs could be made to appear less violent if the degree of violence were escalated while shortening the time devoted to violence in the total program.

Consequently a measure of the *seriousness* of each violent act is also required. An episode's degree of violence could be indicated by providing coders with models or examples of levels of violence against which the aired episode could be matched. Perhaps a first attempt at developing a seriousness indicator might be limited to categorizing violent acts as "serious" versus "not serious." The proportion of serious acts out of all violent episodes could then be reported.

The committee recommends a multiple indicator profile composed of measures of the density, duration, and seriousness of televised violence.[2]

Format of Televised Violence

Another problem to confront in indicator development is the format of portrayal of violence. Must the violence actually be shown? What if the actual act is off screen but actors' reactions are on camera? What if the violence is just described or recounted? How about techniques of presentation such as musical underscoring, zooming in to portray the fear on the face of the victim or the reactions of the aggressor, close-ups showing the "realism" of injuries and the aftereffects of violence? These and other techniques all add to the impact and emotionality of the portrayal of violence.

Although the committee has no specific recommendations on how to confront this type of difficulty, it believes this problem should be carefully considered during the development of a violence profile.

Programs: A Focus on Children

Most people, including the committee, are more concerned over television material to which children are exposed than they are with other programs. Consequently, primary attention should be given to measuring violence in programs with large child audiences and to reporting the violence profile in a manner relevant to this concern.

---

[2]In addition to these measures that are standardized by "per unit time" or by "per program," it may prove useful to report the total number of violent incidences portrayed over time and programs.

Some programs are targeted particularly at children, such as those aired on Saturday mornings and during the pre- and afterschool hours on weekdays. Of course children are not restricted in their viewing to "children's programs." In fact, about six million children between two and eleven years of age watch prime television (7:30 to 11 p.m.) and are exposed to "adult programs." Consequently the committee recommends, on the basis of information from the Surgeon General's Report (1971), that televised violence be measured during these hours:

Weekdays: 7 to 9 a.m. and 3 to 11 p.m.
Weekends 7 a.m. to 2 p.m.

This focus on the hours of heavy viewing by children raises the possibility of weighting a profile by the kind of exposure a program gets. The user of televised violence profiles may in some cases not be interested in just an accounting of violence but might be interested in the size of the audience, exposed to the materials. Some shows that are aired very late in the day, for example, do not attract much of an audience, while others are widely viewed. One might consider differentiating profiles of "violence aired" from profiles of "violence received."

In terms of reporting the profile in a manner relevant to the concern with children, the profile should be reported separately by types of programs. For example, the amount of violence (i.e, its density, duration, and seriousness) should be reported separately for cartoon programs, all "children programming," and for all "adult programming." Reporting categories should be developed in order to differentiate programs in terms of their exposure to children.

Sampling Times During the Broadcasting Year

The developer of indicators of television violence is faced with the question of when to measure television violence during the course of the broadcasting year. Given the logistical problems and costs of measuring and reporting data on a daily basis,[3] some choice has to be made to sample—say, one or two weeks out of 52 weeks. In selecting sample weeks, attention should be given to their importance to the television industry and to their representativeness. The former refers to those weeks during which new programs are aired and/or during which important ratings of audience size are made by such organizations as Nielsen and the FCC. Thus these weeks have important implications for the decision to keep a television program or remove it from the air. The latter refers to how "typical" the sample week is in comparison to the rest of the year's programming fare.

---

[3]To monitor and code 75 hours of television programming, it currently costs Dr. Gerbner and his colleagues $62,594.

With these considerations in mind, the committee recommends that two sample weeks per television season be drawn in which one week is a critical decision-making week (e.g., Nielsen rating week) and the other is a random composite week (i.e., a Monday from one randomly chosen week, a Tuesday from another week, a Wednesday from a third week, etc.).

Priorities in Developing a Profile

The possibilities for increasing the scope, coverage, and detail of indicators of television violence are such that some priorities need to be set. The intended uses of the profiles set some priorities, because too much detail makes the material hard to communicate and unusable. There are also financial limits. Developing a profile for every station or total air time would be prohibitively expensive. The major viewing times of children are proposed as the top priority, and since no accounting can be continuous some time sampling will have to be used. We also propose that the profile be applied to fictional entertainment only and that news and documentaries be excluded.

CONCLUSIONS AND RECOMMENDATIONS

Conclusions and Assumptions

We conclude that there is enough evidence on some negative effects of television violence to warrant the development and use of indicators of televised violence.

In constructing indicators of televised violence, careful consideration should be given to the intended uses of the indicators. The intended uses will influence the format of the indicators and the resources required for their development. The committee assumes that the purposes in such development are those of monitoring and education and not censorship.

After exploring the problems associated with defining "violence" for the purpose of developing television indicators, the committee proposes a modified version of Dr. Gerbner's definition.

The broad nature of this definition shows the unsuitability of a single index of television violence. The evolution of Dr. Gerbner's work indicates the advantages of several indicators of television content. Going beyond his measurements, the committee proposes a profile with three components measuring the density, duration, and seriousness of television violence.

The committee recognizes that any proposal should be feasible. Taken in full, the committee's proposals could become quite costly. We recognize that the degree of their acceptance is dependent on the assessment of the seriousness of the problem and decisions on the resources to be committed to this activity.

Recommendations

In discussing the problems of developing a profile, the committee has already made several recommendations. In this section of our report, these recommendations will be collated and expanded. In making these recommendations, the committee assumes that the public's major concern is with violence appearing in programs to which children are heavily exposed. For these purposes the committee has defined children as being 13 years of age or less.

Aspects of Television Violence to be Measured. The committee proposes the development of a *profile,* whose three components are reported separately, as opposed to a single index figure. The three recommended components are:

(1) Density: Number of violent acts per program (or per unit of time).

(2) Duration: Proportion of total programming time containing violence (i.e., number of minutes of violence divided by number of minutes of total programming, excluding programming time devoted to news, documentaries, and commercials).

(3) Seriousness: Severity of violence reported in terms of number or proportion of "seriously" violent acts out of the total number of violent acts.

Although the committee recommends the separate reporting of each component, it leaves open the eventual advisability of combining the three components into a summary figure. Initially, the aggregation of the components should be avoided and should not be undertaken without prior research into the technical problems involved, the understanding of the profile by its users, the consequences for the intended functions of the violence (i.e., density, duration, and seriousness measures).

The committee suggests that the following other aspects of television violence be measured and reported but not necessarily included in the standard summaries constituting the most commonly reported profile of violence (i.e., density duration, and seriousness measures).

1. Characteristics of harm-doers and victims (separate accountings by harm-doers and victims).
   a) Sex
   b) Age
   c) Race
   d) Nationality and/or ethnicity
   e) Socioeconomic level

(2) Characteristics of violent situations.
   a) Serious versus comic
   b) Effectiveness of violence in achieving harm-doer's goals and/or in thwarting victim's goals.
   c) Location in historic time: past, present, future
   d) Physical location: domestic versus foreign; urban versus rural

The foregoing is not an exhaustive list of the characteristics of actors and situations that might be important in assessing televised violence. It is not necessary to include these other aspects of programming in order to have a profile of violence. However, there is much to be gained by keeping track of other aspects of violent episodes, because they provide insights into the message–content of the aired material to which people, particularly children, are exposed. If the purpose of a set of indicators is to monitor violence on television, it is of interest to know who is perpetrating the violence, who are the victims, and when does violence occur. This information provides the profile-users with a clearer picture of changing patterns in televised violence than do measures of density, duration, and seriousness by themselves.

Programs to be Coded. The committee recommends the coding of the violent content in all regularly scheduled and special programs aired on all of the major network stations and on a sample of independent stations during the time sampling period (specified below) except for (1) news and documentary programs and (2) commercials. Besides measuring violence during network programming on ABC, CBS, and NBC affiliated stations, nonnetwork programming should be coded in a sample of stations selected to represent *network-affiliated stations* (in nonnetwork time) and *independent stations.*

Time Sampling Period. As noted earlier, the committee suggests restricting the coverage of television broadcasting time to those hours in which children are likely to be viewing. On the basis of presently available data on the viewing habits of children (those 13 years old or less), the following time periods should be monitored:

Weekdays: 7 to 9 a.m. and 3 to 11 p.m.
Weekends: 7 a.m. to 2 p.m.

Developing a profile of the total year's programming would be prohibitively expensive both in terms of man-hours and money. Consequently, a sample of weeks should be drawn on the basis of their importance to the television industry's decisions about programming and on their representativeness of the year's programming. The committee proposes that two sample weeks per television season be drawn in which one week is a critical decision-making week (e.g., Nielsen rating week) and the other is a random composite week (i.e., a Monday from one randomly chosen week, a Tuesday from another randomly selected week, etc.).

Reporting Categories. Besides reporting the violence profile based on the total sample of programs, various reporting categories should be developed. The density, duration, and seriousness of televised violence should be reported for programs falling within each category and subcategory. A few suggested reporting categories are:

(1) Types of violence
   a) Intentional versus unintentional
   b) Instrumental versus expressive
(2) Types of programs
   a) Cartoon versus noncartoon
   b) Children's versus adults' programs
   c) Crime, police, and detective programs
   d) Other fiction programs (e.g., including family comedies, adventure shows, etc.)
   e) Nonfiction programs, excluding news and documentaries but including such programs as variety and quiz shows
(3) Types of social actors
   a) Agents of social control (e.g., police, private detectives, etc.)
   b) Criminals and other antisocial characters
   c) Heroes and other "good guys" versus villains or "bad guys" who are not agents of social control or criminals, respectively
(4) Types of television networks
   a) Major national networks (NBC, CBS, and ABC)
   b) Network-affilitated stations during nonnetwork programming
   c) Independent stations

Advisory Committee. The committee proposes that a technical advisory committee be established to serve as an overseer and consultant during the development and implementation of the violence profile. Either the technical advisory committee or a separate body should engage in evaluation research on how the profile is employed by different user groups (e.g., parents, television executives) and with what effects. Once a profile and reporting categories are developed, they should not be considered as immutable. Frequent evaluations of the effectiveness of the profile and reporting categories should be carried out.

Funding. As yet the committee has not specified the needs for research projects and further work in the area of television and social behavior or the types of indicators that might be valuable in such work. It is clear, however, that research-based knowledge existing today is not adequate for answering all of the questions and concerns raised vis-à-vis, in particular, the development of a violence profile or, in general, the effects of television on social behavior. This means that more research is required. Consequently the committee is concerned that the funds for the development of a violence profile not be taken from those available for "basic" research. If a violence profile is developed, it should be funded by a regulator agency, such as the Federal Communications Commission, and not by an agency such as the National Institute for Mental Health.

CONCLUDING STATEMENT

Gerbner's Violence Profile

Obviously, the committee's thinking about the development of a violence profile has been strongly influenced by the work of Dr. George Gerbner and his colleagues at the University of Pennsylvania. Their systematic and longitudinal collection and coding of television content is now entering its seventh year and provides valuable information about the "world of television." In making its recommendations, the committee gratefully acknowledges its debts to Dr. Gerbner and his colleagues.

While there are similarities between the committee's recommendations and Dr. Gerbner's violence profile (e.g., the definition of violence), there are differences in terms of (1) the component indicators of the violence profile, (2) types of programs coded, (3) selection and number of sample weeks, (4) selection and number of monitored hours of programming per week, and (5) reporting categories.

Chart I presents a comparison of Gerbner's and the committee's recommended violence profiles. The entries are self-explanatory, except in the case of "Profile Components." Gerbner's "rate" and the committee's "density" measures are identical; both measure the frequency of violence per program or time unit. Gerbner's "prevalence" measure is the proportion of programs containing any violence, regardless of frequency or seriousness of violence. "Role" refers to the characterization of perpetrators of violence and their victims and is composed of several measures: (1) per cent of perpetrators out of all characters, (2) per cent of victims, (3) per cent of victims or perpetrators, (4) per cent of killers, and (5) per cent of characters killed, and (6) per cent of all characters who are either killed or are killers. Thus to some

CHART I
Comparison of Gerbner's and Committee's Violence Profiles

	Gerbner[a]	Committee
Profile components[b]	Prevalence	Density
	Rate	Duration
	Role	Seriousness
	Program content	Supplementary codes
	Violence index	Characteristics of
	Program score	harm-doers
	Character score	Characteristics of
		violent situations
Programs coded	Network	Network and nonnetwork
	Fiction (dramatic)	Fiction (dramatic)
		Nonfiction, *except* for
		news, documentary,
		sports, and commer-
		cial programs
	Regularly scheduled	Regularly scheduled
		and specials
	Weekend children's	Weekday and weekend
	programs	children's programs
Broadcasting hours monitored	Monday–Sunday: 8–11 p.m.	Monday–Friday 7–9 a.m. and 3–11 p.m.
	Saturday–Sunday: 8 a.m.–2 p.m.	Saturday–Sunday: 7 a.m.–2 p.m.
Sample weeks	One week in fall TV "season"	Two weeks per TV "season"
		One critical week
		One random composite week
Reporting categories	All programs	All programs
	Networks	Networks
	Cartoon versus general	Cartoon versus general
	Program types	Program types
		Types of violence
		Types of social actors

[a]From Gerbner & Gross, 1974.
[b]See text for explanation of terms.

extent "role" corresponds to the concept behind the committee's seriousness measure (i.e., level of violence present), and the information about the prevalence of killers and victims is available from the supplementary codes suggested by the committee. Finally, the Gerbner profile contains a composite "violence index" that is composed of a "program score" and a "character score." The program score is a weighted sum of the prevalence and rate measures, and the character score is a weighted sum of the per cent killers and the per cent killed from the role measures (Gerbner & Gross, 1974). At this time the committee does not recommend the aggregation of measures. Finally, Gerbner's "program content" indicator refers to the characteristics of programs in such terms as locale, time, tone, and thematic structure. Most of these are included in the committee's "supplementary codes."

Beyond the composition of the profiles themselves, Chart I reveals that the committee's suggestions are mainly expansions of the scope of the Gerbner profile. The committee recommends expanding the types of program monitored, enlarging the number of broadcasting hours, selecting different and more sample weeks, and reporting breakdowns of the profile by different categories.

Summary

Expert opinion, based on studies, indicates that there are some negative effects of the violence aired on television. Studies also indicate that there is some concern over this problem on the part of the adult public. Most of the concern and most of the evidence points to effects on children and youth.

People do not agree on what to call violence in a strict definitional sense, but there is probably broad agreement on the bulk of televised episodes that are candidates for such a classification (though people would differ on whether they were bad or good or neither) and fair agreement on rough orderings of the seriousness of violence portrayed.

The nature of indicators of violence on television is heavily dependent on the uses to which such measures are going to be put. There is no clear specification of these uses and purposes, and one can imagine several. We assumed informational uses involved in monitoring program offerings and educating interested parties in the violent content of television fare.

We have proposed a broad definition of violence and several sets of categories to serve as a guide to content coding and reporting. These categories could be fully elaborated for some reporting purposes and reduced or combined for other reporting objectives. In all cases, however, we propose a profile of program characteristics be reported rather than any single index. No single rating system or single index labeling is proposed, but a number of profiles, simple or complex, can be derived from the indicators proposed.

The profile would provide information about the violent content appearing during those hours when children are likely to be viewing television. It is based on monitoring network programming and a highly clustered sample of aired programs from network-affiliated stations and independent stations as sampled by two weeks of broadcasting per television season. Using the same categories, definitions, and methodologies, any group could monitor stations or areas of their choosing and develop profiles that could be compared with those developed and reported by officially contracted organizations.[4] Producers, networks, or stations could also use the profiles in researching and reporting their own products.

Much is still unknown about the consequences of televised violence. Indeed the same can be said about the consequences of any type of television content. More research is required to assess carefully the different effects of television; the differential impact of various aspects of television (e.g., the medium per se vs. its content); the social and psychological circumstances under which particular effects are likely; the characteristics of individuals who are more or less influenced by television; the social and psychological functions that television serves for different people, etc. These are major issues that are not yet adequately understood.

In submitting this report the committee does not necessarily recommend the profile as the *most* effective or even as *an* effective means toward reducing violence in American society. Given the existing gaps in knowledge about television's effects (as just cited above), it is questionable whether the production of a violence profile would contribute to a reduction in actual violence. The committee is well aware that televised violence is but one of many potential contributors to violence. More efforts are required to weigh the relative importance of various potential causes of violence in order to recommend steps toward reducing violence. In addition, attention should be paid to the relative ease (in the sense of cost-effectiveness) of altering various contributors to violence.

If sound social policies vis-à-vis television programming are to be developed, an expanding base of empirical knowledge about the effects of television is required.

REFERENCES

Blumenthal, Kahn, Andrews, & Head. *Justifying violence: Attitudes of American men.* Ann Arbor: Institute for Social Research, University of Michigan, 1972.

Gerbner, G., & Gross, L. *Violence Profile No. 6.* Philadephia: Annenberg School of Communications, University of Pennsylvania, 1974, p. 5.

Hearings before the Subcommittee on Communications of the Committee on Commerce, The United States Senate, 92nd Congress, 2nd Session, March 21–24, 1972. Washington: The U.S. Government Printing Office, p. 374.

Surgeon General's Scientific Advisory Committee on Television and Social Behavior. *Television and growing up: The impact of televised violence.* National Institute of Mental Health, Rockville, Md., 1971, p. 9.

---

[4]The BBC in Great Britain monitored the violent content of six months of all fictional television programming produced by BBC and ITV (excluding news and sports). Volunteers, who had some training, were used, with two volunteers monitoring each program. This project suggests that a coding schema can be used by local community groups to monitor programming in specific localities. *Programme Content and Viewers' Perceptions: A BBC Audience Research Department Report.* BBC Publications, 1972.

Appendix III:
Report of the Study Group on the Entertainment Function of Television, March 4, 1976

TO: SSRC Committee on TV and Social Behavior
FROM: Percy Tannenbaum
RE: Entertainment Project

As we had planned, a committee-sponsored conference on "Television and Entertainment" was held in New York, October 24–25, 1975, immediately on the heels of our regular fall committee meeting. The main agenda for this initial gathering was to assess "the lay of the land," so to speak—to ascertain the current conceptualization and research in the field, what the main research needs seem to present themselves, and the specific interests of each of the participants in the area. The major aim was to solicit their suggestions for further activity, if any, our committee should undertake in this particular arena.

All but two of the original list of invitees were able to attend (Sylvan Tomkins, Rutgers, and Daniel Berlyne, Toronto, were unable to attend for reasons of health):

Paul Ekman, University of California at San Francisco
Seymore Feshbach, University of California at Los Angeles
Gerald Lesser, Harvard University
William McGuire, Yale University
Harold Mendelsohn, University of Denver
Jerome Singer, Yale University
Dolf Zillmann, Indiana University

In addition, the following members of our committee participated:

Ron Abeles
Leo Bogart
Hilde Himmelweit
Aimee Leifer
Jack McLeod
Percy Tannenbaum
Steve Withey

Most of the group assembled for dinner on Friday, October 24, at which time the main interests and concerns of the committee were presented and a conference agenda of sorts was established. The entire group reconvened Saturday morning and met into the latter part of the afternoon, when fatigue seemed to afflict most of the assemblage (especially the SSRC committee members). In retrospect, it would appear that while there are obvious reasons for combining such functions with regular committee meetings, substantial costs are incurred which should be considered in any future such undertakings.

CONFERENCE PROCEEDINGS

We did not deem the conference to be of the sort that a full recording of the proceedings would be useful. We do have some scattered notes, but they mainly indicate that somewhat unstructured nature of the discussion. This was not entirely accidental, since we were trying to come to grips with a rather amorphous concept in focusing on "entertainment." Even with such a relatively (and deliberately) homogeneous group (most label themselves as social psychologists), this served to guarantee a wide-ranging, multifaceted discussion, which is not very easy to summarize into several dominant themes.

As noted above, one of the participants was Bill McGuire, who as many of you know has an uncommon capacity for organizing and conceptually cataloguing seemingly disparate materials. At my urging, Bill prepared such an outline of what the conference "churned up" in him. I have taken considerable liberties with his draft in preparing the following statement but the major contribution is his.

I. Three Dimensions That Block Out the
   Conceptual Space Defined by the
   "Television and Entertainment" Rubric

When a group of social scientists get (or are pulled) together to discuss a topic which is as unconventional for them as this one of television and entertainment, there tend to be three different subgroups pulling in definably different directions. One group focuses on the TV medium and is more independent-variable oriented. A second is preoccupied with entertainment as a behavioral phenomenon and can be considered to more dependent-variable oriented. For the third, the focus is on some other phenomenon (e.g., functioning) which has some relationship to TV and/or entertainment.

A. TV as the Focus. Some of us are primarily interested in the topic as part of our preoccupation with the effects of television, one of which is entertainment. Other effects such as violence, prosocial behavior, political socialization, etc. have been more studied with respect to what impact television has on them than has this topic of entertainment. Yet the main social function of television and its economic basis are its contribution to human entertainment. The members of our group who are primarily interested in the television medium begin to get uncomfortable or impatient when other members who are more dependent-variable oriented start talking about the appeals of the ballet, or of participation in sports, or what makes for a popular novel, on the grounds that these other topics are pulling away from our appropriate point of departure—namely, the impact of television.

B. Entertainment as the Focus. Those who are attracted to the area via its dependent variable are more preoccupied with what constitutes entertainment and what are its motivational properties. Obviously television is a major source of entertainment in our society, but those social scientists with this primary interest tend to feel constrained by confining our thoughts only to those forms of entertainment which are possible through that medium. These members of our group move easily to other forms of entertainment, such as participation in or direct observation of sports, outdoor recreation, music, print, etc. They may feel uncomfortable when the talk drifts into other effects of television (violence, socialization, stimulation of cognitive development, etc.) which have little to do with entertainment and would prefer to grapple with the complexities of entertainment as a fairly constant and almost universal behavioral phenomenon.

C. Focus on Related Phenomena. Still another subgroup approaches this area with primary concerns in other topics—e.g., the development and function of fantasy, humor, arousal, recreation, aesthetics, sex, other prime motives. For some, the dominant paradigm is with any one of these phenomena mediating between TV and entertainment. For others, the given phenomenon is the primary motive system behind entertainment behavior, or TV viewing. At times the phenomenon becomes the main dependent variable, and TV and/or entertainment are merely part of an array of antecedent conditions. Either way, such people are bound to get uneasy when the talk drifts into other effects of TV which do not particularly involve their central concern or into discussions of entertainment processes other than this one are involved.

Together, these three dimensions appear to map our conceptual space of this "Television and entertainment" domain. It is good to keep all three in mind to know where we stand in relationship to one another and to anticipate that preoccupations with different cells of this three-dimensional structure will tend to have several of us pulling in different directions on occasion. While this might be occasionally uncomfortable, it will also keep each of us from getting unduly parochial.

II. Three Conservative Tools for Filling in the Conceptual Space

A. Subdivisions of the Field. The topic is a vast one, so that to avoid being overwhelmed it is useful to break down the field into manageable size by classifying the various entertainment materials available on television. The relevant questions thus generated include: What are the types of entertainment programs and the variants within these types? Do the conventional categories have any meaning (the daytime serial, the situation comedy, etc.) as regards the kind of entertainment offered? Would it be better to approach the topic of types of television entertainment, not from the point of view of the material but from the point of view of the different needs which might be gratified? How would one go about subdividing the type of television entertainment so as to have the division be maximally provocative or optimally useful?

B. Provocative Contrasts. It is often a useful insight into the dynamics of an area and provocative to further thought to search out the seeming contradictions or paradoxes in our conception of what is going on. The following represents a variety of such heuristically provocative dialectics in the television and entertainment area. There is obviously a grain of truth in each term of these oppositions. The questions are obviously raised not for an answer in favor of one or the other alternatives but to stimulate appreciation of the complexities of television entertainment.

1. Does television entertainment involve the individual's getting outside of him- or herself or does it involve intensifying, enriching one's self-consciousness?

2. Does television entertainment involve escaping from one's problems or getting wrapped up in new problems?

3. Does television entertain by calming or exciting?

4. Is a given entertainment experience more intense when the person is isolated or in a social context (for example, is watching football on television versus at the stadium more pleasurable?)?

5. Is television entertainment best explained negatively in terms of getting away from things—escapism—or more positively by the attractiveness of the entertaining material?

6. How does one explain the paradoxical attraction of seemingly negative material, such as being entertained by suspense and puzzles, by horror shows, by tragedies and tear-jerkers, etc. Or outside the television show domain, why we are attracted by the exertion, pain, danger, etc., involved in vigorous sports and dangerous recreation?

7. Is the entertainment selected by choice or is it accidental?

8. Is what is entertaining socially defined or is it determined by the intrinsic needs of the individual?

9. To what extent does entertainment involve novelty versus confirmation of expectedness, as in the ritualized depictions of comedy situations versus the unexpected punchline, etc.?

10. What is the relative appeal of spectator versus participant entertainment?

11. Is television entertainment mostly a matter of momentary diversion or does it involve remote recurrences and fantasies?

12. Does television entertainment involve left hemisphere or right hemisphere functions or both in some mix?

C. Terms and Constructs. A good deal of the discussion tended to be in terms of other social science concepts that are partially but not fully associated with television entertainment. They are included here, in a somewhat incomplete and haphazard fashion, to give some insight into the vast range of constructs that is generated when one pries open the black box represented by the concept of entertainment. The fact that many of these terms are not better defined than that of entertainment, as such, certainly does not help our task any.

aesthetics	fun
affect	gratification
amusement	humor
arousal	interest
art	leisure
content–style	means versus end
ceremony	novelty
diversion	pleasure
empathy	quality of life
enjoyment	recreation
entertainment	ritual
escapism	symbolism
excitement	vicarious experience
fantasy–make-believe	volitional–autotelic–instrumental

III. Methodology Opportunities

In the course of our discussions, a wide variety of research methods were mentioned in passing, some in connection with brief reports of ongoing research, others as potential tools for exploring new problems. The following represent McGuire's culling of those that were mentioned, along with the kind of problem to which a given method might be applied.

1. Propositional inventory of related fields (sports, popular culture, literary criticism, movie producers, etc.) giving the wisdom (some of which is probably quite unwise) of what entertains, as judged by a wide variety of people in the entertainment business who serve gatekeeper roles, and also other people.

2. Multidimensional scaling techniques that would help us to get some insights into how programs should cluster or the dimensions underlying the entertainment programs, in order to get some insight into the dimensions of television entertainment.

3. Open-ended interviews (of producers and consumers of television) trying to get at what entertains, when the respondent himself or herself is probably aware of this and unable to articulate inklings of which he or she may be aware.

4. Thought experiments, involving imagining how entertainment would be in a different world (for example, before television or before print and literature, etc.).

5. Look for the categorizing of entertainment materials by different subpopulations (for example, children as well as adults, men and women separately, different social classes separately, etc.).

6. Use actors, directors, television administrators, etc. as resource people to suggest the entertainment variables and then test these by survey and experimental manipulation.

7. Analyze television entertainment from each of several different areas that have some bearing on it (for example, the psychology of humor, quality of life, factors in best sellers, popular culture, literary criticism, aesthetics, sociology of sports, leisure studies, etc.). Knowledgeable people in each of these areas could be asked to write a paper on television entertainment, perhaps with more specific directive.

8. Analyses of Nielsen ratings with demographic characteristics, program content analysis, etc. Perhaps the Gerbner content analysis (or something similar) would be a useful starting point.

9. Use of "concept testing" approach used by networks and advertisers as a convenient way of studying the hypothetical variables involved in television entertainment.

10. Studying how people choose the programs to which they listen. Studying also the actual patterns of listening (as in "watching children watch television") in order to zero in on precise viewing patterns and get insights into the determinants of these patterns.

11. Study the effort people are willing to expend to watch one versus another type of program—for example, by the cycling apparatus or keeping the switch depressed to avoid static, etc.

12. Study joint choices as regards program preferences to get insights into the meaning of patterns of show preferences within individuals (for example, are they likely to solve different needs or the same needs, etc.).

REACTIONS AND SUGGESTIONS

The diversity represented by the foregoing potpourri of ideas and methods was not totally without its rewards. The conference did serve to provide a useful forum for a once-over-quickly exchange of conceptual approaches, updated reports on current and expected personal lines of research, etc. In the process, it exposed us to different ideas which we could apply in our own work and forced us to address other issues which we may otherwise tend to overlook. Although all agreed on the unstructured nature of the meeting, most felt that was an unavoidable feature at this stage of the game.

It was apparent early (actually, many of us anticipated it before hand) that there would be an inadequate basis for a consensus view to develop with respect to defining entertainment, establishing research priorities, and the like. I nevertheless tried to push in this direction, if for no other reason than to get the group to face that situation directly. Once this was apparent, we were able to address alternative activities from the perspective of the committee and its concern with stimulating researchers in the field, other than those present, to pursue ideas in the general area under consideration.

What emerged from this discussion was an agreement that the field might benefit most by a more detailed presentation and airing of the type and range of ideas and issues we had engaged in at our relatively brief meeting. It was further agreed that the best avenue for such a presentation would be a volume of contributions from the participants at the conference (and possibly several others whose names they were invited to submit). Diversified as such a volume is bound to be, it was felt that this could have a stimulating effect on other researchers. Conversely, it was felt that the imposition of a single model or other such constraints would be artificial and probably counterproductive to the goals of the committee. The meeting adjourned with each participant promising to send me a brief outline of a chapter they might expect to submit to such a volume.

By now, I have received a dozen such outlines, in varying degrees of detail. Several of our guests were obviously stimulated by the meeting and took pains to restructure their thinking to prepare a rather detailed outline. Others focused more on elaborating on the ideas and issues that had been raised at the meeting. All in all, I believe we have the makings of a reasonable volume, not that much more unstructured than most other offerings at this stage of the game and well in keeping with our aim of providing some impetus to a much neglected area of research.

Appendix IV:
Report of the Study Group on Television and Ethnicity, March 2, 1976

Ronald P. Abeles*
Social Science Research Council

Peter Almond
Carnegie Council on Children

Tony Batten
American Broadcasting Companies

Gordon Berry
University of California,
    Los Angeles

James Blackwell
University of Massachusetts,
    Boston

Jane Crowley
National Broadcasting Company

Dennis Doty
American Broadcasting Companies

Sherryl Browne Graves
New York University

Mary Harper
National Institute of
    Mental Health

Anna Lee Hopson
Columbia Broadcasting Company

Irving Janis*
Yale University

Hope Klapper
New York University

Joseph Klapper
Columbia Broadcasting Company

Aimee Dorr Leifer*
Harvard University

Chester M. Pierce*
Harvard University

Morris Rosenberg
University of Maryland

*Conference conveners and members of the Committee on Television and Social Behavior.

CATALOGUE OF IDEAS

For the purpose of report the committee has elected to make five divisions. Hence ideas produced by the conference will be placed in one of these categories; (1) researchable issues, (2) personnel stimulation, (3) dissemination of ideas, (4) public relations between the TV industry and the academy, and (5) ideas from the TV professionals. Save for category (5) any idea will be placed arbitrarily in a given area even though there may be equal reason to place it in one or more other categories.

RESEARCHABLE ISSUES

A. Studies on Ethnic Identification.
1. How do people socialize to ethnic identification?
2. What modulates ethnic identification?
3. What competes with ethnic identification?
4. What is the relation between social class and ethnicity?
5. What is the effect of stereotypes on the majority and on minorities?
6. What is the effect of positive as well as negative presentation of blacks on both whites and blacks?
7. Who makes the decisions and how are they made concerning atypical stereotypes—for example, a black professional on a daytime serial?

B. Studies on Perception.
1. Perception of roles as they relate to mass media.
2. Cross cultural perceptions—for example, use of Osgood's semantic differentials to see how minorities are represented or could be represented.
3. Positive and negative differences about how programs and ideas are perceived by different cohorts or people.
4. Study the disparity between what is intended and what is perceived.
5. What is the effect of "contrasts" on black and white populations—for example, seeing a black judge decide the fate of a white.

C. Studies on Cognitive Styles.
1. Recall and decay studies, especially about black-related issues.
2. Racial and class differences in accepting ideas presented by TV.

D. Search for Subtle Predictable Indicators of Racism.
1. Qualitative and quantitative content analyses for possible differences in how the races are presented in such areas as handling of time–space, amount of freedom and mobility, evidences of dependent status, depiction of sanity and stability, etc.

E. Studies on Psychophysiology.
1. Racial differences in terms of body systems in response to TV presentations.

In the course of its deliberations about what might be researchable the committee reviewed many possibilities that provoked thoughtful commentary. For instance, there were spirited discussions relative to studying the effect of using "Black English" on TV and scrutinizing which groups would accept or oppose its use. Another area of animated exchange involved the possible value of assigning roles in TV programs in a randomized manner, in terms of skin color.

Although no conclusions were reached in either of these examples (the consensus in both being more negative than positive), they represent some inquiry areas in which political, social, and emotional bias might run so high that the research benefits–results would stir up well-nigh irresistable and insoluble controversy. Thus such areas might be better avoided until the field of ethnic research on TV can command a wider acceptance and can be seen by all parties as being not unduly threatening.

PERSONNEL STIMULATION

A. Appeals to Relevant Professional Groups.
1. Place issue of need for personnel development on programs, in journals, in newsletters, etc.

B. Increase public awareness about possible opportunities for blacks.
1. Use of media, clergy, schools, civic groups, to inform black youth of the satisfactions in doing TV research.

There seemed to be no end of ways to increase the numbers of black researchers. If a mechanism was found to provide coordinating services on an ongoing basis over a period of time, much could be accomplished.

In addition, the conferees beliefed that SSRC in its suasive role could exert a highly salutary effect. For instance, SSRC could suggest that it would be advantageous (as well as fair) for minority people to be on research advisory review panels used by the networks, public, and private funding agencies. Similarly, SSRC could ask the networks or other communications-related industries to set up special fellowships for minority trainees. SSRC could remind critical facilities that there is an existing reservoir of relevant research skill in the black population, even if it is argued that there is a paucity of expertise in communication research. Or SSRC could indicate to ongoing programs such as American Association for the Advancement of Science (AAAS) that any mass media fellowships should include minority recipients.

DISSEMINATION OF IDEAS

A. Follow-through Efforts.
1. The results of this conference will be circularized to a large number of critical and potential researchers via the NIMH Minority Coalitions and Minority Research and Development Centers.
2. The participants of this meeting will have "coffee meetings" for interested, selected students in an attempt to make a ripple effect.
3. One participant has volunteered to seek funding for an expanded version of this conference so that more ideas can be generated, more people stimulated, more products exchanged.
4. It is suggested that ongoing, viable black TV research groups be in contact with the committee. There are a couple of organizations working full time. It is suggested that Dr. Leifer visit the one in San Francisco in order to set up active collaboration.

In discussing idea dissemination it was apparent that any efforts undertaken under the section on personnel stimulation or public relations would disseminate the ideas in a way to meet the charge of the conference. Likewise the meeting with network representatives was an important idea dissemination site.

Other concerns came up for which we seek the guidance of the full committee. Dr. Bogart, for example, might be able to tell us with whom "creative editors" (a name we learned was more acceptable than "censors") consult in terms of ethnic issues. Also we need to know how is the consultation performed.

Our total committee might have to give over some consideration for possible informed consent by individuals and communities if certain studies are actualized. In promulgating ideas it might be that this sensitivity should be underlined in order to make our efforts seem more believable.

PUBLIC RELATIONS BETWEEN INDUSTRY AND THE ACADEMY

Not surprisingly the conferees went over much the same material that the full committee has addressed in this regard. The overarching question was how can the TV industry be brought to change its values, so that it will take seriously any relevant social science research. The logical subquestion was how can a role be created for ethnic research in the TV industry.

The pariticipants would like for researchers to be informed about many areas so that they could be more useful. For example, they needed to know how issues are selected, what areas are acknowledged to be guesswork, what research data is there about blacks and how was it collected, reduced, and analyzed. The conferees believed it was important to avoid an adversary stance. They hoped to operate on the basis that the industry, like all good citizens, would be motivated to make a better life quality for everyone, if possible.

Most of the discussion related to the need to have a sufficiently trusting relationship with key personnel that researchers could learn what the industry believed research could do, what factors are believed to make ethnic programs fail, how bad seasons modify future risks about ethnic programs, how a franchise (specialty- or discipline-depicted) is chosen, what seems to be the difference between audience expectation and audience acceptance.

Thus the obvious lack of informed opinion obliged a modest if not conciliatory posture toward the industry. This could have massive sequelae positive and negative if there is mobilized an intense effort to find new black researchers and to have such researchers impact on the industry.

There was agreement, however, that the racism inherent in the society demands that those with more scientific understanding of it make their views known to black and white creative and business people, who are in the TV industry. Hence directors, producers should be instructed about possible damage to whites and blacks if blacks always are undercut in presentations in gratuitous ways, such as never being shown as true winners.

IDEAS FROM TV PROFESSIONALS

The TV professionals were generous with their time and their views. They provided much counsel. It is our opinion that such meetings can and should be sustained somehow, so that dialogue can continue. The list of research points that the TV experts listed include:

1) Need to judge effect and intent differently.

2) The need to specify in advance the intent and the target audience. In considering the effect on the target audience, the effect on the general population needs attention, too.

3) There is a need for the researcher to know (and specify) who are the consumers of this research.

4) The research must be "dramatic."

5) It would be advantageous to begin with a producing entity as a focal point—for example, who, how, what, why, are the plans and goals in the Lear organization?

6) It was suggested that researchers watch carefully TV developments in England since often they presage what happens in the United States.

7) In ethnic research, what makes people like or dislike a program? What people are in each group.

8) Ethnic research should be moving toward a cumulative body of knowledge and should be building itself in that manner.

9) A mechanism must be found whereby thought-through products of research can be offered.

10) A caveat: Content analysis cannot be confused with effects. Content must relate to what the audience, in fact, sees.

11) Analyze the number of "ethnics" at the celebrated communication schools.

12) Social Scientists must be concerned with far more than "the networks,"—for example, affilate stations, package program producers.

13) What does the ethnic community want to see?

14) Study public service advertisements.

15) Follow through from the beginning of a script through the airing to observe the mesh between the views of the broadcast standards people and the outcome of their interventions.

16) Work with production groups before and after something has been aired.

17) Remember that TV is an educational medium as well as an entertainment medium.

18) How much of a market is composed of ethnic groups and what per cent of the advertising dollar do they represent?

MISCELLANY

There remain a few remarks that should be stated. The conference was informed by printout from NIMH that only two of fifteen funded TV projects could be construed to have any possible ethnic emphasis.

Many of the ideas suggested in this conference would profit by being thought of in terms of longitudinal as well as cross-sectional studies.

There was constant emphasis on the specificity of black culture and its ability to enrich the total culture.

In the view of the subcommittee the conferees attended to the task with admirable zeal, dedication, and usefulness. We are grateful for their cooperation and productivity. We are indebted, too, for the excellent staff work from which we were beneficiary. We have every hope that some of the results of this meeting will have far-flung and continuous meaningful influence.

Appendix V:
Description of Television
Programs Cited

DESCRIPTIONS OF TELEVISION SERIES

All in the Family

A popular, long-running, network, prime time situation comedy series about a working class white family. The father, Archie Bunker, is a lovable, bigoted man who endorses traditional roles, values, and activities. The mother, Edith Bunker is something of a nervous fussbudget, but she cares lovingly and consistently for people and evidences considerable independence. The Bunker's daughter, Gloria, and her husband, Mike—a feisty, liberal student—live with them. In various seasons the Bunker's neighbors or friends have included a black family, the Jeffersons, and a couple who evidence nontraditional sex roles. Episodes usually center around the clashes in values that are nearly inevitable among such a group of people.

Bionic Woman

A network, prime time adventure series, especially popular with children. The white heroine, Jamie Sommers, a sometime schoolteacher, spends most of each episode using her bionically produced superhuman powers of hearing, speed, and strength and her unusual human intelligence and concern for others to save those in extreme danger anywhere in the world. She reports to Oscar Goldman, the head of an American intelligence agency, and receives care from Rudy, the doctor–scientist who gave her bionic powers.

Brady Bunch

Originally a network, prime time family situation comedy, the series is now in syndication and broadcast for children at a variety of times by stations not affiliated with the three major networks. The white middle-class family includes a mother, father, and numerous male and female children spanning the ages of approximately 6 to 16. Episodes center around various minor comedies and tradegies in the lives of the family members.

Captain Kangaroo

A long-running, network weekday morning series designed for preschool-age children. Captain Kangaroo serves as the host throughout, interacting regularly with Mr. Greenjeans, an ersatz farmer and animal lover; with Beth, a young woman; and with Bunny Rabbit, a puppet. At times he talks directly to children viewing at home. Various filmed, animated, and studio segments present material traditionally of interest to or instructionally appropriate for preschoolers. All of the regular cast is white.

Charlie's Angels

A popular, network, prime time crime drama series featuring three gorgeous detectives. In the first season or two, Sabrina, Kelly, and Jill (Farrah Fawcett-Majors), taking directions from Charlie—only a voice to them—and receiving support from Bosley—a lovable, somewhat bumbling man—rescue people, solve mysteries, and catch criminals in situations that always include danger and violence. In later seasons Jill was replaced by her "sister" Chris, but otherwise the series remains unchanged. All the regular cast is white.

Columbo

A network, prime time crime drama series featuring a dishevelled, self-effacing, apparently inept but actually quite brilliant big city white detective. In each episode a crime is committed and Columbo proceeds to investigate it. He always solves it in an hour, through clever questioning and deduction.

Eight is Enough

A network, prime time low-key situation comedy featuring a white middle-class family with eight children ranging from approximately 7 to 19. In the first season the family members cope humorously with their minor comedies and tragedies, but major tragedy rears its head when we learn the mother will die (as the actress did). In the second season the father, Tom Bradford, dates, then remarries, and the material for each episode returns to the more usual family show.

Good Times

A network, prime time situation comedy featuring a black working-class family. During the seasons in which this series has been broadcast the family has changed from an intact one with mother, father, and children to one with a mother and children and from there to one in which the children are alone with guidance provided by a neighbor. The episodes often center around the minor comedies and tragedies of family life, but they sometimes focus on issues that are more pertinent to a black than to a white family (or to a white's stereotype of a black family) or on issues such as intelligence testing, child abuse, or unemployment, which are not as often dealt with in white family situation comedies.

Happy Days

A popular, network, prime time situation comedy that is now also broadcast in the daytime, featuring Arthur Fonzarelli (The Fonz) and his teenage peers. Set in the 1950s, episodes usually center around dancing, dating, work, school performance, social clubs, and other such teenage concerns. All of the regular cast is white and roughly middle class.

Hardy Boys/Nancy Drew Mysteries

A network, prime time adventure series loosely based on the books by the same names. In the first season episodes alternated between the Hardy Boys and Nancy Drew, but the format for each involved establishing a mystery (usually not murder or other violent crime) and then having it solved through the investigative work of the white middle-class teenage amateur detectives. In the second season the Hardy Boys gradually became the focus of each episode, with Nancy Drew occasionally joining them, and violence became more evident.

Ironside

Originally a network, prime time crime drama series that is not in syndication and usually broadcast in the evenings by stations that are not affiliated with the three major networks. It features a white "Chief" of police who is now permanently confined to a wheelchair because he was shot by a criminal bearing a grudge. He is aided by two police officers, a white man and woman, and by a black law student. Episodes usually center around the solving of a crime or the prevention of a crime.

The Jeffersons

A network, prime time situation comedy featuring an upper-middle-class black family. George Jefferson, the father of the family, is a bigoted nouveau riche buffoon. Sensibility is supplied by his wife, mother, maid, son, or friends—including an interracial couple. Episodes usually center around George's pomposity, stupidity, avarice, or prejudice.

Lou Grant

A network, prime time dramatic series set in the offices of a big city newspaper, which Lou Grant has a primary role in running. The rest of the all white middle-class regular cast have various jobs at the paper. Episodes usually center on the various moral, legal, and personal problems that are likely to arise in such a business, but they may also focus on someone facing the possibility of a terminal illness, on a father's relationship with an "errant" son, and other such more personal dramas.

Mary Tyler Moore Show

A popular, long-running, network, prime time situation comedy that is no longer in production but is now syndicated on stations not affiliated with the three major networks. The series featured Mary Tyler Moore, a young careerist on a television news staff. In the early seasons she had a good friend, Rhoda, and some episodes centered around their friendship or Rhoda's dilemmas. When Rhoda moved to her own series, episodes most often focused on comedic dilemmas among the station news staff. Lou Grant was Mary's boss throughout the series. The regular cast is all white and middle class.

McCloud

A prime time, network crime drama series featuring a cowboy-like, modern white detective with a big city police force. Episodes usually involved McCloud chasing and catching a dangerous criminal in any of a variety of places in the United States.

Misterogers' Neighborhood

A long-running PBS series for preschool-children, which is no longer in production. Fred Rogers serves as the host throughout, interacting regularly with various members of his neighborhood—the mailman and his wife, a woman who runs a hobby shop, and the like. Most of the studio segments are set in his home, and at times he talks directly to children viewing at home. There is always one segment in a fantasy land peopled by puppets (people and animals) and by live actors. Most of each episode deals with children's feelings, hopes, fears, wishes, and the like with occasional instruction about cognitive material. Originally the regular cast was all white, but black characters have appeared in later seasons.

Perry Mason

Originally prime time, network crime drama series now in syndication and broadcast in the late afternoon, early evening, or late night by stations that are not affiliated with the major networks. It features Perry Mason, a criminal lawyer; his investigator, Paul; and his secretary, Della. Each episode usually begins with a crime having been committed that Perry and his crew then solve, most often in court and most often freeing someone wrongly accused of it. All of the cast is white and middle class.

Rebop

A PBS series for children from approximately the age of 8 through mid-adolescence. Designed to present positively the variety of ethnic groups in the United States, each program features film portraits of three adolescents who do interesting things or live in interesting places.

Rhoda

A prime time network situation comedy that originally focused on the adventures and misadventures of Rhoda and her husband Joe as they lived together and worked at their separate jobs, with occasional ventures into the lives of her sister, her mother, or their friends. In a subsequent season Rhoda and Joe get a divorce, and the series turns to events in Rhoda's life as a divorcee. All of the regular cast is white and middle class.

Sesame Street

A long-running PBS series for preschool-age children. It has a regular cast of black, white, and Hispanic children and adults in a New York City ghetto setting and a variety of adult guests, usually entertainers or sportspeople. Filmed, animated, and studio segments present basic cognitive information and skills, with lesser emphasis given to feelings, self-concept, and social interaction. Over the years greater emphasis has been given to these latter topics, to ecology, and to physical disabilities.

Shazam/Isis

A Saturday morning network dramatic series designed for children in the latter years of elementary school. Originally episodes alternated between Shazam and Isis, both young and white, but the format for each involved establishing some human dilemma that was resolved through the use of superpowers that each was able to draw upon with the intercession of some higher power. At the end the hero or heroine of the episode always delivered the moral message that the story had illustrated. Subsequent seasons have not kept such strict alternation, and most recently only the Isis episodes have been broadcast.

Six Million Dollar Man

A long-running network prime time adventure series, especially popular with children. The white hero, Steve Austin, spends most of each episode using his bionically produced superhuman powers of sight, hearing, speed, and strength and his unusual human intelligence and concern for others to save those in extreme danger anywhere in the world. He reports to Oscar Goldman, the head of an American intelligence agency, and receives care from Rudy, the doctor–scientist who gave him bionic powers.

Superman

At various times in the recent past this has been broadcast as a syndicated adventure series with live actors on stations not affiliated with the major networks or as an animated adventure series on Saturday morning on network stations. In both cases the series is intended for a child audience. Each episode involves a danger, a crime, or an intended crime that Clark Kent, newspaper reporter turned superhuman by removing his reporter's disguise, successfully resolves. He has superhuman strength and the ability to fly. All of the cast is white.

Switch

A prime time, network crime drama series featuring two white men—an ex-con and an ex-policeman or ex-detective. Together they redress grievances, right wrongs, stop crime, or catch criminals, with the first two more frequent that the latter two. The means to success is always some kind of con game.

Welcome Back Kotter

A prime time network situation comedy featuring Mr. Kotter, a white teacher who returns to the inner-city high school he attended. He and his wife—and now their childen—live in his old neighborhood. Most of his interactions, when not with his wife and children, are with the crusty school principal or with the Sweat Hogs, an exuberant, racially mixed group of male high school students. The comedy usually turns on the down-and-outers besting the others, with Mr. Kotter's connivance and support.

Author Index

Italics denote pages with bibliographic information.

A

Abeles, R. P., 295, *301*
Abelson, H., 282, *285*
Abelson, R. P., 14, *16*, 165, *187, 189*
Achelpohl, C., 177, *187*
Adler, R. P., 48, *51*, 92, 93, *101*, 172, *187*, 192, *226*, 231, *247*
Ajzen, I., 155, *159*
Akers, R., 177, *187*
Allsopp, R., 262, *285*
Alt, J., 156, *159*
American Association of Advertising Agencies, 127
Andison, F. S., 115, *131*
Andrews, 312, *324*
Andrews, L., 206, 217, *228*
Annan, Lord, 136, 137, *159*
Aptheker, H., 232, *247*
Arons, S., 167, *187*
Aronson, E., 202, *226*
Atkin, C. K., 29, *53*, 179, *187*, 262, 278, 279, *285*
Austin, L. J., 271, *285*

B

Baer, D. M., 191, *226*
Bagdikian, B., 92, *101*

Baker, R. K., 114, *131*
Baldwin, A. L., 214, *226*
Baldwin, C. P., 214, *226*
Baldwin, T., 72, 74, *80*
Ball, S. J., 114, *131*, 192, *226, 227*, 266, 273, 282, 283, 284, *285*
Ball-Rokeach, S. J., 298, *300*
Balsley, D. F., 262, *285*
Bandura, A., 272, *285, 295, 300*
Banks, J., 234, *247*
Banks, S., 91, *101*
Banks, W. C., 271, *285*
Barcus, F. E., 262, *285*
Barenboim, D., 218, *227*
Barnouw, E., 136, *159*
Batten, Barton, Durstine & Osborn, Inc., *132*
Bauer, A., 35, *51*, 70, *80*
Bauer, R. A., 35, *51*, 70, *80*, 297, *300*
Bayton, J. A., 271, *285*
BBC Publications, 323, *323*
Becker, L. B., 19, 26, 28, 30, 32, *51, 53*
Bechtel, R. B., 177, *187*
Belson, W., 121, *132*
Bennett, J. W., 12, *16*, 97, *101*
Bennett, R., 238, *248*
Berelson, B., 26, 34, 35, *51, 53*
Berkowitz, L., 19, *51*, 219, 220, *226*
Berndt, T. J., 206, 216, 217, *227*
Bijou, S. W., 191, *226*

347

Birt, J., 140, *159*
Blackmer, E. R., 206, 217, *228*
Bloomquist, L. E., 236, *248*
Blumenthal, 312, *324*
Blumler, J. G., 24, 26, 49, *51*, 56, 62, *80*, 92, *101*, 298, *301*
Bogart, L., 27, *51*, 75, *80*, 91, 92, *101*, 104, 115, *132*, 200, *226*, 263, 264, *285*
Bogatz, G. A., 192, *226*, *227*, 266, 273, 282, 283, 284, *285*
Bogle, D., 233, *247*
Boorstin, D. J., 291, *301*
Bower, R. T., 263, *285*
Brandt, M. T., 48, *51*
Breed, W., 97, *101*
British Broadcasting Corporation, 92, *101*
Broadcasting, 137, *159*
Bronfenbrenner, U., 244, *247*, 271, *285*
Brooks, V., 194, *228*
Broom, G. M., 48, *51*
Brown, A., 204, *227*
Brown, J. D., 29, *53*
Brown, L., 264, *285*
Bullock, P., 242
Burke, K. R., 271, *285*
Burns, T., 140, 149, 158, *159*
Butcher, M. J., 233, *247*
Bybee, C. R., 20, 30, 33, *53*
Byrne, D. F., 193, 213, 214, *229*
Byrnes, J. E., 19, *53*

C

Cantor, J. R., 25, *51*, 71, 72, 74, *81*
Carey, J. W., 263, *286*
Casey, B., 242
Cater, D., 79, *80*, 92, 93, *101*, 104, *132*
Cawelti, J., 97, *101*
Chaffee, S. H., 26, 29, 31, 40, *51*, *52*, *53*, 179, *187*
Chandler, M., 218, *227*
Charters, W. W., 265, *286*
Clark, C., 238, *247*, 262, 273, 274, *286*
Clark, K. B., 271, *286*
Clark, M. P., 271, *286*
Coates, B., 259, *286*
Cohen, B. C., 19, *52*
Cohen, L. B., 194, *227*
Collins, M. A., 27, *52*, 91, 92, 93, 95, *101*
Collins, W. A., 206, 216, 217, *227*, *228*

Comstock, G. A., 24, 37, *52*, 100, *101*, 104, 115, *132*, 161, 172, 176, 177, 178, 179, *187*, *188*, 259, 260, 265. *286*, 294, *301*
Congressional Black Caucus, 260, *286*
Coward, E. W., 11, *16*
Cramond, J., 31, *52*
Crewe, I., 156, *159*
Cripps, T., 234, *248*
Crisis, 246, *248*
Culley, J. D., 238, *248*
Cyclops, 240, *248*

D

Daily News (New York), 264, *286*
Davidson, E. S., 115, 121, *132*, 191, 192, *228*, 259, 263, 265, *287*
Davis, D., 31, *52*
DeFleur, L. B., 168, *188*
DeFleur, M. L., 168, *188*, 298, *300*
DeLoache, J. S., 194, *227*, *230*
Dervin, B., 264, 284, *286*
De Sola Pool, I., 294, *301*
Dienstbier, P. A., 130, *132*
Dirks, J., 194, *227*
Dominick, J. R., 29, *52*, 85, *101*, 166, *188*, 238, *248*, 262, 264, 279, 284, *286*, *287*, 289
Donohue, G. A., 18, *54*
Doolittle, J. C., 32, *51*
Dorr, A., 199, *227*
Downing, M., 262, 278, *286*
Durall, J. A., 20, 30, 33, *53*

E

Efron, E., 264, *286*
Ehrenberg, A. A. C., 94, *101*
Ehrenberg, A. S. C., 27, *52*, 91, 92, 93, 95, *101*
Eisenstein, E., 93, *101*
Ekman, P., 31, *52*
Emmerich, W., 192, *227*
Epstein, E. J., 100, *101*

F

Faber, R., 179, *189*
Fagan, J. F., 194, *227*

Fantz, R. L., 194, *227*
Feingold, M., 122, *132*
Feldman, J., 218, *227*
Feshbach, S., 20, 29, *52,* 118, *132,* 201, *227*
Festinger, L., 26, *52*
Field, J., 194, *227*
Fife, M. D., 238, 240, *248, 262, 286*
Firth, R., 12, *16*
Fischer, C. S., 295, *301*
Fishbein, M., 155, *159*
Fisher, M., 100, *101,* 115, *132,* 294, *301*
Fiske, M., 35, *54*
Flapan, D., 206, *227*
Flavell, J. H., 192, *227*
Fletcher, A. D., 263, 264, 284, *286*
Fontes, B. F., 236, *248*
Fortis, J. G., 209, *228*
Frank, R. S., 178, *188*
Freidson, E., 67, *81*
Friedlander, B. Z., 192, *226*
Friedman, L., 92, *101*
Friedrich, L. K., 192, *229*
Friedrick, L. K., 265, *289*
Frueh, T., 192, *227*
Fryrear, J. L., 272, *289*
Furu, T., 31, *52,* 152, *159,* 259, *286*

G

Gans, H. J., 36, *52,* 63, 64, 68, 69, 70, 72,
    81, 99, *101,* 110, *132*
Gaudet, H., 26, 34, 35, *53*
Gensch, D., 92, *101*
Gerbner, G., 88, 96, *101,* 116, 124, *132,* 151,
    159, 162, 176, *188,* 209, *227,* 262, 278,
    279, *286,* 299, *301,* 313, 322, 323, *324*
Gibson, E., 194, *227*
The Glasgow Media Project, 147, *159*
Glenn, C., 204, *230*
Glidewell, J. C., 85, *102*
Goffman, E., 13, *16,* 97, *101*
Goldberg, M. E., 266, 268, 279, *286*
Golden, B., 202, *226*
Goldman, K. S., 192, *227*
Goodhardt, G. J., 27, *52,* 91, 92, 93, 95, *101*
Goodman, I., 259, *286*
Goodman, P., 85, *101*
Goranson, R. E., 192, *227*
Gordon, N. J., 191, 198, *228,* 259, 265, 281,
    287
Gormley, W. T., 18, *52*

Gorn, G. I., 266, 268, 279, *286*
Grambs, J., 234, *247*
Grant, G., 271, *287*
Graves, S. B., 191, 192, 198, 199, *227, 229,*
    259, 265, 268, 269, 273, 274, 279, 281,
    282, 283, *286, 287*
Gray, L. N., 209, *228*
Greaves, W., 260, *286*
Greeley, B., 264, 284, *286*
Greenberg, B. S., 20, 29, *52,* 116, *132,* 179,
    188, 210, *227,* 238, *248, 262,* 263, 264,
    267, 269, 270, 272, 274, 277, 278, 279,
    282, 284, *285, 286, 287*
Greendale, S., 20, *53*
Greenspan, S., 218, *227*
Greenwald, H. J., 271, *287*
Gregor, A. J., 271, *287*
Gross, B. M., 206, 217, *228*
Gross, L., 88, 96, *101,* 124, *132,* 151, *159,*
    176, *188,* 209, *227,* 262, 278, 279, *286,*
    299, *301,* 313, 322, 323, *324*
Gurevitch, M., 49, *51,* 56, *81,* 298, *301*

H

Haggerty, S., 238, *248*
Hale, G. A., 216, *228*
Hall, S., 135, *159*
Hanneman, G. J., 264, 270, 274, 282, *286*
Harrison, R., 31, *52*
Hatano, G., 156, 157, *159*
Haupt, D. L., 194, *229*
Hawkins, R. P., 20, 29, *52,* 216, *228*
Head, 312, *324*
Heckel, R. V., 272, *287, 288*
Heller, M. S., 120, *132*
Henderson, E. H., 267, 275, 283, 285, *288*
Herold, C., 266, *288*
Hess, V. L., 206, 216, 217, *227*
Hetherington, E. M., 272, *289*
Hicks, D. J., 179, *188*
Higgins, P. B., 122, *133*
Himmelweit, H. T., 25, 31, *52,* 58, 59, *81,*
    99, *101,* 151, 152, 153, 155, 156, *159,*
    166, 167, *188,* 259, *287*
Hinton, J. L., 236, 237, 241, *248*
Hirsch, P., 70, 71, *81,* 83, *101,* 149, *159,*
    203, 215, *228*
Hochberg, J. E., 194, *228*
Hoffman, H. R., 25, 29, *53,* 177, *188,* 194,
    204, *228,* 263, 264, 284. *287, 288*

Hollenbeck, A. R., 194, *229*
Hornby, M., 266, *288*
Hornick, R., 147, *160*
Horton, D., 64, *81,* 197, *228*
Howard, B., 117
Hraba, J., 271, *287*
Hughes, H. M., 26, *54*
Humphreys, P., 156, *159*
Huston-Stein, A., 213, *228*
Hyman, H. H., 26, *52,* 284, *287*

I

Inhelder, B., 191, *229*
Insko, C. A., 202, *228*
Israel, H., 264, *287*
Ives, J., 49, *51*
Izcaray, F., 31, *51*

J

Jackson-Beeck, M., 116, 124, *132*
Jaeger, M. E., 99, *101*
Jaeger, M. J., 153, 156, *159*
James, L., 106, *132*
Janis, I. L., 163, 170, 177, 183, 184, 186, *188*
Jeffries-Fox, S., 116, 124, *132*
Johnson, G. T., 122, *132*
Johnson, N., 204, *229*
Jones, J. M., 271, *287*
Junger Witt, P., 129

K

Kahn, 312, *324*
Kahneman, D., 164, *189*
Kanungo, R. N., 266, 268, 279, *286*
Katsch, E., 167, *187*
Katz, D., 298, *301*
Katz, E., 24, 34, *51, 52,* 56, 60, 62, 63, *80, 92, 101,* 140, 151, 156, *159, 160,* 298, *301*
Kazarow, K., 192, *226*
Kelley, H. H., 15, *16*
Kendall, P. L., 281, *287*
Kenny, D. A., 29, *52*
Kerlinger, F., 20, 30, *52*
Kernan, K. T., 205, *228*

King, S., 266, *288*
Kirsch, A., 91, *101*
Klapper, J. T., 34, *52,* 282, *287*
Klein, P., 63, 65, 71, *81,* 92, *101*
Klemesrud, J., 239, *248*
Kline, F. G., 26, *52*
Kline, S., 154, *160*
Klineberg, O., 265, *287*
Knowlton, A., 127
Kohlberg, L., 191, 193, 214, *228*
Kohler, C. J., 194, *229*
Kraus, S., 31, *52,* 268, 279, *287*
Krugman, H., 59, *81,* 94, 98, *102*
Krull, R., 25, *54*
Kunce, M., 272, *287*

L

Labov, W., 204, 205, *228*
Lambert, W. E., 265, *287*
Landry, R., 77, *81*
Lang, G. E., 20, 31, 40, *52,* 156, *160*
Lang, K., 20, 31, 40, *52,* 165, *160*
Larsen, O. N., 209, *228*
Lazarsfeld, P. F., 26, 34, 35, *51, 52, 53,* 60, *81*
Lear, N., 125, 270, *287*
Lechenby, J. D., 264, 270, 274, 277, *287*
Leichtman, H. M., 272, *288* –
Leifer, A. D., 167, 168, *188,* 191, 198, 206, 210, 217, 218, 219, 220, *228,* 259, 265, 269, 274, 281, 283, *287*
Lelchuk, H., 127, *132*
Lemon, J., 279, *287*
Lerner, D., 37, *53*
Lesser, G. S., 50, *53,* 167, 168, 179, *188, 189,* 192, *226*
Levy, M., 62, *81*
Lewis, C., 72, 74, *80*
Lieberman Research, Inc., 120, *132*
Liebert, R. M., 115, 121, *132,* 191, 192, *228,* 259, 263, 265, *287*
Lindsey, G., 100, *101,* 115, *132,* 172, 178, *188*
Locke, J., 191, *228*
Lourenco, S., 85, *102*
Lund, R., 127
Lyle, J., 25, 29, 31, *53, 54,* 152, *160,* 177, *188,* 194, 204, *228,* 259, 263, 264, 284, *287, 288*

M

MacCorquodale, K., 19, *53*
MacDonald, D., 78, *81*
Magnuson, W. B., 306
Mandler, J. M., 204, *229*
Mann, L., 163, 170, 183, 184, 186, *188*
Marcus, S., 77, *81*
Marcuse, H., 78, *81*
Margulies, L., 240, *248*
Maynard, J., 194, *230*
Mays, L., 267, 275, 283, *288*
Mazingo, S., 238, *248*
McAdoo, H. P., 271, *288*
McClure, R., 79, *81,* 92, *102*
McCombs, M. E., 19, 26, *51, 53*
McDonald, D. L., 259, *288*
McGhee, P. E., 192, *227*
McGowan, J., 84, *102*
McGuire, W. J., 26, *53,* 161, *188,* 202, *229,*
    277, *288,* 298, *301*
McLeod, J. M., 20, 26, 28, 29, 30, 33, *51,*
    53, 179, *187,* 296, *301*
McLuhan, M., 25, *53,* 211, *229*
McQuail, D., 62, *80,* 91, *102*
McPhee, W. N., 26, 34, 35, *51,* 99, *102*
McPherson, D. A., 271, *287*
Meehl, P. E., 19, *53*
Mendelson, G., 262, *288*
Meringoff, L., 192, *226*
Milavsky, J. R., 119, 172, *188,* 299, *301*
Milgram, S., 118, 122, *132*
Miller, L. K., 216, *228*
Miller, M. M., 44, *53*
Miller, P. V., 26, *52*
Miranda, S. B., 194, *227*
Mitchell-Kernan, C., 244, *248*
Morland, J. K., 271, *288*
Morrison, A. J., 20, 26, *52, 53*
Moscovici, S., 135, *160*
Murray, J. P., 104, *132,* 263, *288*

N

National Advisory Commission on Civil
    Disorders, 260, *288*
National Citizens Committee for
    Broadcasting, 116, *132*
Neale, J. M., 115, 121, *132,* 191, 192, *228,*
    259, 263, 265, *287*
Neeley, J. J., 272, *288*

Neier, A., 129
Nelson, B., 267, *288*
Newcomb, H., 92, 93, 97, *102*
Newsweek, 272, *288*
The New York Times, 125, 129
Nie, N. H., 36, *53*
Noble, G., 197, 206, *229*
Noble, P., 235, *248*
Noll, R., 84, *102*
Northcott, H. C., 236, 237, 241, *248*

O

Obatala, J. K., 239, *248*
O'Connor, J. J., 239, *248*
Odum, E. P., 11, *16*
Office of Social Research, 269, *288*
O'Keefe, G. J., 296, *301*
O'Kelley, C. G., 236, *248*
Olien, C. N., 18, *54*
Oppenheim, A. N., 25, 31, *52,* 58, 59, *81,*
    152, 155, *159,* 167, *188,* 259, *287*
Oppenheim, D. B., 271, *287*
Ormiston, L. H., II, 262, 279, *288*

P

Paisley, M. B., 115, *132*
Paisley, W., 38, *53*
Papazian, E., 128
Parke, R. D., 272, *289*
Parker, E. B., 25, 31, *53, 54,* 152, *160,* 194,
    229, 259, 263, *288*
Pastore, J. O., 306
Patterson, T., 79, *81,* 92, *102*
Paulson, F. L., 259, *288*
Pearl, R. A., 194, *227*
Peck, M., 84, *102*
Pedhazur, E., 20, 30, *52*
Pekowsky, B., 172, *188,* 299, *301*
Peterson, R. C., 265, *288*
Petrocik, J. R., 36, *53*
Phelps, E. M., 202, *229,* 269, 274, 283, *287,*
    288
Piaget, J., 191, *229*
Pierce, C., 246, *248*
Pierce, F. S., 118
Pierce, M., 85, *101*
Pingree, S., 192, 201, *229*
Pinkey, A., 232, *248*

Platt, J., 162, *188*
Polsky, S., 120, *132*
Pool, I., 170, *188*
Powell, G. J., 271, *288*
Powell, R. M., 112
Powers, W. T., 14, *16*
Preston, I. L., 48, *51*
Pusser, H. E., 259, *286*

R

Ranganathan, B., 92, *101*
Rapport, D. J., 11, *16*
Rawlings, E., 219, 220, *226*
Ray, M., 94, *102*
Reeves, B., 20, 24, 29, 44, *52, 53,* 179, *188,*
    201, 210, *227, 229,* 269, 277, 282, *287*
Riley, J., 67, *81*
Riley, M., 67, *81*
Roberts, C., 237, 240, *248,* 279, *288*
Roberts, D. F., 202, 217, 218, 219, 220, *228,*
    229, 266, *288*
Robertson, T. S., 192, 223, *226*
Robinson, J. P., 18, *53,* 100, *102,* 131, *132,*
    264, *287*
Rodenkirchen, J. M., 18, *54*
Rogers, E., 94, *102*
Rogers, E. M., 37, *53*
Rogers, R. S., 154, *160*
Rokeach, M., 269, 270, 277, 281, 284, *289*
Rosekrans, M. A., 272, *288*
Rossiter, J. R., 192, 223, *226*
Rothenberg, M., 122, *133*
Rousseau, J.-J., 191, *229*
Rubin, D. E., 192, *226*
Rubinstein, E. A., 24, 37, *52,* 104, 110, *132*
Ruff, H. A., 194, *229*
Rumelhart, D. E., 204, *229*
Rushton, J. P., 153, *160*

S

Saint, W. S., 11, *16*
Salomon, G., 156, *160,* 213, *229*
Särlvik, B., 156, *159*
Scherer, K. R., 295, *301*
Schneider, A. R., 116
Schramm, W. M., 25, 31, 32, 37, *53, 54,*
    152, *160,* 194, *229,* 259, 263, *288*
Scriven, M., 50, *54*

Seggar, J. F., 168, *188,* 236, 237, 241, *248,*
    262, 279, *288*
Seidman, S. K., 267, 275, 283, 285, *288*
Selman, R. L., 193, 213, 214, *229*
Senate Subcommittee on Communications,
    233, 234, *248,* 308, *324*
Serbin, L. A., 262, *289*
Shank, R. C., 165, *189*
Shantz, C. V., 191, 213, *229*
Shaw, D. C., 19, *53*
Sheatsley, P. B., 26, *52,* 284, *287*
Shirley, K., 259, *289*
Shore, R. E., 192, *227*
Shotland, R. L., 118, 122, *132*
Showalter, S. W., 117, *133*
Siebold, D. T., 26, *54*
Sigal, L. V., 100, *102*
Signorielli, N., 116, 124, *132*
Silverman, F., 125
Silverman, T., 266, *288*
Simon, H. A., 11, *16*
Singer, D. G., 123, *133,* 181, *189*
Singer, J. L., 123, *133,* 166, 177, 181, *189*
Singer, R. D., 118, *132*
Sklar, R., 76, 78, *81*
Slaby, R. G., 194, *229*
Smiley, S., 204, *227*
Smith, G. C., 48, *51*
Smythe, D. W., 262, *289*
Somers, A. R., 122, *133*
Spock, B., 191, *229*
Spurlock, J., 271, *289*
Star, S. A., 26, *54*
Steeper, F., 20, *53*
Stein, A. H., 191, 192, *229, 265, 289*
Stein, B., 178
Stein, N. L., 204, *229, 230*
Steiner, G. A., 35, *51*
Steiner, V. S., 267, 275, 283, 285, *288*
Sterne, D., 266, *288*
Sternglanz, S. H., 262, *289*
Stevenson, H. W., 216, *228*
Stipp, H., 172, *188*
Stouwie, R. J., 272, *289*
Strauss, M. S., 194, *230*
Strickland, S., 79, *80,* 104, *132*
Stripp, H., 299, *301*
Stutzman, B., 129, *133*
Surber, C. F., 214, *230*
Surgeon General's Scientific Advisory
    Committee on Television and Social
    Behavior, 104, *133, 259, 289,* 307, 308,
    311, 316, *324*

Surlin, S. H., 264, 270, 274, 277, 284, *287,* *289*
Sussman, M., 49, *54*
Svennevig, 148
Swafford, T., 118
Swift, B., 99, *101,* 151, 152, 153, *159*

T

Tankard, J. W., 117, *133*
Tannenbaum, P. H., 26, *54,* 61, *81*
Taylor, P. H., 206, 217, *228*
Teplin, L. A., 271, *289*
Thelen, M. H., 272, *287*
Thurstone, L. L., 265, *288*
Tichenor, P. J., 18, *54*
Time, 239, *247*
Tower, J., 125
Turner, J. E., 11, *16*
Tversky, A., 164, *189*

U

U.S. Commission on Civil Rights, 209, *230*

V

Vane, E. T., 129
Verba, S., 36, *53*
Vidmar, N., 269, 270, 277, 281, 284, *289*
Vince, P., 25, 31, *52,* 58, 59, 81, 152, 155, *159,* 167, *188,* 259, *287*

W

Wackman, D. B., 179, *189*
Waletzky, J., 205, *228*

Wall Street Journal, 114, 127
Ward, S., 179, *189,* 192, *226*
Warner, W. L., 95, 97, *102*
Watt, J. H., 25, *54*
Wedell, C., 151, *160*
Welles, C., 86, *102*
Wells, W., 118, *133*
Werner, A., 31, *54*
Westby, S. D., 206, *227*
Wheeler, P., 168, *188,* 262, 279, *289*
Wheldon, H., 150, *160*
White, H., 92, *102*
Whiteley, S., 266, *288*
Whittemore, S. L., 259, *288*
Williams, R., 89, *102*
Williams, S., 262, 279, *288*
Wilson, D., 31, *52*
Wilson, H., 148, *160*
Wilson, M., 122, *133*
Winick, C., 112, *133*
Wohl, R. R., 64, *81,* 197, *228*
Wolf, K. M., 281, *287*
Wolfe, K. M., 35, *54*
Wood, P. H., 89, *102*
Wright, J., 213, *228*
Wright, W., 77, *81*

Y

Yoder, J. T., 243, *248*
Young, M., 262, *288*

Z

Ziemke, D. A., 20, 30, 33, *53*
Zillmann, D., 24, 25, *51, 54,* 61, *81*
Zimbardo, P., 277, 281, 282, 284, *289*

Subject Index

A

Agenda-setting, 19, 148
Aggression, 24–25, 58–59, 219–221, 295 (*see also* Violence)
Annan Committee, 136–141
Attitudes:
  and behavior, 25–26, 57, 59–60, 94, 297–298
  and information gain, 153–154
  and script models, 14–16

B

Black Americans:
  and ethnic identity, 244–245
  history of their media portrayals, 233–234, 238–242, 262
  history in America, 232–234
  research issues, 261, 333–338
  stereotypes of, 235, 239, 250, 260
  and Study Group on Ethnicity, 3–4, 333–338
  and televised violence, 58, 235–236
  as TV characters, 65–66, 236–238, 251–257, 262

C

Catharsis theory, 118–119
Children:
  learning to understand TV, 156–157, 172, 192–226
  TV effects on, 20, 58, 156–157, 167–169, 177–180, 192–226, 297
Committee on Television and Social Behavior, vii–ix, 1–8
Communications research, history of, 151–153
Construction of reality, 162–163, 170–171, 194–226, 299 (*see also* Research paradigms, script models)
Content analysis, 27, 56–57, 93, 97, 176–177, 262–263 *(see also* Violence)
  of portrayal of Black Americans, 236–238
Content effects, 19
  of recurrent themes, 173–177
  and viewer involvement in TV content, 55–56, 59–60, 62, 91, 155, 211

E, I

Entertainment functions of TV, 2–3, 325–332
Involvement of audience in TV content (*see* Content effects)

M, N

Market research, 89–98, 110–112
Motion picture industry, 111–114
Needs-and-gratifications (*see* Research
  paradigms)

P

Parasocial interaction, 64–65, 197–198
Perceived reality of TV content, 20–23, 58,
  66, 96–97, 201
Politics and political behavior, 60–62, 67,
  94–95, 148–149, 156, 291
Pseudo-events, 291

R

Research methodology:
  difficulties in, 7–8, 17–51, 56–57, 69–70,
    77, 293–300
  field research, 69–70
Research paradigms, 91–100, 293
  cultural, 11–12, 95–98, 103–104, 110–111
  concepts of media effects, 17–33, 292–293
  ecological, 10–11
  "effects research," 9, 17, 35–36, 40–41,
    57–62, 296–297
  medical models, 249–257

Research paradigms *(contd.)*
  needs-and-gratifications, 26, 62–66, 76–77,
    298
  script models, 13–16, 164–187

S, T

Social Science Research Council, viii–ix
Stereotypes (*see* Black Americans)
Television industry:
  Committee on Television and Social
    Behavior, 4–6
  economics of, 11, 75, 84–89, 108
  networks, 84–89, 144–145, 149, 157–158
  organization and structure, 4–6, 70–75,
    107
  and relation to society, 138–141, 296
  selection of programs, 70–75, 87–89,
    107–108, 128–130, 145–146, 196–197
  social research by, 118–122

V

Viewing behavior, 62–63, 87, 95
Violence index, 114–117, 124, 307–324
Violence, televised:
  effects of, 58–59, 61–62, 94, 104–106,
    118–122, 152–153, 201, 219–221
  measures of (*see* Violence index)
  and script model, 13–15
  as social issue, 75–76, 124–130